DAVID LIVINGSTONE (1813-1873)

The titles Missionary, Doctor, Explorer all describe David Livingstone, but they do not begin to breathe life into a man so dedicated to the Lord. At the age of twenty, Livingstone committed his life to Christ and became one of the most influential missionaries to ever serve the African people. Noted not only for his missionary accomplishments, but known also for his efforts to stop the slave trade and for his extensive exploration of the mysterious African continent.

Reading Livingstone's incredible life story one begins to understand the heart and soul of a man who felt "that the salvation of man ought to be the chief desire and aim of every Christian."

HEROES OF THE FAITH has been designed and produced for the discerning book lover. These classics of the Christian faith have been printed and bound with beauty, readability and longevity in mind.

Greatest care has gone into the selection of these volumes, with the hope that you will not only find books that are a joy to read, but books that will stir your faith and enlighten your daily walk with the Lord.

Titles available include:

Fanny Crosby, Bernard Ruffin

David Livingstone, W. Garden Blaikie, D.D., LL.D.

Dwight L. Moody, W.R. Moody

George Müller, William Henry Harding

Mary Slessor, W.P. Livingstone

John Wesley, C.E. Vulliamy

The Personal Life

OF

David Livingstone

CHIEFLY FROM HIS UNPUBLISHED
JOURNALS AND CORRESPONDENCE
IN THE POSSESSION OF HIS FAMILY

W. Garden Blaikie, D.D., LL.D.

Barbour and Company, Inc.
164 Mill Street
Westwood, New Jersey

ISBN 0916441-482

Published by: **BARBOUR AND COMPANY, INC.**
164 Mill Street
Westwood, New Jersey 07675

(In Canada, THE CHRISTIAN LIBRARY,
960 The Gateway, Burlington, Ontario L7L 5K7)

EVANGELICAL CHRISTIAN PUBLISHERS ASSOCIATION ᴄᴇᴘᴀ MEMBER

Printed in the United States of America

DAVID LIVINGSTONE.

CHAPTER I.

EARLY YEARS.

A.D. 1813-1936.

Ulva—The Livingstones—Traditions of Ulva life—The "baughting-time"—
"Kirsty's Rock"—Removal of Livingstone's grandfather to Blantyre—
Highland blood—Neil Livingstone—His marriage to Agnes Hunter—Her
grandfather and father—Monument to Neil and Agnes Livingstone in
Hamilton Cemetery—David Livingstone, born 19th March, 1813—Boyhood
—At home—In school—David goes into Blantyre Mill—First Earnings—
Night-school—His habits of reading—Natural-history expeditions—Great
spiritual change in his twentieth year—Dick's *Philosophy of a Future State*
—He resolves to be a missionary—Influence of occupation at Blantyre—
Sympathy with the people—Thomas Burk and David Hogg—Practical
character of his religion.

THE family of David Livingstone sprang, as he has
himself recorded, from the island of Ulva, on the west
coast of Mull, in Argyllshire. Ulva, "the island of
wolves," is of the same group as Staffa, and, like it,
remarkable for its basaltic columns, which, according to
MacCulloch, are more deserving of admiration than those
of the Giant's Causeway, and have missed being famous
only from being eclipsed by the greater glory of Staffa.
The island belonged for many generations to the Mac-
quaires, a name distinguished in our home annals, as well
as in those of Australia. The Celtic name of the Livingstones was M'Leay, which, according to Dr. Livingstone's
own idea, means "son of the gray-headed," but according
to another derivation, "son of the physician." It has been

(17)

surmised that the name may have been given to some son of the famous Beatoun, who held the post of physician to the Lord of the Isles. Probably Dr. Livingstone never heard of this derivation; if he had, he would have shown it some favor, for he had a singularly high opinion of the physician's office.

The Saxon name of the family was originally spelt Livingstone, but the Doctor's father had shortened it by the omission of the final "e." David wrote it for many years in the abbreviated form, but about 1857, at his father's request, he restored the original spelling.[1] The significance of the original form of the name was not without its influence on him. He used to refer with great pleasure to a note from an old friend and fellow-student, the late Professor George Wilson, of Edinburgh, acknowledging a copy of his book in 1857: "Meanwhile, may your name be propitious; in all your long and weary journeys may the *Living* half of your title outweigh the other; till after long and blessed labors, the white *stone* is given you in the happy land."

Livingstone has told us most that is known of his forefathers; how his great-grandfather fell at Culloden, fighting for the old line of kings; how his grandfather could go back for six generations of his family before him, giving the particulars of each; and how the only tradition he himself felt proud of was that of the old man who had never heard of any person in the family being guilty of dishonesty, and who charged his children never to introduce the vice. He used also to tell his children, when spurring them to diligence at school, that neither had he ever heard of a Livingstone who was a donkey. He has also recorded a tradition that the people of the island were converted from being Roman Catholics "by the laird coming round with a man having a yellow staff, which would seem to have attracted more attention than his

[1] See Journal of Geographical Society, 1857, p. clxviii.

teaching, for the new religion went long afterward— perhaps it does so still—by the name of the religion of the yellow stick." The same story is told of perhaps a dozen other places in the Highlands; the "yellow stick" seems to have done duty on a considerable scale.

There were traditions of Ulva life that must have been very congenial to the temperament of David Livingstone. In the "Statistical Account" of the parish to which it belongs[1] we read of an old custom among the inhabitants, to remove with their flocks in the beginning of each summer to the upland pastures, and bivouac there till they were obliged to descend in the month of August. The open-air life, the free intercourse of families, the roaming frolics of the young men, the songs and merriment of young and old, seem to have made this a singularly happy time. The writer of the account (Mr. Clark, of Ulva) says that he had frequently listened with delight to the tales of pastoral life led by the people on these occasions; it was indeed a relic of Arcadia. There were tragic traditions, too, of Ulva; notably that of Kirsty's Rock, an awful place where the islanders are said to have administered Lynch law to a woman who had unwittingly killed a girl she meant only to frighten, for the alleged crime—denied by the girl— of stealing a cheese. The poor woman was broken-hearted when she saw what she had done; but the neighbors, filled with horror, and deaf to her remonstrances, placed her in a sack, which they laid upon a rock covered by the sea at high water, where the rising tide slowly terminated her existence. Livingstone quotes Macaulay's remark on the extreme savagery of the Highlanders of those days, like the Cape Caffres, as he says; and the tradition of Kirsty's Rock would seem to confirm it. But the stories of the "baughting-time" presented a fairer aspect of Ulva life, and no doubt left happier impressions on his mind. His

---

[1] Kilninian and Kilmore. See *New Statistical Account of Scotland,* Argyll-shire, p. 345.

grandfather, as he tells us, had an almost unlimited stock of such stories, which he was wont to rehearse to his grandchildren and other rapt listeners.

When, for the first and last time in his life, David Livingstone visited Ulva, in 1864, in a friend's yacht, he could hear little or nothing of his relatives. In 1792, his grandfather, as he tells us, left it for Blantyre, in Lanarkshire, about seven miles from Glasgow, on the banks of the Clyde, where he found employment in a cotton factory. The dying charge of the unnamed ancestor must have sunk into the heart of his descendant, for, being a God-fearing man and of sterling honesty, he was employed in the conveyance of large sums of money from Glasgow to the works, and in his old age was pensioned off, so as to spend his declining years in ease and comfort. There is a tradition in the family, showing his sense of the value of education, that he was complimented by the Blantyre school-master for never grudging the price of a school-book for any of his children—a compliment, we fear, not often won at the present day. The other near relations of Livingstone seem to have left the island at the same time, and settled in Canada, Prince Edward's Isle, and the United States.

The influence of his Highland blood was apparent in many ways in David Livingstone's character. It modified the democratic influences of his earlier years, when he lived among the cotton spinners of Lanarkshire. It enabled him to enter more readily into the relation of the African tribes to their chiefs, which, unlike some other missionaries, he sought to conserve, while purifying it by Christian influence. It showed itself in the dash and daring which were so remarkbly combined in him with Saxon forethought and perseverance. We are not sure but it gave a tinge to his affections, intensifying his likes, and some of his dislikes too. His attachment to Sir Roderick Murchison was quite that of a Highlander, and hardly less so was his feeling toward the Duke of Argyll,—a man whom he had no doubt many grounds for esteeming highly, but of whom, after visiting him at Inveraray, he spoke with all the enthusiasm of a Highlander for his chief.

The Ulva emigrant had several sons, all of whom but one eventually entered the King's service during the French war, either as soldiers or sailors. The old man was somewhat disheartened by this circumstance, and especially by the fate of Charles, head-clerk in the office of Mr. Henry Monteith, in Glasgow, who was pressed on board a man-of-war, and died soon after in the Mediterranean. Only one son remained at home, Neil, the father of David, who eventually became a tea-dealer, and spent his life at Blantyre and Hamilton. David Livingstone has told us that his father was of the high type of character portrayed in the *Cottar's Saturday Night*. There are friends still alive who remember him well, and on whom he made a deep impression. He was a great reader from his youth upward, especially of religious works. His reading and his religion refined his character, and made him a most pleasant and instructive companion. His conversational powers were remarkable, and he could pour out in a most interesting way the stores of his reading and observation.

Neil Livingstone was a man of great spiritual earnestness, and his whole life was consecrated to duty and the fear of God. In many ways he was remarkable, being in some things before his time. In his boyhood he had seen the evil effects of convivial habits in his immediate circle, and in order to fortify others by his example he became a strict teetotaler, suffering not a little ridicule and opposition from the firmness with which he carried out his resolution. He was a Sunday-school teacher, an ardent member of a missionary society, and a promoter of meetings for prayer and fellowship, before such things had ceased to be regarded as badges of fanaticism. While traveling through the neighboring parishes in his voca-

tion of tea-merchant, he acted also as colporteur, distribu-
ting tracts and encouraging the reading of useful books.
He took suitable opportunities when they came to him of
speaking to young men and others on the most important
of all subjects, and not without effect. He learned Gaelic
that he might be able to read the Bible to his mother, who
knew that language best. He had indeed the very soul of
a missionary. Withal he was kindly and affable, though
very particular in enforcing what he believed to be right.
He was quick of temper, but of tender heart and gentle
ways; anything that had the look of sternness was the
result not of harshness but of high principle. By this
means he commanded the affection as well as the respect
of his family. It was a great blow to his distinguished
son, to whom in his character and ways he bore a great
resemblance, to get news of his death, on his way home
after his great journey, dissipating the cherished pleasure
of sitting at the fireside and telling him all his adventures
in Africa.

The wife of Neil Livingstone was Agnes Hunter, a
member of a family of the same humble rank and the
same estimable character as his own. Her grandfather,
Gavin Hunter, of the parish of Shotts, was a doughty
Covenanter, who might have sat for the portrait of
David Deans. His son David (after whom the traveler
was named) was a man of the same type, who got his first
religious impressions in his eighteenth year, at an open-
air service conducted by one of the Secession Erskines.
Snow was falling at the time, and before the end of the
sermon the people were standing in snow up to the
ankles; but David Hunter used to say he had no feeling
of cold that day. He married Janet Moffat, and lived at
first in comfortable circumstances at Airdrie, where he
owned a cottage and a croft. Mrs. Hunter died, when her
daughter Agnes, afterward Mrs. Neil Livingstone, was but
fifteen. Agnes was her mother's only nurse during a long

illness, and attended so carefully to her wants that the minister of the family laid his hand on her head, and said, " A blessing will follow you, my lassie, for your duty to your mother." Soon after Mrs. Hunter's death a reverse of fortune overtook her husband, who had been too good-natured in accommodating his neighbors. He removed to Blantyre, where he worked as a tailor. Neil Livingstone was apprenticed to him by his father, much against his will; but it was by this means that he became acquainted with Agnes Hunter, his future wife. David Hunter, whose devout and intelligent character procured for him great respect, died at Blantyre in 1834, at the age of eighty-seven. He was a great favorite with his grandchildren, to whom he was always kind, and whom he allowed to rummage freely among his books, of which he had a considerable collection, chiefly theological.

Neil Livingstone and Agnes Hunter were married in 1810, and took up house at first in Glasgow. The furnishing of their house indicated the frugal character and self-respect of the occupants; it included a handsome chest of drawers, and other traditional marks of respectability. Not liking Glasgow, they returned to Blantyre. In a humble home there, five sons and two daughters were born. Two of the sons died in infancy, to the great sorrow of the parents. Mrs. Livingstone's family spoke and speak of her as a very loving mother, one who contributed to their home a remarkable element of brightness and serenity Active, orderly, and of thorough cleanliness, she trained her family in the same virtues, exemplifying their value in their own home. She was a delicate little woman, with a wonderful flow of good spirits, and remarkable for the beauty of her eyes, to which those of her son David bore a strong resemblance. She was most careful of household duties, and attentive to her children. Her love had no crust to penetrate, but came beaming out freely like the light of the sun. Her son loved her, and in many ways followed her. It was the genial, gentle influences that had moved him under his mother's training that enabled him to move the savages of Africa.

She, too, had a great store of family traditions, and, like the mother of Sir Walter Scott, she retained the power of telling them with the utmost accuracy to a very old age. In one of Livingstone's private journals, written in 1864, during his second visit home, he gives at full length one of his mother's stories, which some future Macaulay may find useful as an illustration of the social condition of Scotland in the early part of the eighteenth century:

"Mother told me stories of her youth : they seem to come back to her in her eighty-second year very vividly. Her grandfather, Gavin Hunter, could write, while most common people were ignorant of the art. A poor woman got him to write a petition to the minister of Shotts parish to augment her monthly allowance of sixpence, as she could not live on it. He was taken to Hamilton jail for this, and having a wife and three children at home, who without him would certainly starve, he thought of David's feigning madness before the Philistines, and beslabbered his beard with saliva. All who were found guilty were sent to the army in America, or the plantations. A sergeant had compassion on him, and said, 'Tell me, gudeman, if you are really out of your mind. I'll befriend you.' He confessed that he only feigned insanity, because he had a wife and three bairns at home who would starve if he were sent to the army. 'Dinna say onything mair to ony body,' said the kind-hearted sergeant. He then said to the commanding officer, 'They have given us a man clean out of his mind : I can do nothing with the like o' him.' The officer went to him and gave him three shillings, saying, 'Tak' that, gudeman, and gang awa' hame to your wife and weans.' 'Ay,' said mother, 'mony a prayer went up for that sergeant, for my grandfather was an unco godly man. He had never had so much money in his life before, for his wages were only threepence a day.'"

Mrs. Livingstone, to whom David had always been a ost dutiful son, died on the 18th June, 1865, after a lingering illness which had confined her to bed for several years. A telegram received by him at Oxford announced her death; that telegram had been stowed away in one of his traveling cases, for a year after (19th June, 1866), in

his *Last Journals,* he wrote this entry: "I lighted on a telegram to-day:
'Your mother died at noon on the 18th June.
This was in 1865; it affected me not a little."[1]

The home in which David Livingstone grew up was bright and happy, and presented a remarkable example of all the domestic virtues. It was ruled by an industry that never lost an hour of the six days, and that welcomed and honored the day of rest; a thrift that made the most of everything, though it never got far beyond the bare necessaries of life; a self-restraint that admitted no stimu·lant within the door, and that faced bravely and steadily all the burdens of life; a love of books that showed the presence of a cultivated taste, with a fear of God that dignified the life which it moulded and controlled. To the last David Livingstone was proud of the class from which he sprang. When the highest in the land were showering compliments on him, he was writing to his old friends of "my own order, the honest poor," and trying, by schemes of colonization and otherwise, to promote their benefit. He never had the least hankering for any title or distinction that would have seemed to lift him out of his own class; and it was with perfect sincerity that on the tombstone which he placed over the resting-place of his parents in the cemetery of Hamilton, he expressed his feelings in these words, deliberately refusing to change the "and" of the last line into "but":

TO SHOW THE RESTING-PLACE OF

NEIL LIVINGSTONE,

AND AGNES HUNTER, HIS WIFE,

AND TO EXPRESS THE THANKFULNESS TO GOD

OF THEIR CHILDREN,

JOHN, DAVID, JANET, CHARLES, AND AGNES,

FOR POOR AND PIOUS PARENTS.

---

[1] *Last Journals,* vol. i. p. 55.

David Livingstone's birthday was the 19th March, 1813. Of his early boyhood there is little to say, except that he was a favorite at home. The children's games were merrier when he was among them, and the fireside brighter. He contributed constantly to the happiness of the family. Anything of interest that happened to him he was always ready to tell them. The habit was kept up in after-years. When he went to study in Glasgow, returning on the Saturday evenings, he would take his place by the fireside and tell them all that had occurred during the week, thus sharing his life with them. His sisters still remember how they longed for these Saturday evenings. At the village school he received his early education. He seems from his earliest childhood to have been of a calm, self-reliant nature. It was his father's habit to lock the door at dusk, by which time all the children were expected to be in the house. One evening David had infringed this rule, and when he reached the door it was barred. He made no cry nor disturbance, but having procured a piece of bread, sat down contentedly to pass the night on the doorstep. There, on looking out, his mother found him. It was an early application of the rule which did him such service in later days, to make the best of the least pleasant situations. But no one could yet have thought how the rule was to be afterward applied. Looking back to this period, Livingstone might have said, in the words of the old Scotch ballad:

> "O little knew my mother,
>   The day she cradled me,
> The lands that I should wander o'er,
>   The death that I should dee."

At the age of nine he got a New Testament from his Sunday-school teacher for repeating the 119th Psalm on two successive evenings with only five errors, a proof that perseverance was bred in his very bone.

His parents were poor, and at the age of ten he was put

to work in the factory as a piecer, that his earnings might aid his mother in the struggle with the wolf which had followed the family from the island that bore its name. After serving a number of years as a piecer, he was promoted to be a spinner. Greatly to his mother's delight, the first half crown he ever earned was laid by him in her lap. Livingstone has told us that with a part of his first week's wages he purchased Ruddiman's Rudiments of Latin, and pursued the study of that language with unabated ardor for many years afterward at an evening class which had been opened between the hours of eight and ten. "The dictionary part of my labors was followed up till twelve o'clock, or later, if my mother did not interfere by jumping up and snatching the books out of my hands. I had to be back in the factory by six in the morning, and continue my work, with intervals for breakfast and dinner, till eight o'clock at night. I read in this way many of the classical authors, and knew Virgil and Horace better at sixteen than I do now." [1]

In his reading, he tells us that he devoured all the books that came into his hands but novels, and that his plan was to place the book on a portion of the spinning-jenny, so that he could catch sentence after sentence as he passed at his work. The labor of attending to the wheels ·was great, for the improvements in spinning machinery that have made it self-acting had not then been introduced. The utmost interval that Livingstone could have for reading at one time was less than a minute.

The thirst for reading so early shown was greatly stimulated by his father's example. Neil Livingstone, while fond of the old Scottish theology, was deeply interested in the enterprise of the nineteenth century, or, as he called it, "the progress of the world," and endeavored to interest his family in it too. Any books of travel, and especially of missionary enterprise, that he could lay his hands on,

---

[1] *Missionary Travels*, p. 3.

he eagerly read. Some publications of the Tract Society, called the *Weekly Visitor,* the *Child's Companion and Teacher's Offering,* were taken in, and were much enjoyed by his son David, especially the papers of "Old Humphrey." Novels were not admitted into the house, in accordance with the feeling prevalent in religious circles. Neil Livingstone had also a fear of books of science, deeming them unfriendly to Christianity; his son instinctively repudiated that feeling, though it was some time before the works of Thomas Dick, of Broughty-Ferry, enabled him to see clearly, what to him was of vital significance, that religion and science were not necessarily hostile, but rather friendly to each other.

The many-sidedness of his character showed itself early; for not content with reading, he used to scour the country, accompanied by his brothers, in search of botanical, geological, and zoological specimens. Culpepper's *Herbal* was a favorite book, and it set him to look in every direction for as many of the plants described in it as the countryside could supply. A story has been circulated that on these occasions he did not always confine his researches in zoology to fossil animals. That Livingstone was a poacher in the grosser sense of the term seems hardly credible, though with the Radical opinions which he held at the time it may readily be believed that he had no respect for the sanctity of game. If a salmon came in his way while he was fishing for trout, he made no scruple of bagging it. The bag on such occasions was not always made for the purpose, for there is a story that once when he had captured a fish in the "salmon pool," and was not prepared to transport such a prize, he deposited it in the leg of his brother Charles's trousers, creating no little sympathy for the boy as he passed through the village with his sadly swollen leg!

It was about his twentieth year that the great spiritual change took place which determined the course of Livingstone's future life. But before this time he had earnest thoughts on religion. "Great pains," he says in his first book, "had been taken by my parents to instill the doctrines of Christianity into my mind, and I had no difficulty in understanding the theory of a free salvation by the atonement of our Saviour; but it was only about this time that I began to feel the necessity and value of a personal application of the provisions of that atonement to my own case."[1] Some light is thrown on this brief account in a paper submitted by him to the Directors of the London Missionary Society in 1838, in answer to a schedule of queries sent down by them when he offered himself as a missionary for their service. He says that about his twelfth year he began to reflect on his state as a sinner, and became anxious to realize the state of mind that flows from the reception of the truth into the heart. He was deterred, however, from embracing the free offer of mercy in the gospel, by a sense of unworthiness to receive so great a blessing, till a supernatural change should be effected in him by the Holy Spirit. Conceiving it to be his duty to wait for this, he continued expecting a ground of hope within, rejecting meanwhile the only true hope of the sinner, the finished work of Christ, till at length his convictions were effaced, and his feelings blunted. Still his heart was not at rest; an unappeased hunger remained, which no other pursuit could satisfy.

In these circumstances he fell in with Dick's *Philosophy of a Future State*. The book corrected his error, and showed him the truth. "I saw the duty and inestimable privilege *immediately* to accept salvation by Christ. Humbly believing that through sovereign mercy and grace I have been enabled so to do, and having felt in some measure its effects on my still depraved and deceitful heart, it is my desire to show my attachment to the cause of Him who died for me by devoting my life to his service."

---

[1] *Missionary Travels*, p. 4.

There can be no doubt that David Livingstone's heart was very thoroughly penetrated by the new life that now flowed into it. He did not merely apprehend the truth— the truth laid hold of him. The divine blessing flowed into him as it flowed into the heart of St. Paul, St. Augustine, and others of that type, subduing all earthly desires and wishes. What he says in his book about the freeness of God's grace drawing forth feelings of affectionate love to Him who bought him with his blood, and the sense of deep obligation to Him for his mercy, that had influenced, in some small measure, his conduct ever since, is from him most significant. Accustomed to suppress all spiritual emotion in his public writings, he would not have used these words if they had not been very real. They give us the secret of his life. Acts of self-denial that are very hard to do under the iron law of conscience, become a willing service under the glow of divine love. It was the glow of divine love as well as the power of conscience that moved Livingstone. Though he seldom revealed his inner feelings, and hardly ever in the language of ecstasy, it is plain that he was moved by a calm but mighty inward power to the very end of his life. The love that began to stir his heart in his father's house continued to move him all through his dreary African journeys, and was still in full play on that lonely midnight when he knelt at his bedside in the hut in Ilala, and his spirit returned to his God and, Saviour

At first he had no thought of being himself a missionary. Feeling "that the salvation of men ought to be the chief desire and aim of every Christian," he had made a resolution "that he would give to the cause of missions all that he might earn beyond what was required for his subsistence."[1] The resolution to give himself came from his reading an Appeal by Mr. Gutzlaff to the Churches of Britain and America on behalf of China. It was "the

---

[1] Statement to Directors of London Missionary Society.

claims of so many millions of his fellow-creatures, and the complaints of the scarcity, of the want of qualified missionaries," that led him to aspire to the office. From that time—apparently his twenty-first year—his "efforts were constantly directed toward that object without any fluctuation."

The years of monotonous toil spent in the factory were never regretted by Livingstone. On the contrary, he regarded his experience there as an important part of his education, and had it been possible, he would have liked "to begin life over again in the same lowly style, and to pass through the same hardy training."[1] The fellow-feeling he acquired for the children of labor was invaluable for enabling him to gain influence with the same class, whether in Scotland or in Africa. As we have already seen, he was essentially a man of the people. Not that he looked unkindly on the richer classes,—he used to say in his later years, that he liked to see people in comfort and at leisure, enjoying the good things of life,—but he felt that the burden-bearing multitude claimed his sympathy most. How quick the people are, whether in England or in Africa, to find out this sympathetic spirit, and how powerful is the hold of their hearts which those who have it gain! In poetic feeling, or at least in the power of expressing it, as in many other things, David Livingstone and Robert Burns were a great contrast; but in sympathy with the people they were alike, and in both cases the people felt it. Away and alone, in the heart of Africa, when mourning "the pride and avarice that make man a wolf to man," Livingstone would welcome the "good time coming," humming the words of Burns:

> " When man to man, the world o'er,
>   Shall brothers be for a' that."

In all the toils and trials of his life, he found the good of

---

[1] *Missionary Travels*, p. 6.

that early Blantyre discipline, which had forced him to bear irksome toil with patience, until the toil ceased to be irksome, and even became a pleasure.

Livingstone has told us that the village of Blantyre, with its population of two thousand souls, contained some characters of sterling worth and ability, who exerted a most beneficial influence on the children and youth of the place by imparting gratuitous religious instruction. The names of two of the worthiest of these are given, probably because they stood highest in his esteem, and he owed most to them, Thomas Burke and David Hogg. Essentially alike, they seem to have been outwardly very different. Thomas Burke, a somewhat wild youth, had enlisted early in the army. His adventures and hairbreadth escapes in the Forty-second, during the Peninsular and other wars, were marvelous, and used to be told in after-years to crowds of wondering listeners. But most marvelous was the change of heart that brought him back an intense Christian evangelist, who, in season and out of season, never ceased to beseech the people of Blantyre to yield themselves to God. Early on Sunday mornings he would go through the village ringing a bell to rouse the people that they might attend an early prayer-meeting which he had established. His temperament was far too high for most even of the well-disposed people of Blantyre, but Neil Livingstone appreciated his genuine worth, and so did his son. David says of him that "for about forty years he had been incessant and never weary in good works, and that such men were an honor to their country and their profession." Yet it was not after the model of Thomas Burke that Livingstone's own religious life was fashioned. It had a greater resemblance to that of David Hogg, the other of the two Blantyre patriarchs of whom he makes special mention, under whose instructions he had sat in the Sunday-school, and whose spirit may be gathered from his death-bed advice to him: "Now, lad, make religion the every-day business of your life, and not a thing of fits and starts; for if you do, temptation and other things will get the better of you." It would hardly be possible to give a better account of Livingstone's religion than that he did make it quietly, but very really, the every-day business of his life. From the first he disliked men of much profession and little performance; the aversion grew as he advanced in years; and by the end of his life, in judging of men, he had come to make somewhat light both of profession and of formal creed, retaining and cherishing more and more firmly the one great test of the Saviour—"By their fruits ye shall know them."

CHAPTER II.

MISSIONARY PREPARATION.

A.D. 1836—1840.

His desire to be a missionary to China—Medical missions—He studies at Glasgow—Classmates and teachers—He applies to London Missionary Society—His ideas of mission work—He is accepted provisionally—He goes to London—to Ongar—Reminiscences by Rev. Joseph Moore—by Mrs. Gilbert—by Rev. Isaac Taylor—Nearly rejected by the Directors—Returns to Ongar—to London—Letter to his sister—Reminiscences by Dr. Risdon Bennett—Promise to Professor Owen—Impression of his character on his friends and fellow-students—Rev. R. Moffat in England—Livingstone interested—Could not be sent to China—Is appointed to Africa—Providential links in his history—Illness—Last visits to his home—Receives Medical diploma—Parts from his family.

IT was the appeal of Gutzlaff for China, as we have seen, that inspired Livingstone with the desire to be a missionary; and China was the country to which his heart turned. The noble faith and dauntless enterprise of Gutzlaff, pressing into China over obstacles apparently insurmountable, aided by his medical skill and other unusual qualifications, must have served to shape Livingstone's ideal of a missionary, as well as to attract him to the country where Gutzlaff labored. It was so ordered, however, that in consequence of the opium war shutting China, as it seemed, to the English, his lot was not cast there; but throughout his whole life he had a peculiarly lively interest in the country that had been the object of his first love. Afterward, when his brother Charles, then in America, wrote to him that he, too, felt called to the missionary office, China was the sphere which David pointed out to him, in the hope that the door which had been closed to the one brother might be opened to the other.

When he determined to be a missionary, the only persons to whom he communicated his purpose were his minister and his parents, from all of whom he received great encouragement.[1] He hoped that he would be able to go through the necessary preparation without help from any quarter. This was the more commendable, because in addition to the theological qualifications of a missionary, he determined to aquire those of a medical practitioner. The idea of medical missions was at that time comparatively new. It had been started in connection with missions to China, and it was in the prospect of going to that country that Livingstone resolved to obtain a medical education. It would have been comparatively easy for him, in a financial sense, to get the theological training, but the medical education was a costly affair. To a man of ordinary ideas, it would have seemed impossible to make the wages earned during the six months of summer avail not merely for his support then, but for winter too, and for lodgings, fees, and books besides. Scotch students have often done wonders in this way, notably the late Dr. John Henderson, a medical missionary to China, who actually lived on half-a-crown a week, while attending medical classes in Edinburgh. Livingstone followed the same self-denying course. If we had a note of his housekeeping in his Glasgow lodging, we should wonder less at his ability to live on the fare to which he was often reduced

---

[1] Livingstone's minister at this time was the Rev. John Moir, of the Congregational church, Hamilton, who afterward joined the Free Church of Scotland, and is now Presbyterian minister in Wellington, New Zealand. Mr. Moir has furnished us with some recollections of Livingstone, which reached us after the completion of this narrative. He particularly notes that when Livingstone expressed his desire to be a missionary, it was a missionary out and out, a missionary to the heathen, not the minister of a congregation. Mr Moir kindly lent him some books when he went to London, all of which were conscientiously returned before he left the country. A Greek Lexicon, with only cloth boards when lent, was returned in substantial calf. He was ever careful, conscientious, and honorable in all his dealings, as his father had been before him.

in Africa. But the importance of the medical qualification had taken a firm hold of his mind, and he persevered in spite of difficulties. Though it was never his lot to exercise the healing art in China, his medical training was of the highest use in Africa, and it developed wonderfully his strong scientific turn.

It was in the winter of 1836–37 that he spent his first session in Glasgow. Furnished by a friend with a list of lodgings, Livingstone and his father set out from Blantyre one wintry day, while the snow was on the ground, and walked to Glasgow. The lodgings were all too expensive. All day they searched for a cheaper apartment, and at last in Rotten Row they found a room at two shillings a week. Next evening David wrote to his friends that he had entered in the various classes, and spent twelve pounds in fees; that he felt very lonely after his father left, but would put " a stout heart to a stey brae," and " either mak' a spune or spoil a horn." At Rotten Row he found that his landlady held rather communistic views in regard to his tea and sugar; so another search had to be made, and this time he found a room in the High street, where he was very comfortable, at half-a-crown a week.

At the close of the session in April he returned to Blantyre and resumed work at the mill. He was unable to save quite enough for his second session, and found it necessary to borrow a little from his elder brother.[1] The classes he attended during these two sessions were the Greek class in Anderson's College, the theological classes of Rev. Dr. Wardlaw, who trained students for the Independent Churches, and the medical classes in Anderson's. In the Greek class he seems to have been entered as a private

---

[1] The readiness of elder brothers to advance part of their hard-won earnings, or otherwise encourage a younger brother to attend college, is a pleasant feature of family life in the humbler classes of Scotland. The case of James Beattie, the poet, assisted by his brother David, and that of Sir James Simpson, who owed so much to his brother Alexander, will be remembered in this connection.

student exciting little notice.[1] In the same capacity he attended the lectures of Dr. Wardlaw. He had a great admiration for that divine, and accepted generally his theological views. But Livingstone was not much of a scientific theologian.

His chief work in Glasgow was the prosecution of medical study. Of his teachers, two attracted him beyond the rest—the late Dr. Thomas Graham, the very distinguished Professor of Chemistry, and Dr. Andrew Buchanan, Professor of the Institutes of Medicine, his life-long and much-attached friend. While attending Dr. Graham's class he was brought into frequent contact with the assistant to the Professor, Mr. James Young. Originally bred to a mechanical employment, this young man had attended the evening course of Dr. Graham, and having attracted his attention, and done various pieces of work for him, he became his assistant. The students used to gather round him, and several met in his room, where there was a bench, a turning-lathe, and other conveniences for mechanical work. Livingstone took an interest in the turning-lathe, and increased his knowledge of tools—a knowledge which proved of the highest service to him when—as he used to say all missionaries should be ready to do—he had to become a Jack-of-all-trades in Africa.

Livingstone was not the only man of mark who frequented that room, and got lessons from Mr. Young "how to use his hands." The Right Hon. Lyon Playfair, who has had so distinguished a scientific career, was another of its habitués. A galvanic battery constructed by two young men on a new principle, under Mr. Young's instructions, became an object of great attraction, and among

---

[1] A very sensational and foolish reminiscence was once published of a raw country youth coming into the class with his clothes stained with grease and whitened by cotton-wool. This was Livingstone. The fact is, nothing could possibly have been more unlike him. At this time Livingstone was not working at the mill; and, in regard to dress, however plainly he might be clad, he was never careless, far less offensive.

4

those who came to see it and its effects were two sons of the Professor of Mathematics in the University. Although but boys, both were fired at this interview with enthusiasm for electric science. Both have been for many years Professors in the University of Glasgow. The elder, Professor James Thomson, is well known for his useful inventions and ingenious papers on many branches of science. The younger, Sir William Thomson, ranks over the world as prince of electricians, and second to no living man in scientific reputation.

Dr. Graham's assistant devoted himself to practical chemistry, and made for himself a brilliant name by the purification of petroleum, adapting it for use in private houses, and by the manufacture of paraffin and paraffin-oil. Few men have made the art to which they devoted themselves more subservient to the use of man than he whom Livingstone first knew as Graham's assistant, and afterward used to call playfully "Sir Paraffin." "I have been obliged to knight him," he used to say, "to distinguish him from the other Young." The "other" Young was Mr. E. D. Young, of the Search Expedition, and subsequently the very successful leader of the Scotch Mission at Lake Nyassa. The assistant to Dr. Graham still survives, and is well known as Mr. Young, of Kelly, LL.D. and F.R.S.

When Livingstone returned from his first journey his acquaintance with Mr. Young was resumed, and their friendship continued through life. It is no slight testimony from one who knew him so long and so intimately, that, in his judgment, Livingstone was the best man he ever knew, had more than any other man of true filial trust in God, more of the spirit of Christ, more of integrity, purity, and simplicity of character, and of self-denying love for his fellow-men. Livingstone named after him a river which he supposed might be one of the sources of the Nile, and used ever to speak with great respect of the chief achievement of Mr. Young's life,—filling houses with a clear white light at a fraction of the cost of the smoky article which it displaced.

Beyond their own department, men of science are often as lax and illogical as any; but when scientific training is duly applied, it genders a habit of thorough accuracy, inasmuch as in scientific inquiry the slightest deviation from truth breeds endless mischief. Other influences had already disposed Livingstone to great exactness of statement, but along with these his scientific training may be held to have contributed to that dread of exaggeration and of all inaccuracy which was so marked a feature of his character through life.

It happened that Livingstone did not part company with Professor Graham and Mr. Young when he left Glasgow. The same year, Dr. Graham went to London as Professor in University College, and Livingstone, who also went to London, had the opportunity of paying occasional visits to his class. In this way, too, he became acquainted with the late Dr. George Wilson, afterward Professor of Technology in the University of Edinburgh, who was then acting as unsalaried assistant in Dr. Graham's laboratory. Frank, genial, and chivalrous, Wilson and Livingstone had much in common, and more in after-years, when Wilson, too, became an earnest Christian. In the simplicity and purity of their character, and in their devotion to science, not only for its own sake, but as a department of the kingdom of God, they were brothers indeed. Livingstone showed his friendship in after-years by collecting and transmitting to Wilson whatever he could find in Africa worthy of a place in the Edinburgh Museum of Science and Art, of which his friend was the first Director.

In the course of his second session in Glasgow (1837–38) Livingstone applied to the London Missionary Society, offering his services to them as a missionary. He had learned that that Society had for its sole object to send the gospel to the heathen; that it accepted missionaries from different Churches, and that it did not set up any particular form of Church, but left it to the converts to choose the form they considered most in accordance with the Word of God. This agreed with Livingstone's own notion of what a Missionary Society should do. He had already connected himself with the Independent communion, but this preference for it was founded chiefly on his greater regard for the *personnel* of the body, and for the spirit in which it was administered, as compared with the Presbyterian Churches of Scotland. He had very strong views of the spirituality of the Church of Christ, and the need of a profound spiritual change as the only true basis of Christian life and character. He thought that the Presbyterian Churches were too lax in their communion, and particularly the Established Church. He was at this time a decided Voluntary, chiefly on the ground maintained by such men as Vinet, that the connection of Church and State was hurtful to the spirituality of the Church; and he had a particular abhorrence of what he called "geographical Christianity,"—which gave every man within a certain area a right to the sacraments. We shall see that in his later years Dr. Livingstone saw reason to modify some of these opinions; surveying the Evangelical Churches from the heart of Africa, he came to think that, established or non-established, they did not differ so very much from each other, and that there was much good and considerable evil in them all.

In his application to the London Missionary Society, Livingstone stated his ideas of missionary work in comprehensive terms: "The missionary's object is to endeavor by every means in his power to make known the gospel by preaching, exhortation, conversation, instruction of the young; improving, so far as in his power, the temporal condition of those among whom he labors, by introducing the arts and sciences of civilization, and doing everything to commend Christianity to their hearts and consciences. He will be exposed to great trials of his faith and patience from the indifference, distrust, and even direct opposition and scorn of those for whose good he is laboring; he may be tempted to despondency from the little apparent fruit of his exertions, and exposed to all the contaminating influence of heathenism." He was not about to undertake this work without counting the cost. "The hardships and dangers of missionary life, so far as I have had the means of ascertaining their nature and extent, have been the subject of serious reflection, and in dependence on the promised assistance of the Holy Spirit, I have no hesitation in saying that I would willingly submit to them, considering my constitution capable of enduring any ordinary share of hardship or fatigue." On one point he was able to give the Directors very explicit information: he was not married, nor under any engagement of marriage, nor had he ever made proposals of marriage, nor indeed been in love! He would prefer to go out unmarried, that he might, like the great apostle, be without family cares, and give himself entirely to the work.

His application to the London Missionary Society was provisionally accepted, and in September, 1838, he was summoned to London to meet the Directors. A young Englishman came to London on the same errand at the same time, and a friendship naturally arose between the two. Livingstone's young friend was the Rev. Joseph Moore, afterwards missionary at Tahiti; now of Congleton, in Cheshire. Nine years later, Livingstone, writing to Mr. Moore from Africa, said: "Of all those I have met since we parted, I have seen no one I can compare to you for sincere, hearty friendship." Livingstone's family used to speak of them as Jonathan and David. Mr. Moore has kindly furnished us with his recollections of Livingstone at this time:—

" I met with Livingstone first in September, 1838, at 57 Aldersgate street, London. On the same day we had received a letter from the Secretary informing us severally that our applications had been received, and that we must appear in London to be examined by the Mission Board there. On the same day, he from Scotland, and I from the south of England, arrived in town. On that night we simply accosted each other, as those who meet at a lodging house might do. After breakfast on the following day we fell into conversation, and finding that the same object had brought us to the metropolis, and that the same trial awaited us, naturally enough we were drawn to each other. Every day, as we had not been in town before, we visited places of renown in the great city, and had many a chat about our prospects.

" On Sunday, in the morning, we heard Dr. Leifchild, who was then in his prime, and in the evening Mr. Sherman, who preached with all his accustomed persuasiveness and mellifluousness. In the afternoon we worshiped at St. Paul's, and heard Prebendary Dale.

" On Monday we passed our first examination. On Tuesday we went to Westminster Abbey. Who that had seen those two young men passing from monument to monument could have divined that one of them would one day be buried with a nation's—rather with the civilized world's—lament, in that sacred shrine ? The wildest fancy could not have pictured that such an honor awaited David Livingstone. I grew daily more attached to him. If I were asked why, I should be rather at a loss to reply. There was truly an indescribable charm about him, which, with all his rather ungainly ways, and by no means winning face, attracted almost every one, and which helped him so much in his after-wanderings in Africa.

" He won those who came near him by a kind of spell. There happened to be in the boarding-house at that time a young M.D., a a saddler from Hants, and a bookseller from Scotland. To this hour they all speak of him in rapturous terms.

" After passing two examinations, we were both so far accepted by the Society that we were sent to the Rev. Richard Cecil, who resided at Chipping Ongar, in Essex. Most missionary students were sent to him for three months' probation, and if a favorable opinion was sent to the Board of Directors, they went to one of the Independent colleges. The students did not for the most part live with Mr. Cecil, but took lodgings in the town, and went to his house for meals and instruction in classics and theology. Livingstone and I lodged together. We read Latin and Greek, and began Hebrew together. Every day we took walks, and visited all the spots of interest in the neighborhood, among them the country churchyard which was the burial-place of John Locke. In a place so quiet, and a life so ordinary as that of a student, there did not occur many events worthy of recital. I will, however, mention one or two things, because they give an insight—a kind of prophetic glance—into Livingstone's after-career.

"One foggy November morning, at three o'clock, he set out from Ongar to walk to London to see a relative of his father's.[1] It was about twenty-seven miles to the house he sought. After spending a few hours with his relation, he set out to return on foot to Ongar. Just out of London, near Edmonton, a lady had been thrown out of a gig. She lay stunned on the road. Livingston immediately went to her, helped to carry her into a house close by, and having examined her and found no bones broken, and recommending a doctor to be called, he resumed his weary tramp. Weary and footsore, when he reached Stanford Rivers he missed his way, and finding after some time that he was wrong, he felt so dead-beat that he was inclined to lie down and sleep; but finding a directing-post he climbed it, and by the light of the stars deciphered enough to know his whereabouts. About twelve that Saturday night he reached Ongar, white as a sheet, and so tired he could hardly utter a word. I gave him a basin of bread and milk, and I am not exaggerating when I say I put him to bed. He fell at once asleep, and did not awake till noonday had passed on Sunday.

"Total abstinence at that time began to be spoken of, and Livingstone and I, and a Mr. Taylor, who went to India, took a pledge together to abstain.[2] Of that trio, two, I am sorry to say (*heu me miserum!*), enfeebled health, after many years, compelled to take a little wine for our stomachs' sake. Livingstone was one of the two.

"One part of our duties was to prepare sermons, which were submitted to Mr. Cecil, and, when corrected, were committed to memory, and then repeated to our village congregations. Livingstone prepared one, and one Sunday the minister of Stamford Rivers, where the celebrated Isaac Taylor resided, having fallen sick after the morning service, Livingstone was sent for to preach in the evening. He took his text, read it out very deliberately, and then—then—his sermon had fled! Midnight darkness came upon him, and he abruptly said: 'Friends, I have forgotten all I had to say,' and hurrying out of the pulpit, he left the chapel.

---

[1] We learn from the family that the precise object of the visit was to transact some business for his eldest brother, who had begun to deal in lace. In the darkness of the morning Livingstone fell into a ditch, smearing his clothes, and not improving his appearance for smart business purposes. The day was spent in going about in London from shop to shop, greatly increasing Livingstone's fatigue.

[2] Livingstone had always practiced total abstinence, according to the invariable custom of his father's house. The third of the trio was the Rev. Joseph V. S. Taylor, now of the Irish Presbyterian Mission, Gujerat, Bombay.

"He never became a preacher" [we shall see that this does not apply to his preaching in the Sichuana language], "and in the first letter I received from him from Elizabeth Town, in Africa, he says: 'I am a very poor preacher, having a bad delivery, and some of them said if they knew I was to preach again they would not enter the chapel. Whether this was all on account of my manner I don't know; but the truth which I uttered seemed to plague very much the person who supplies the missionaries with wagons and oxen. (They were bad ones.) My subject was the necessity of adopting the benevolent spirit of the Son of God, and abandoning the selfishness of the world.' Each student at Ongar had also to conduct family worship in rotation. I was much impressed by the fact that Livingstone never prayed without the petition that we might imitate Christ in all his imitable perfections."[1]

In the Autobiography of Mrs. Gilbert, an eminent member of the family of the Taylors of Ongar, there occur some reminiscenses of Livingstone, corresponding to those here given by Mr. Moore.[2]

The Rev. Isaac Taylor, LL.D., now rector of Settringham, York, son of the celebrated author of *The Natural History of Enthusiasm*, and himself author of *Words and Places, Etruscan Researches*, etc., has kindly furnished us with the following recollection: "I well remember as a boy taking country rambles with Livingstone when he was studying at Ongar. Mr. Cecil had several missionary students, but Livingstone was the only one whose personality made any impression on my boyish imagination. I might sum up my impression of him in two words— simplicity and resolution. Now, after nearly forty years, I remember his step, the characteristic forward tread, firm,

---

[1] In connection with this prayer, it is interesting to note the impression made by Livingstone nearly twenty years afterward on one who saw him but twice— once at a public breakfast in Edinburgh, and again at the British Association in Dublin in 1857. We refer to Mrs. Sime, sister of Livingstone's early friend, Professor George Wilson, of Edinburgh. Mrs. Sime writes: "I never knew any one who gave me more the idea of power over other men, such power as our Saviour showed while on earth, the power of love and purity combined."

[2] Page 386, third edition.

simple, resolute, neither fast nor slow, no hurry and no dawdle, but which evidently meant—getting there."[1] We resume Mr. Moore's reminiscences:

" When three months had elapsed, Mr. Cecil sent in his report to the Board. Judging from Livingstone's hesitating manner in conducting family worship, and while praying on the week-days in the chapel, and also from his failure so complete in preaching, an unfavorable report was given in. . . . Happily, when it was read, and a decision was about to be given against him, some one pleaded hard that his probation should be extended, and so he had several months' additional trial granted. I sailed in the same boat, and was also sent back to Ongar as a naughty boy. . . . At last we had so improved that both were fully accepted. Livingstone went to London to pursue his medical studies, and I went to Cheshunt College. A day or two after reaching college, I sent to Livingstone, asking him to purchase a second-hand carpet for my room. He was quite scandalized at such an exhibition of effeminacy, and positively refused to gratify my wish. . . . In the spring of 1840 I met Livingstone at London in Exeter Hall, when Prince Albert delivered his maiden speech in England. I remember how nearly he was brought to silence when the speech, which he had lodged on the brim of his hat, fell into it, as deafening cheers made it vibrate. A day or two after, we heard Binney deliver his masterly missionary sermon, ' Christ seeing of the travail of his soul and being satisfied.' "

The meeting at Exeter Hall was held to inaugurate the Niger Expedition. It was on this occasion that Samuel Wilberforce became known as a great platform orator.[2] It must have been pleasant to Livingstone in after-years to

---

[1] On one occasion, in conversation with his former pastor, the Rev. John Moir, Livingstone spoke of Mr. Isaac Taylor, who had shown him much kindness, and often invited him to dine in his house. He said that though Mr. Taylor was connected with the Independents, he was attached to the principles of the Church of England. Mr. Taylor used to lay very great stress on acquaintance with the writings of the Fathers as necessary for meeting the claims of the Tractarians, and did not think that that study was sufficiently encouraged by the Nonconformists. Any one who has been in Mr. Taylor's study at Stanford Rivers, and who remembers the top-heavy row of patristic folios that crowned his collection of books, and the glance of pride he cast on them as he asked his visitor whether many men in his Church were well read in the Fathers, will be at no loss to verify this reminiscence. Certainly Livingstone had no such qualification, and undoubtedly he never missed it.

[2] *Life of Bishop Wilberforce,* vol. i. p. 160.

recall the circumstance when he became a friend and correspondent of the Bishop of Oxford.

Notwithstanding the dear postage of the time, Livingstone wrote regularly to his friends, but few of his letters have survived. One of the few, dated 5th May, 1839, is addressed to his sister, and in it he says that there had been some intention of sending him abroad at once, but that he was very desirous of getting more education. The letter contains very little news, but is full of the most devout aspirations for himself and exhortations to his sister. Alluding to the remark of a friend that they should seek to be " uncommon Christians, that is, eminently holy and devoted servants of the Most High," he urges:

" Let us seek—and with the conviction that we cannot do without it —that all selfishness be extirpated, pride banished, unbelief driven from the mind, every idol dethroned, and everything hostile to holiness and opposed to the divine will crucified; that ' holiness to the Lord ' may be engraven on the heart, and evermore characterize our whole conduct. This is what we ought to strive after; this is the way to be happy ; this is what our Saviour loves—entire surrender of the heart. May He enable us by his Spirit to persevere till we attain it! All comes from Him, the disposition to ask as well as the blessing itself.

" I hope you improve the talents committed to you whenever there is an opportunity. You have a class with whom you have some influence. It requires prudence in the way of managing it; seek wisdom from above to direct you; *persevere*—don't be content with once or twice recommending the Saviour to them—again and again, in as kind a manner as possible, familiarly, individually, and privately, exhibit to them the fountain of happiness and joy, never forgetting to implore divine energy to accompany your endeavors, and you need not fear that your labor will be unfruitful. If you have the willing mind, that is accepted; nothing is accepted if that be wanting. God desires that. He can do all the rest. After all, He is the sole agent, for the ' willing mind ' comes alone from Him. This is comforting, for when we think of the feebleness and littleness of all we do, we might despair of having our services accepted, were we not assured that it is not these God looks to, except in so far as they are indications of the state of the heart."

Dr. Livingstone's sisters have a distinct recollection that the field to which the Directors intended to send him was the West Indies, and that he remonstrated on the ground that he had spent two years in medical study, but in the West Indies, where there were regular practitioners, his medical knowledge would be of little or no avail. He pleaded with the Directors, therefore, that he might be allowed to complete his medical studies, and it was then that Africa was provisionally fixed on as his destination. It appears, however, that he had not quite abandoned the thought of China. Mr. Moir, his former pastor, writes that being in London in May, 1839, he called at the Mission House to make inquiries about him. He asked whether the Directors did not intend to send him to the East Indies, where the field was so large and the demand so urgent, but he was told that though they esteemed him highly, they did not think that his gifts fitted him for India, and that Africa would be a more suitable field.

On returning to London, Livingstone devoted himself with special ardor to medical and scientific study. The church with which he was connected was that of the late Rev. Dr. Bennett, in Falcon Square. This led to his becoming intimate with Dr. Bennett's son, now the well-known J. Risdon Bennett, M.D., LL.D., F.R.S., and President of the Royal College of Physicians, London. The friendship continued during the whole of Dr. Livingstone's life. From some recollections with which Dr. Bennett has kindly furnished us we take the following:

"My acquaintance with David Livingstone was through the London Missionary Society, when, having offered himself to that Society, he came to London to carry on those medical and other studies which he had commenced in Glasgow. From the first, I became deeply interested in his character, and ever after maintained a close friendship with him. I entertained toward him a sincere affection, and had the highest admiration of his endowments, both of mind and heart, and of his pure and noble devotion of all his powers to the highest purposes of life. One could not fail to be impressed with his simple, loving, Christian spirit, and the combined modest, unassuming, and self-reliant character of the man.

"He placed himself under my guidance in reference to his medical studies, and I was struck with the amount of knowledge that he had already acquired of those subjects which constitute the foundation of medical science. He had, however, little or no acquaintance with the practical departments of medicine, and had had no opportunities of studying the nature and aspects of disease. Of these deficiencies he was quite aware, and felt the importance of acquiring as much practical knowledge as possible during his stay in London. I was at that time physician to the Aldersgate Street Dispensary, and was lecturing at the Charing Cross Hospital on the practice of medicine, and thus was able to obtain for him free admission to hospital practice as well as attendance on my lectures and my practice at the dispensary. I think that I also obtained for him admission to the opthalmic hospital in Moorfields. With these sources of information open to him, he obtained a considerable acquaintance with the more ordinary forms of disease, both surgical and medical, and an amount of scientific and practical knowledge that could not fail to be of the greatest advantage to him in the distant regions to which he was going, away from all the resources of civilization. His letters to me, and indeed all the records of his eventful life, demonstrate how great to him was the value of the medical knowledge with which he entered on missionary life. There is abundant evidence that on various occasions his own life was preserved through his courageous and sagacious application of his scientific knowledge to his own needs; and the benefits which he conferred on the natives to whose welfare he devoted himself, and the wonderful influence which he exercised over them, were in no small degree due to the humane and skilled assistance which he was able to render as a healer of bodily disease. The account which he gave me of his perilous encounter with the lion, and the means he adopted for the repair of the serious injuries which he received, excited the astonishment and admiration of all the medical friends to whom I related it, as evincing an amount of courage, sagacity, skill, and endurance that have scarcely been surpassed in the annals of heroism."

Another distinguished man of science with whom Livingstone became acquainted in London, and on whom he made an impression similar to that made on Dr Bennett, was Professor Owen. Part of the little time at his disposal was devoted to studying the series of comparative anatomy in the Hunterian Museum, under Professor Owen's charge Mr. Owen was interested to find that the Lanarkshire student was born in the same neighborhood as Hunter,[1] but still more interested in the youth himself and his great love of natural history. On taking leave, Livingstone promised to bear his instructor in mind if any curiosity fell in his way. Years passed, and as no communication reached him, Mr. Owen was disposed to class the promise with too many others made in the like circumstances. But on his first return to this country Livingstone presented himself, bearing the tusk of an elephant with a spiral curve. He had found it in the heart of Africa, and it was not easy of transport. " You may recall," said Professor Owen, at the Farewell Festival in 1858, "the difficulties of the progress of the weary sick traveler on the bullock's back. Every pound weight was of moment; but Livingstone said, ' Owen shall have this tusk,' and he placed it in my hands in London." Professor Owen recorded this as a proof of Livingstone's inflexible adherence to his word. With equal justice we may quote it as a proof of his undying gratitude to any one that had shown him kindness.

On all his fellow-students and acquaintances the simplicity, frankness, and kindliness of Livingstone's character made a deep impression. Mr. J. S. Cook, now of London, who spent three months with him at Ongar, writes: " He was so kind and gentle in word and deed to all about him that all loved him. He had always words of sympathy at command, and was ready to perform acts of sympathy for those who were suffering." The Rev. G. D. Watt, a brother Scotchman, who went as a missionary to India, has a vivid remembrance of Livingstone's mode of discussion; he showed great simplicity of view, along with a certain roughness or bluntness of manner; great kindliness, and yet great persistence in holding to his own ideas. But none of his friends seem to have had any foresight of

---

[1] Not in the same *parish*, as stated afterward by Professor Owen. Hunter was born in East Kilbride, and Livingstone in Blantyre. The error is repeated in notices of Livingstone in some other quarters.

5

the eminence he was destined to attain. The Directors of
the Society did not even rank him among their ablest
men. It is interesting to contrast the opinion entertained
of him then with that expressed by Sir Bartle Frere, after
much personal intercourse, many years afterward. "Of
his intellectual force and energy," wrote Sir Bartle, "he
has given such proof as few men could afford. Any five
years of his life might in any other occupation have established a character and raised for him a fortune such as
none but the most energetic of our race can realize."[1]

But his early friends were not so much at fault.
Livingstone was somewhat slow of maturing. If we may
say so, his intellect hung fire up to this very time, and it
was only during his last year in England that he came to
his intellectual manhood, and showed his real power. His
very handwriting shows the change; from being cramped
and feeble it suddenly becomes clear, firm, and upright, very
neat, but quite the hand of a vigorous, independent man.

Livingstone's prospects of getting to China had been
damaged by the Opium War; while it continued, no new
appointments could be made, even had the Directors
wished to send him there. It was in these circumstances
that he came into contact with his countryman, Mr. (now
Dr.) Moffat, who was then in England, creating much
interest in his South African mission. The idea of his
going to Africa became a settled thing, and was soon
carried into effect.

"I had occasion" (Dr. Moffat has informed us) "to call for some one
at Mrs. Sewell's, a boarding-house for young missionaries in Aldersgate
street, where Livingstone lived. I observed soon that this young man
was interested in my story, that he would sometimes come quietly and
ask me a question or two, and that he was always desirous to know
where I was to speak in public, and attended on these occasions. By
and by he asked me whether I thought he would do for Africa. I said
I believed he would, if he would not go to an old station, but would
advance to unoccupied ground, specifying the vast plain to the north,
where I had sometimes seen, in the morning sun, the smoke of a thou-

---

[1] *Good Words,* 1874, p. 285.

and villages, where no missionary had ever been. At last Livingstone said: 'What is the use of my waiting for the end of this abominable opium war? I will go at once to Africa.' The Directors concurred, and Africa became his sphere."

It is no wonder that all his life Livingstone had a very strong faith in Providence, for at every turn of his career up to this point, some unlooked-for circumstance had come in to give a new direction to his history. First, his reading Dick's *Philosophy of a Future State*, which led him to Christ, but did not lead him away from science; then his falling in with Gutzlaff's *Appeal*, which induced him to become a medical missionary; the Opium War, which closed China against him; the friendly word of the Director who procured for him another trial; Mr. Moffat's visit, which deepened his interest in Africa; and finally, the issue of a dangerous illness that attacked him in London—all indicated the unseen hand that was preparing him for his great work.

The meeting of Livingstone with Moffat is far too important an event to be passed over without remark. Both directly and indirectly Mr. Moffat's influence on his young brother, afterward to become his son-in-law, was remarkable. In after-life they had a thorough appreciation of each other. No family on the face of the globe could have been so helpful to Livingstone in connection with the great work to which he gave himself. If the old Roman fashion of surnames still prevailed, there is no household of which all the members would have been better entitled to put AFRICANUS after their name. The interests of the great continent were dear to them all. In 1872, when one of the Search Expeditions for Livingstone was fitted out, a grandson of Dr. Moffat, another Robert Moffat, was among those who set out in the hope of relieving him; cut off at the very beginning, in the flower of his youth, he left his bones to moulder in African soil.

The illness to which we have alluded was an attack of

congestion of the liver, with an affection of the lungs. It seemed likely to prove fatal, and the only chance of recovery appeared to be a visit to his home, and return to his native air. In accompanying him to the steamer, Mr. Moore found him so weak that he could scarcely walk on board. He parted from him in tears, fearing that he had but a few days to live. But the voyage and the visit had a wonderful effect, and very soon Livingstone was in his usual health. The parting with his father and mother, as they afterward told Mr. Moore, was very affecting. It happened, however, that they met once more. It was felt that the possession of a medical diploma would be of service, and Livingstone returned to Scotland in November, 1840, and passed at Glasgow as Licentiate of the Faculty of Physicians and Surgeons. It was on this occasion he found it so inconvenient to have opinions of his own and the knack of sticking to them. It seemed as if he was going to be rejected for obstinately maintaining his views in regard to the stethoscope; but he pulled through. A single night was all that he could spend with his family, and they had so much to speak of that David proposed they should sit up all night. This, however, his mother would not hear of. "I remember my father and him," writes his sister, "talking over the prospects of Christian missions. They agreed that the time would come when rich men and great men would think it an honor to support whole stations of missionaries, instead of spending their money on hounds and horses. On the morning of 17th November we got up at five o'clock. My mother made coffee. David read the 121st and 135th Psalms, and prayed. My father and he walked to Glasgow to catch the Liverpool steamer." On the Broomielaw, father and son looked for the last time on earth on each other's faces. The old man walked back slowly to Blantyre, with a lonely heart no doubt, yet praising God. David's face was now set in earnest toward the Dark Continent.

CHAPTER III.

FIRST TWO YEARS IN AFRICA.

A.D. 1841–1843.

His ordination—Voyage out—At Rio de Janeiro—At the Cape—He proceeds to Kuruman—Letters—Journey of 700 miles to Bechuana country—Selection of site for new station—Second excursion to Bechuana country—Letter to his sister—Influence with chiefs—Bubi—Construction of a water-dam—Sekomi—Woman seized by a lion—The Bakaa—Sebehwe—Letter to Dr. Risdon Bennett—Detention at Kuruman—He visits Sebehwe's village—Bakhatlas—Sechéle, chief of Bakwains—Livingstone translates hymns—Travels 400 miles on oxback—Returns to Kuruman—Is authorized to form new station—Receives contributions for native missionary—Letters to Directors on their Mission policy—He goes to new station—Fellow-travelers—Purchase of site—Letter to Dr. Bennett—Desiccation of South Africa—Death of a servant, Sehamy—Letter to his parents.

ON the 20th November, 1840, Livingstone was ordained a missionary in Albion Street Chapel, along with the Rev. William Ross, the service being conducted by the Rev. J. J. Freeman and the Rev. R. Cecil. On the 8th of December he embarked on board the ship "George," under Captain Donaldson, and proceeded to the Cape, and thence to Algoa Bay. On the way the ship had to put in at Rio de Janeiro, and he had a glance at Brazil, with which he was greatly charmed. It was the only glimpse he ever got of any part of the great continent of America. Writing to the Rev. G. D. Watt, with whom he had become intimate in London, and who was preparing to go as a missionary to India, he says:

"It is certainly the finest place I ever saw; everything delighted me except man. . . . We lived in the home of an American Episcopal Methodist minister—the only Protestant missionary in Brazil. . . . Tracts and Bibles are circulated, and some effects might be expected, were a most injurious influence not exerted by European visitors.

These alike disgrace themselves and the religion they profess by drunk-
enness. All other vices are common in Rio. When will the rays of
Divine light dispel the darkness in this beautiful empire? The climate
is delightful. I wonder if disabled Indian missionaries could not make
themselves useful there."

During the voyage his chief friend was the captain of
the ship. "He was very obliging to me," says Livingstone, "and gave me all the information respecting the
use of the quadrant in his power, frequently sitting up
till twelve o'clock at night for the purpose of taking lunar
observations with me." Thus another qualification was
acquired for his very peculiar life-work. Sundays were
not times of refreshing, at least not beyond his closet.
"The captain rigged out the church on Sundays, and we
had service; but I being a poor preacher, and the chaplain
addressing them all as Christians already, no moral influence was exerted, and even had there been on Sabbath, it
would have been neutralized by the week-day conduct.
In fact, no good was done." Neither at Rio, nor on board
ship, nor anywhere, could good be done without the element of personal character. This was Livingstone's strong
conviction to the end of his life.

In his first letter to the Directors of the London Missionary Society he tells them that he had spent most of
his time at sea in the study of theology, and that he was
deeply grieved to say that he knew of no spiritual good
having been done in the case of any one on board the
ship. His characteristic honesty thus showed itself in his
very first dispatch.

Arriving at the Cape, where the ship was detained a
month, he spent some time with Dr. Philip, then acting as
agent for the Society, with informal powers as superinten-
dent. Dr. Philip was desirous of returning home for a
time, and very anxious to find some one to take his place
as minister of the congregation of Cape Town, in his
absence. The office was offered to Livingstone, who

rejected it with no little emphasis—not for a moment
would he think of it, nor would he preach the gospel
within any other man's line. He had not been long at
the Cape when he found to his surprise and sorrow that
the missionaries were not all at one, either as to the
general policy of the mission, or in the matter of social
intercourse and confidence. The shock was a severe one;
it was not lessened by what he came to know of the spirit
and life of a few—happily only a few—of his brethren
afterward; and undoubtedly it had an influence on his
future life. It showed him that there were missionaries
whose profession was not supported by a life of consistent
well-doing, although it did not shake his confidence in the
character and the work of missionaries on the whole. He
saw that in the mission there was what might be called a
colonial side and a native side; some sympathizing with
the colonists and some with the natives. He had no difficulty in making up his mind between them; he drew
instinctively to the party that were for protecting the
natives against the unrighteous encroachments of the
settlers.

On leaving the ship at Algoa Bay, he proceeded by land
to Kuruman or Lattakoo, in the Bechuana country, the
most northerly station of the Society in South Africa, and
the usual residence of Mr. Moffat, who was still absent in
England. In this his first African journey the germ of
the future traveler was apparent. "Crossing the Orange
River," he says, "I got my vehicle aground, and my oxen
got out of order, some with their heads where their tails
should be, and others with their heads twisted round in
the yoke so far that they appeared bent on committing
suicide, or overturning the wagon. . . . I like travel-
ling very much indeed. There is so much freedom connected with our African manners. We pitch our tent,
make our fire, etc., wherever we choose, walk, ride, or
shoot at abundance of all sorts of game as our inclination

leads us; but there is a great drawback: we can't study or
read when we please. I feel this very much. I have
made but very little progress in the language (can speak
a little Dutch), but I long for the time when I shall give
my undivided attention to it, and then be furnished with
the means of making known the truth of the gospel."
While at the Cape, Livingstone had heard something of
a fresh-water lake ('Ngami) which all the missionaries
were eager to see. If only they would give him a month
or two to learn the colloquial language, he said they might
spare themselves the pains of being "the first in at the
death." It is interesting to remark further that, in this
first journey, science had begun to receive its share of
attention. He is already bent on making a collection for
the use of Professor Owen,[1] and is enthusiastic in describing some agatized trees and other curiosities which he met
with.

Writing to his parents from Port Elizabeth, 19th May,
1841, he gives his first impressions of Africa. He had
been at a station called Hankey:

"The scenery was very fine. The white sand in some places near
the beach drifted up in large wreaths exactly like snow. One might
imagine himself in Scotland were there not a hot sun overhead. The
woods present an aspect of strangeness, for everywhere the eye meets
the foreign-looking tree from which the bitter aloes is extracted, popping
up its head among the mimosa bushes and stunted acacias. Beautiful
humming-birds fly about in great numbers, sucking the nectar from the
flowers, which are in great abundance and very beautiful. I was much
pleased with my visit to Hankey. . . . The state of the people
presents so many features of interest, that one may talk about it and
convey some idea of what the Gospel has done. The full extent of the
benefit received can, however, be understood only by those who witness
it in contrast with other places that have not been so highly favored.
My expectations have been far exceeded. Everything I witnessed surpassed my hopes, and if this one station is a fair sample of the whole,
the statements of the missionaries with regard to their success are far
within the mark. The Hottentots of Hankey appear to be in a state

---

[1] This collection never reached its destination.

similar to that of our forefathers in the days immediately preceding the times of the Covenanters. They have a prayer-meeting every morning at four o'clock, *and well attended.* They began it during a visitation of measles among them, and liked it so much that they still continue."

He goes on to say that as the natives had no clocks or watches, mistakes sometimes occurred about ringing the bell for this meeting, and sometimes the people found themselves assembled at twelve or one o'clock instead of four. The welcome to the missionaries (their own missionary was returning from the Cape with Livingstone) was wonderful. Muskets were fired at their approach, then big guns; and then men, women, and children rushed at the top of their speed to shake hands and welcome them, The missionary had lost a little boy, and out of respect each of the people had something black on his head. Both public worship and family worship were very interesting, the singing of hymns being very beautiful. The bearing of these Christianized Hottentots was in complete contrast to that of a Dutch family whom he visited as a medical man one Sunday. There was no Sunday; the man's wife and daughters were dancing before the house, while a black played the fiddle.

His instructions from the Directors were to go to Kuruman, remain there till Mr. Moffat should return from England, and turn his attention to the formation of a new station farther north, awaiting more specific instructions. He arrived at Kuruman on the 31st July, 1841, but no instuctions had come from the Directors; his sphere of work was quite undetermined, and he began to entertain the idea of going to Abyssinia. There could be no doubt that a Christian missionary was needed there, for the country had none; but if he should go, he felt that probably he would never return. In writing of this to his friend Watt, he used words almost prophetic: " Whatever way my life may be spent so as but to promote the glory

of our gracious God, I feel anxious to do it. . . . *My
life may be spent as profitably as a pioneer as in any other
way.*"

In his next letter to the London Missionary Society,
dated Kuruman, 23d September, 1841, he gives his impressions of the field, and unfolds an idea which took hold of
him at the very beginning, and never lost its grip. It
was, that there was not population enough about the
South to justify a concentration of missionary labor there,
and that the policy of the Society ought to be one of
expansion, moving out far and wide wherever there was
an opening, and making the utmost possible use of native
agency, in order to cultivate so wide a field. In England
ne had thought that Kuruman might be made a great
missionary institute, whence the beams of divine truth
might diverge in every direction, through native agents
supplied from among the converts; but since he came to
the spot he had been obliged to abandon that notion; not
that the Kuruman mission had not been successful, or that
the attendance at public worship was small, but simply
because the population was meagre, and seemed more likely
to become smaller than larger. The field from which
native agents might be drawn was thus too small. Farther
north there was a denser population. It was therefore his
purpose, along with a brother missionary, to make an early
journey to the interior, and bury himself among the
natives, to learn their language, and slip into their modes
of thinking and feeling. He purposed to take with him
two of the best qualified native Christians of Kuruman, to
plant them as teachers in some promising locality; and in
case any difficulty should arise about their maintenance,
he offered, with characteristic generosity, to defray the cost
of one of them from his own resources.

Accordingly, in company with a brother missionary from
Kuruman, a journey of seven hundred miles was performed
before the end of the year, leading chiefly to two results:

in the first place, a strong confirmation of his views on the subject of native agency; and in the second place, the selection of a station, two hundred and fifty miles north of Kuruman, as the most suitable for missionary operations. Seven hundred miles traveled over *more Africano* seemed to indicate a vast territory; but on looking at it on the map, it was a mere speck on the continent of heathenism. How was that continent ever to be evangelized? He could think of no method except an extensive method of native agency. And the natives, when qualified, were admirably qualified. Their warm, affectionate manner of dealing with their fellow-men, their ability to present the truth to their minds freed from the strangeness of which foreigners could not divest it, and the eminent success of those employed by the brethren of Griqua Town, were greatly in their favor. Two natives had likewise been employed recently by the Kuruman Mission, and these had been highly efficient and successful. If the Directors would allow him to employ more of these, conversions would increase in a compound ratio, and regions not yet explored by Europeans would soon be supplied with the bread of life.

In regard to the spot selected for a mission, there were many considerations in its favor. In the immediate neighborhood of Kuruman the chiefs hated the gospel, because it deprived them of their supernumerary wives. In the region farther north, this feeling had not yet established itself; on the contrary, there was an impression favorable to Europeans, and a desire for their alliance. These Bechuana tribes had suffered much from the marauding invasions of their neighbors; and recently, the most terrible marauder of the country, Mosilikatse, after being driven westward by the Dutch Boers, had taken up his abode on the banks of a central lake, and resumed his raids, which were keeping the whole country in alarm. The more peaceful tribes had heard of the value of the white man, and of the weapons by which a mere handful of whites had repulsed hordes of marauders. They were therefore disposed to welcome the stranger, although this state of feeling could not be relied on as sure to continue, for Griqua hunters and individuals from tribes hostile to the gospel were moving northward, and not only circulating rumors unfavorable to missionaries, but by their wicked lives introducing diseases previously unknown. If these regions, therefore, were to be taken possession of by the gospel, no time was to be lost. For himself, Livingstone had no hesitation in going to reside in the midst of these savages, hundreds of miles away from civilization, not merely for a visit, but, if necessary, for the whole of his life.

In writing to his sisters after this journey (8th December, 1841), he gives a graphic account of the country, and some interesting notices of the people:

"Janet, I suppose, will feel anxious to know what our dinner was. We boiled a piece of the flesh of a rhinoceros which was toughness itself, the night before. The meat was our supper, and porridge made of Indian corn-meal and gravy of the meat made a very good dinner next day. When about 150 miles from home we came to a large village. The chief had sore eyes; I doctored them, and he fed us pretty well with milk and beans, and sent a fine buck after me as a present. When we had got about ten or twelve miles on the way, a little girl about eleven or twelve years of age came up and sat down under my wagon, having run away for the purpose of coming with us to Kuruman. She had lived with a sister whom she had lately lost by death. Another family took possession of her for the purpose of selling her as soon as she was old enough for a wife. But not liking this, she determined to run away from them and come to some friends near Kuruman. With this intention she came, and thought of walking all the way behind my wagon. I was pleased with the determination of the little creature, and gave her some food. But before we had remained long there, I heard her sobbing violently, as if her heart would break. On looking round, I observed the cause. A man with a gun had been sent after her, and he had just arrived. I did not know well what to do now, but I was not in perplexity long, for Pomare, a native convert who accompanied us, started up and defended her cause. He being the son of a chief, and possessed of some little authority, managed the matter nicely. She had been loaded with beads to render her more attractive, and fetch a higher price. These she stripped off and gave to the man, and desired him to go away. I afterward took measures for hiding her, and though fifty men had come for her, they would not have got her."

The story reads like an allegory or a prophecy. In the person of the little maid, oppressed and enslaved Africa comes to the good Doctor for protection; instinctively she knows she may trust him; his heart opens at once, his ingenuity contrives a way of protection and deliverance, and he will never give her up. It is a little picture of Livingstone's life.

In fulfillment of a promise made to the natives in the interior that he would return to them, Livingstone set out on a second tour into the interior of the Bechuana country on 10th February, 1842. His objects were, first, to acquire the native language more perfectly, and second, by suspending his medical practice, which had become inconveniently large at Kuruman, to give his undivided attention to the subject of native agents. He took with him two native members of the Kuruman church, and two other natives for the management of the wagon.

The first person that specially engaged his interest in this journey was a chief of the name of Bubi, whose people were Bakwains. With him he stationed one of the native agents as a teacher, the chief himself collecting the children and supplying them with food. The honesty of the people was shown in their leaving untouched all the contents of his wagon, though crowds of them visited it. Livingstone was already acquiring a powerful influence, both with chiefs and people, the result of his considerate and conciliatory treatment of both. He had already observed the failure of some of his brethren to influence them, and his sagacity had discerned the cause. His success in inducing Bubi's people to dig a canal was contrasted in a characteristic passage of a private letter, with the experience of others ·

6

"The doctor and the rainmaker among these people are one and the same person. As I did not like to be behind my professional brethren, I declared I could make rain too, not, however, by enchantments like them, but by leading out their river for irrigation. The idea pleased mightily, and to work we went instanter. Even the chief's own doctor is at it, and works like a good fellow, laughing heartily at the cunning of the 'foreigner' who can make rain so. We have only one spade, and this is without a handle; and yet by means of sticks sharpened to a point we have performed all the digging of a pretty long canal. The earth was lifted out in 'gowpens' and carried to the huge dam we have built in karosses (skin cloaks), tortoise-shells, or wooden bowls. We intended nothing of the ornamental in it, but when we came to a huge stone, we were forced to search for a way round it. The consequence is, it has assumed a beautifully serpentine appearance. This is, I believe, the first instance in which Bechuanas have been got to work without wages. It was with the utmost difficulty the earlier missionaries got them to do anything. The missionaries solicited their permission to do what they did, and this was the very way to make them show off their airs, for they are so disobliging; if they perceive any one in the least dependent upon them, they immediately begin to tyrannize. A more mean and selfish vice certainly does not exist in the world. I am trying a different plan with them. I make my presence with any of them a favor, and when they show any impudence, I threaten to leave them, and if they don't amend, I put my threat into execution. By a bold, free course among them I have had not the least difficulty in managing the most fierce. They are in one sense fierce, and in another the greatest cowards in the world. A kick would, I am persuaded, quell the courage of the bravest of them. Add to this the report which many of them verily believe, that I am a great wizard, and you will understand how I can with ease visit any of them. Those who do not love, fear me, and so truly in their eyes am I possessed of supernatural power, some have not hesitated to affirm I am capable of even raising the dead! The people of a village visited by a French brother actually believed it. Their belief of my powers, I suppose, accounts, too, for the fact that I have not missed a single article either from the house or wagon since I came among them, and this, although all my things lay scattered about the room, while crammed with patients."

It was unfortunate that the teacher whom Livingstone stationed with Bubi's people was seized with a violent fever, so that he was obliged to bring him away. As for Bubi himself, he was afterward burned to death by an explosion of gunpowder, which one of his sorcerers was trying, by means of burnt roots, to *un*-bewitch.

In advancing, Livingstone had occasion to pass through a part of the great Kalahari desert, and here he met with Sekomi, a chief of the Bamangwato, from whom also he received a most friendly reception. The ignorance of this tribe he found to be exceedingly great:

" Their conceptions of the Deity are of the most vague and contradictory nature, and the name of God conveys no more to their understanding than the idea of superiority. Hence they do not hesitate to apply the name to their chiefs. I was every day shocked by being addressed by that title, and though it as often furnished me with a text from which to tell them of the only true God and Jesus Christ, whom he has sent, yet it deeply pained me, and I never felt so fully convinced of the lamentable detoriation of our species. It is indeed a mournful truth that man has become like the beasts that perish."

The place was greatly infested by lions, and during Livingstone's visit an awful occurrence took place that made a great impression on him:

" A woman was actually devoured in her garden during my visit, and that so near the town that I had frequently walked past it. It was most affecting to hear the cries of the orphan children of this woman. During the whole day after her death the surrounding rocks and valleys rang and re-echoed with their bitter cries. I frequently thought as I listened to the loud sobs, painfully indicative of the sorrows of those who have no hope, that if some of our churches could have heard their sad wailings, it would have awakened the firm resolution to do more for the heathen than they have done."

Poor Sekomi advanced a new theory of regeneration which Livingstone was unable to work out:

" On one occasion Sekomi, having sat by me in the hut for some time in deep thought, at length addressing me by a pompous title said, 'I wish you would change my heart. Give me medicine to change it, for it is proud, proud and angry, angry always.' I lifted up the Testament and was about to tell him of the only way in which the heart can be changed, but he interrupted me by saying, 'Nay, I wish to have it changed by medicine, to drink and have it changed at once, for it is always very proud and very uneasy, and continually angry with some one.' He then rose and went away."

A third tribe visited at this time was the Bakaa, and here, too, Livingstone was able to put in force his wonderful powers of management. Shortly before, the Bakaa had murdered a trader and his company. When Livingstone appeared their consciences smote them, and, with the exception of the chief and two attendants, the whole of the people fled from his presence. Nothing could allay their terror, till, a dish of porridge having been prepared, they saw Livingstone partake of it along with themselves without distrust. When they saw him lie down and fall asleep they were quite at their ease. Thereafter he began to speak to them:

" I had more than ordinary pleasure in telling these murderers of the precious blood which cleanseth from all sin. I bless God that He has conferred on one so worthless the distinguished privilege and honor of being the first messenger of mercy that ever trod these regions. Its being also the first occasion on which I had ventured to address a number of Bechuanas in their own tongue without reading it, renders it to myself one of peculiar interest. I felt more freedom than I had anticipated, but I have an immense amount of labor still before me, ere I can call myself a master of Sichuana. This journey discloses to me that when I have acquired the Batlapi, there is another and perhaps more arduous task to be accomplished in the other dialects, but by the Divine assistance I hope I shall be enabled to conquer. When I left the Bakaa, the chief sent his son with a number of his people to see me safe part of the way to the Makalaka."

On his way home, in passing through Bubi's country, he was visited by sixteen of the people of Sebehwe, a chief who had successfully withstood Mosilikatse, but whose cowardly neighbors, under the influence of jealousy, had banded together to deprive him of what they had not had the courage to defend. Consequently he had been driven into the sandy desert, and his object in sending to Livingstone was to solicit his advice and protection, as he wished to come out, in order that his people might grow corn, etc. Sebehwe, like many of the other people of the country, had the notion that if he got a single white man to live with him, he would be quite secure. It was no wonder that Livingstone early acquired the strong conviction that if missions could only be scattered over Africa, their immediate effect in promoting the tranquillity of the continent could hardly be over-estimated.

We have given these details somewhat fully, because they show that before he had been a year in the country Livingstone had learned how to rule the Africans. From the very first, his genial address, simple and fearless manner, and transparent kindliness formed a spell which rarely failed. He had great faith in the power of humor. He was never afraid of a man who had a hearty laugh. By a playful way of dealing with the people, he made them feel at ease with him, and afterward he could be solemn enough when the occasion required. His medical knowledge helped him greatly; but for permanent influence all would have been in vain if he had not uniformly observed the rules of justice, good feeling, and good manners. Often he would say that the true road to influence was patient continuance in well-doing. It is remarkable that, from the very first, he should have seen the charm of that method which he employed so successfully to the end.

In the course of this journey, Livingstone was within ten days of Lake 'Ngami, the lake of which he had heard at the Cape, and which he actually discovered in 1849; and he might have discovered it now, had discovery alone been his object. Part of his journey was performed on foot, in consequence of the draught oxen having become sick:

"Some of my companions," he says in his first book, "who had recently joined us, and did not know that I understood a little of their speech, were overheard by me discussing my appearance and powers: 'He is not strong, he is quite slim, and only appears stout because he puts himself in those bags (trousers); he will soon knock up.' This caused my Highland blood to rise, and made me despise the fatigue of keeping them all at the top of their speed for days together, and until I heard them expressing proper opinions of my pedestrian powers."

We have seen how full Livingstone's heart was of the missionary spirit; how intent he was on making friends of the natives, and 'how he could already preach in one dialect, and was learning another. But the activity of his mind enabled him to give attention at the same time to other matters. He was already pondering the structure of the great African Continent, and carefully investigating the process of desiccation that had been going on for a long time, and had left much uncomfortable evidence of its activity in many parts. In the desert, he informs his friend Watt that no fewer than thirty-two edible roots and forty-three fruits grew without cultivation. He had the rare faculty of directing his mind at the full stretch of its power to one great object, and yet, apparently without effort, giving minute and most careful attention to many other matters,—all bearing, however, on the same great end.

A very interesting letter to Dr. Risdon Bennett, dated Kuruman, 18th December, 1841, gives an account of his first year's work from the medical and scientific point of view. First, he gives an amusing picture of the Bechuana chiefs, and then some details of his medical practice:

The people are all under the feudal system of government, the chieftainship is hereditary, and although the chief is usually the greatest ass, and the most insignificant of the tribe in appearance, the people pay a deference to him which is truly astonishing. . . . I feel the benefit often of your instructions, and of those I got through your kindness. Here I have an immense practice. I have patients now under treatment who have walked 130 miles for my advice; and when these go home, others will come for the same purpose. This is the country for a medical man if he wants a large practice, but he must leave fees out of the ques-tion! The Bechuanas have a great deal more disease than I expected to find among a savage nation; but little else can be expected, for they are nearly naked, and endure the scorching heat of the day and the chills of the night in that condition. Add to this that they are abso-lutely omnivorous. Indigestion, rheumatism, opthalmia are the pre-vailing diseases. . . . Many very bad cases were brought to me, and sometimes, when traveling, my wagon was quite besieged by their blind and halt and lame. What a mighty effect would be produced if one of the seventy disciples were among them to heal them all by a word! The Bechuanas resort to the Bushmen and the poor people that live in the desert for doctors. The fact of my dealing in that line a little is so strange, and now my fame has spread far and wide. But if one of Christ's apostles were here, I should think he would be very soon known all over the continent to Abyssinia. The great deal of work I have had to do in attending to the sick has proved beneficial to me, for they make me speak the language perpetually, and if I were inclined to be lazy in learning it, they would prevent me indulging the propensity. And they are excellent patients, too, besides. There is no wincing; everything prescribed is done *instanter.* Their only failing is that they become tired of a long course. But in any operation, even the women sit unmoved. I have been quite astonished again and again at their calmness. In cutting out a tumor, an inch in diameter, they sit and talk as if they felt nothing. ' A man like me never cries,' they say, 'they are children that cry.' And it is a fact that the men never cry. But when the Spirit of God works on their minds they cry most piteously. Sometimes in church they endeavor to screen themselves from the eyes of the preacher by hiding under the forms or covering their heads with their karosses as a remedy against their convictions. And when they find that won't do, they rush out of the church and run with all their might, crying as if the hand of death were behind them. One would think, when they got away, there they would remain; but no, there they are in their places at the very next meeting. It is not to be wondered at that they should exhibit agitations of body when the mind is affected, as they are quite unaccustomed to restrain their feelings. But that the hardened beings should be moved mentally at all is wonderful indeed. If you saw them in their savage state you would feel the force of this more. . . . *N.B.*—I have got for Professor Owen specimens of the incubated ostrich in abundance, and am waiting for an opportunity to transmit the box to the college. I tried to keep for you some of the fine birds of the interior, but the weather was so horribly hot they were putrid in a few hours.

When he returned to Kuruman in June, 1842, he found that no instructions had as yet come from the Directors as to his permanent quarters. He was preparing for another journey when news arrived that, contrary to his advice, Sebehwe had left the desert where he was encamped, had been treacherously attacked by the chief Mahura, and that many of his people, including women and children, had been savagely murdered. What aggravated the case was that several native Christians from Kuruman had been at the time with Sebehwe, and that these were accused of having acted treacherously by him. But now no native would expose himself to the expected rage of Sebehwe, so that for want of attendants Livingstone could not go to him. He was obliged to remain for some months about Kuruman, itinerating to the neighboring tribes, and taking part in the routine work of the station: that is to say preaching, printing, building a chapel at an out-station, prescribing for the sick, and many things else that would have been intolerable, he said, to a man of " clerical dignity."

He was able to give his father a very encouraging report of the mission work (July 13, 1842): "The work of God goes on here notwithstanding all our infirmities. Souls are gathered in continually, and sometimes from among those you would never have expected to see turning to the Lord. Twenty-four were added to the Church last month, and there are several inquirers. At Motito, a French station about thirty-three miles northeast of this, there has been an awakening, and I hope much good will result. I have good news, too, from Rio de Janeiro. The Bibles that have been distributed are beginning to cause a stir."

The state of the country continued so disturbed that it was not till February, 1843, that he was able to set out for the village where Sebehwe had taken up his residence with the remains of his tribe. This visit he undertook at great personal risk. Though looking at first very ill-pleased, Sebehwe treated him in a short time in a most friendly way, and on the Sunday after his arrival, sent a herald to proclaim that on that day nothing should be done but pray to God and listen to the words of the foreigner. He himself listened with great attention while Livingstone told him of Jesus and the resurrection, and the missionary was often interrupted by the questions of the chief. Here, then, was another chief pacified, and brought under the preaching of the gospel.

Livingstone then passed on to the country of the Ba-khatla, where he had purposed to erect his mission-station. The country was fertile, and the people industrious, and among other industries was an iron manufactory, to which as a bachelor he got admission, whereas married men were wont to be excluded, through fear that they would bewitch the iron! When he asked the chief if he would like him to come and be his missionary, he held up his hands and said, "Oh, I shall dance if you do; I shall collect all my people to hoe for you a garden, and you will get more sweet reed and corn than myself." The cautious Directors at home, however, had sent no instructions as to Living-stone's station, and he could only say to the chief that he would tell them of his desire for a missionary.

At a distance of five days' journey beyond the Bakhatla was situated the village of Sechéle, chief of the Bakwains, afterward one of Livingstone's greatest friends. Sechéle had been enraged at him for not visiting him the year before, and threatened him with mischief. It happened that his only child was ill when the missionary arrived, and also the child of one of his principal men. Living-stone's treatment of both was successful, and Sechéle had not an angry word. Some of his questions struck the heart of the missionary:

" 'Since it is true that all who die unforgiven are lost forever, why did your nation not come to tell us of it before now? My ancestors are all gone, and none of them knew anything of what you tell me. How is this?' I thought immediately," says Livingstone, "of the guilt of the Church, but did not confess. I told him multitudes in our own country were like himself, so much in love with their sins. My ances-tors had spent a great deal of time in trying to persuade them, and yet after all many of them by refusing were lost. We now wish to tell all the world about a Saviour, and if men did not believe, the guilt would be entirely theirs. Sechéle has been driven from another part of his country from that in which he was located last year. and so has Bubi, so that the prospects I had of benefiting them by native teachers are for the present darkened."

Among other things that Livingstone found time for in these wanderings among strange people, was translating hymns into the Sichuana language. Writing to his father (Bakwain Country, 21st March, 1843), he says:

"Janet may be pleased to learn that I am become a poet, or rather a poetaster, in Sichuana. Half a dozen of my hymns were lately printed in a collection of the French brethren. One of them is a translation of "There is a fountain filled with blood;' another, 'Jesus shall reign where'er the sun;' others are on 'The earth being filled with the glory of the Lord,' 'Self-dedication,' 'Invitation to Sinners,' 'The soul that loves God finds him everywhere.' Janet may try to make English ones on these latter subjects if she can, and Agnes will doubtless set them to music on the same condition. I do not boast of having done this, but only mention it to let you know that I am getting a little better fitted for the great work of a missionary, that your hearts may be drawn out to more prayer for the success of the gospel proclaimed by my feeble lips."

Livingstone was bent on advancing in the direction of the country of the Matebele and their chief Mosilikatse, but the dread of that terrible warrior prevented him from getting Bakwains to accompany him, and being thus unable to rig out a wagon, he was obliged to travel on oxback. In a letter to Dr. Risdon Bennett (30th June, 1843), he gives a lively description of this mode of traveling: "It is rough traveling, as you can conceive. The skin is so loose there is no getting one's great-coat, which has to serve both as saddle and blanket, to stick on; and then the long horns in front, with which he can give one a punch in the abdomen if he likes, make us sit as bolt upright as dragoons. In this manner I traveled more than 400 miles." Visits to some of the villages of the Bakalahari gave him much pleasure. He was listened to with great attention, and while sitting by their fires and listening to their traditionary tales, he intermingled the story of the Cross with their conversation, and it was by far the happiest portion of his journey. The people were a poor, degraded, enslaved race, who hunted for other tribes to procure them skins; they were far from wells, and had their gardens far from their houses, in order to have their produce safe from the chiefs who visited them. Coming on to his old friends the Bakaa, he found them out of humor with him, accusing him of having given poison to a native who had been seized with fever on occasion of his former visit. Consequently he could get little or nothing to eat, and had to content himself, as he wrote to his friends, with the sumptuous feasts of his imagination. With his usual habit of discovering good in all his troubles, however, he found cause for thankfulness at their stinginess, for in coming down a steep pass, absorbed with the questions which the people were putting to him, he forgot where he was, lost his footing, and, striking his hand between a rock and his Bible which he was carrying, he suffered a compound fracture of his finger. His involuntary low diet saved him from taking fever, and the finger was healing favorably, when a sudden visit in the middle of the night from a lion, that threw them all into consternation, made him, without thinking, discharge his revolver at the visitor, and the recoil hurt him more than the shot did the lion. It rebroke his finger, and the second fracture was worse than the first. "The Bakwains," he says, "who were most attentive to my wants during the whole journey of more than 400 miles, tried to comfort me when they saw the blood again flowing, by saying, 'You have hurt yourself, but you have redeemed us: henceforth we will only swear by you.' Poor creatures," he writes to Dr. Bennett, "I wished they had felt gratitude for the blood that was shed for their precious souls."

Returning to Kuruman from this journey, in June, 1843, Livingstone was delighted to find at length a letter from the Directors of the Society authorizing the formation of a settlement in the regions beyond. He found another letter that greatly cheered him, from a Mrs. M'Robert, the wife of an Independent minister at Cambuslang (near Blantyre), who had collected and now sent him £12 for a native agent, and was willing, on the part of some young friends, to send presents of clothing for the converts. In acknowledging this letter, Livingstone poured out his very heart, so full was he of gratitude and delight. He entreated the givers to consider Mebalwe as their own agent, and to concentrate their prayers upon him, for prayer, he thought, was always more efficacious when it could be said, "One thing have I desired of the Lord." As to the present of clothing, he simply entreated his friends to send nothing of the kind; such things demoralized the recipients, and bred endless jealousies. If he were allowed to charge something for the clothes, he would be pleased to have them, but on no other terms.

Writing to the Secretary of the Society, Rev. A. Tidman (24th June, 1843), and referring to the past success of the Mission in the nearer localities, he says: "If you could realize this fact as fully as those on the spot can, you would be able to enter into the feelings of irrepressible delight with which I hail the decision of the Directors that we go forward to the dark interior. May the Lord enable me to consecrate my whole being to the glorious work!"

In this communication to the Directors Livingstone modestly, but frankly and firmly, gives them his mind on some points touched on in their letter to him. In regard to his favorite measure—native agency—he is glad that a friend has remitted money for the employment of one agent, and that others have promised the means of employing other two. On another subject he had a communication to make to them which evidently cost him no ordinary effort. In his more private letters to his friends, from an early period after entering Africa, he had expressed himself very freely, almost contemptuously, on the distribution of the laborers. There was far too much cluster-ing about the Cape Colony, and the district immediately beyond it, and a woeful slowness to strike out with the fearless chivalry that became missionaries of the Cross, and take possession of the vast continent beyond. All his letters reveal the chafing of his spirit with this confine-ment of evangelistic energy in the face of so vast a field—this huddling together of laborers in sparsely peopled dis-tricts, instead of sending them forth over the whole of Africa, India, and China, to preach the gospel to every creature. He felt deeply that both the Church at home, and many of the missionaries on the spot, had a poor conception of missionary duty, out of which came little faith, little effort, little expectation, with a miserable ten-dency to exaggerate their own evils and grievances, and fall into paltry squabbles which would not have been possible if they had been fired with the ambition to win the world for Christ.

But what it was a positive relief for him to whisper in the ear of an intimate friend, it demanded the courage of a hero to proclaim to the Directors of a great Society. It was like impugning their whole policy and arraigning their wisdom. But Livingstone could not say one thing in private and another in public. Frankly and fearlessly he proclaimed his views:

" The conviction to which I refer is that a much larger share of the benevolence of the Church and of missionary exertion is directed into this country than the amount of population, as compared with other countries, and the success attending those efforts, seem to call for. This conviction has been forced upon me, both by a personal inspection, more extensive than that which has fallen to the lot of any other, either missionary or trader, and by the sentiments of other missionaries who have investigated the subject according to their opportunities. In reference to the population, I may mention that I was led in England to believe that the population of the interior was dense, and now since I have come to this country I have conversed with many, both of our Society and of the French, and none of them would reckon up the number of 30,000 Bechuanas."

7

He then proceeds to details in a most characteristic way, giving the number of huts in every village, and being careful in every case, as his argument proceeded on there being a small population, rather to overstate than understate the number:

"In view of these facts and the confirmation of them I have received from both French and English brethren, computing the population much below what I have stated, I confess I feel grieved to hear of the arrival of new missionaries. Nor am I the only one who deplores their appointment to this country. Again and again have I been pained at heart to hear the question put, Where will these new brethren find fields of labor in this country? Because I know that in India or China there are fields large enough for all their energies. I am very far from undervaluing the success which has attended the labors of missionaries in this land. No! I gratefully acknowledge the wonders God hath wrought, and I feel that the salvation of one soul is of more value than all the effort that has been expended; but we are to seek the field where there is a possibility that most souls will be converted, and it is this consideration which makes me earnestly call the attention of the Directors to the subject of statistics. If these were actually returned—and there would be very little difficulty in doing so—it might, perhaps, be found that there is not a country better supplied with missionaries in the world, and that in proportion to the number of agents compared to the amount of population, the success may be inferior to most other countries where efforts have been made."

Finding that a brother missionary was willing to accompany him to the station he had fixed on among the Bakhatlas, and enable him to set to work with the necessary arrangements, Livingstone set out with him in the beginning of August, 1843, and arrived at his destination after a fortnight's journey. Writing to his family, "in sight of the hills of Bakhatla," August 21st, 1843, he says: "We are in company with a party of three hunters: one of them from the West Indies, and two from India—Mr. Pringle from Tinnevelly, and Captain Steel of the Coldstream Guards, aide-de-camp to the Governor of Madras. . . . The Captain is the politest of the whole, well versed in the classics, and possessed of much general knowledge." Captain Steele, now General Sir Thomas Steele, proved one of Livingstone's best and most constant friends. In one respect the society of gentlemen who came to hunt would not have been sought by Livingstone, their aims and pursuits being so different from his; but he got on with them wonderfully. In some instances these strangers were thoroughly sympathetic, but not in all. When they were not sympathetic on religion, he had a strong conviction that his first duty as a servant of Christ was to commend his religion by his life and spirit—by integrity, civility, kindness, and constant readiness to deny himself in obliging others; having thus secured their esteem and confidence, he would take such quiet opportunities as presented themselves to get near their consciences on his Master's behalf. He took care that there should be no moving about on the day of rest, and that the outward demeanor of all should be befitting a Christian company. For himself, while he abhorred the indiscriminate slaughter of animals for mere slaughter's sake, he thought well of the chase as a means of developing courage, promptness of action in time of danger, protracted endurance of hunger and thirst, determination in the pursuit of an object, and other qualities befitting brave and powerful men. The respect and affection with which he inspired the gentlemen who were thus associated with him was very remarkable. Doubtless, with his quick apprehension, he learned a good deal from their society of the ways and feelings of a class with whom hitherto he had hardly ever been in contact. The large resources with which they were furnished, in contrast to his own, excited no feeling of envy, nor even a desire to possess their ample means, unless he could have used them to extend missionary operations; and the gentlemen themselves would sometimes remark that the missionaries were more comfortable than they. Though they might at times spend thousands of pounds where Livingstone did not spend as many pence, and would be provided with horses, servants, tents, and stores, enough to secure comfort under almost any conditions, they had not that key to the native heart and that power to command the willing services of native attendants which belonged so remarkably to the missionary. " When we arrive at a spot where we intend to spend the night," writes Livingstone to his family, " all hands immediately unyoke the oxen. Then one or two of the company collect wood; one of us strikes up a fire, another gets out the water-bucket and fills the kettle; a piece of meat is thrown on the fire, and if we have biscuits, we are at our coffee in less than half an hour after arriving. Our friends, perhaps, sit or stand shivering at their fire for two or three hours before they get their things ready, and are glad occasionally of a cup of coffee from us."

The first act of the missionaries on arriving at their destination was to have an interview with the chief, and ask whether he desired a missionary. Having an eye to the beads, guns, and other things, of which white men seemed always to have an ample store, the chief and his men gave them a cordial welcome, and Livingstone next proceeded to make a purchase of land. This, like Abraham with the sons of Heth, he insisted should be done in legal form, and for this purpose he drew up a written contract to which, after it was fully explained to them, both parties attached their signatures or marks. They then proceeded to the erection of a hut fifty feet by eighteen, not getting much help from the Bakhatlas, who devolved such labors on the women, but being greatly helped by the native deacon, Mebalwe. All this Livingstone and his companion had done on their own responsibility, and in the hope that the Directors would approve of it. But if they did not, he told them that he was at their disposal " to go anywhere—*provided it be* FORWARD."

The progress of medical and scientific work during this

period is noted in a letter to Dr. Risdon Bennett, dated 30th June, 1843. In addition to full details of the missionary work, this letter enters largely into the state of disease in South Africa, and records some interesting cases, medical and surgical. Still more interesting, perhaps, is the evidence it affords of the place in Livingstone's attention which began to be occupied by three great subjects of which we shall hear much anon—Fever, Tsetse, and "the Lake." Fever he considered the greatest barrier to the evangelization of Africa. Tsetse, an insect like a common fly, destroyed horses and oxen, so that many traders lost literally every ox in their team. As for the Lake, it lay somewhat beyond the outskirts of his new district, and was reported terrible for fever. He heard that Mr. Moffat intended to visit it, but he was somewhat alarmed lest his friend should suffer. It was not Moffat, but Livingstone, however, that first braved the risks of that fever swamp.

A subject of special scientific interest to the missionary during this period was—the desiccation of Africa. On this topic he addressed a long letter to Dr. Buckland in 1843, of which, considerably to his regret, no public notice appears to have been taken, and perhaps the letter never reached him. The substance of this paper may, however, be gathered from a communication subsequently made to the Royal Geographical Society[1] after his first impression had been confirmed by enlarged observation and discovery. Around, and north of Kuruman, he had found many indications of a much larger supply of water in a former age. He ascribed the desiccation to the gradual elevation of the western part of the country. He found traces of a very large ancient river which flowed nearly north and south to a large lake, including the bed of the present Orange River; in fact, he believed that the whole country south of Lake 'Ngami presented in ancient times very much the same appearance as the basin north of that lake does now,

---

[1] See *Journal*, vol. xxvii. p. 356.

and that the southern lake disappeared when a fissure was
made in the ridge through which the Orange River now
proceeds to the sea. He could even indicate the spot
where the river and the lake met, for some hills there had
caused an eddy in which was found a mound of calcareous
tufa and travertine, full of fossil bones. These fossils he
was most eager to examine, in order to determine the time
of the change; but on his first visit he had no time, and
when he returned, he was suddenly called away to visit a
missionary's child, a hundred miles off. It happened that
he was never in the same locality again, and had therefore
no opportunity to complete his investigation.

Dr. Livingstone's mind had that wonderful power which
belongs to some men of the highest gifts, of passing with
the utmost rapidity, not only from subject to subject, but
from one mood or key to another entirely different. In a
letter to his family, written about this time, we have a
characteristic instance. On one side of the sheet is a pro-
longed outburst of tender Christian love and lamentation
over a young attendant who had died of fever suddenly;
on the other side, he gives a map of the Bakhatla country
with its rivers and mountains, and is quite at home in the
geographical details, crowning his description with some
sentimental and half-ludicrous lines of poetry. No reasonable man will fancy that in the wailings of his heart there
was any levity or want of sincerity. What we are about
to copy merits careful consideration: first, as evincing the
depth and tenderness of his love for these black savages;
next, as showing that it was pre-eminently Christian love,
intensified by his vivid view of the eternal world, and
belief in Christ as the only Saviour; and, lastly, as reveal-
ing the secret of the affection which these poor fellows
bore to him in return. The intensity of the scrutiny
which he directs on his heart, and the severity of the
judgment which he seems to pass on himself, as if he had
not done all he might have done for the spiritual good of

this young man, show with what intense conscientiousness
he tried to discharge his missionary duty:

"Poor Sehamy, where art thou now? Where lodges thy soul
to-night? Didst thou think of what I told thee as thou turnedst from
side to side in distress? I could now do anything for thee. I could
weep for thy soul. But now nothing can be done. Thy fate is fixed.
Oh, am I guilty of the blood of thy soul, my poor dear Sehamy? If
so, how shall I look upon thee in the judgment? But I told thee of a
Saviour; didst thou think of Him, and did He lead thee through the
dark valley? Did He comfort as He only can? Help me, O Lord
Jesus, to be faithful to every one. Remember me, and let me not be
guilty of the blood of souls. This poor young man was the leader of
the party. He governed the others, and most attentive he was to me.
He anticipated my every want. He kept the water-calabash at his
head at night, and if I awoke, he was ready to give me a draught
immediately. When the meat was boiled he secured the best portion
for me, the best place for sleeping, the best of everything. Oh, where
is he now? He became ill after leaving a certain tribe, and believed he
had been poisoned. Another of the party and he ate of a certain dish
given them by a woman whom they had displeased, and having met this
man yesterday he said, 'Sehamy is gone to heaven, and I am almost
dead by the poison given us by that woman.' I don't believe they took
any poison, but they do, and their imaginations are dreadfully excited
when they entertain that belief."

The same letter intimates that in case his family should
have arranged to emigrate to America, as he had formerly
advised them to do, he had sent home a bill of which £10
was to aid the emigration, and £10 to be spent on clothes
for himself. In regard to the latter sum, he now wished
them to add it to the other, so that his help might be more
substantial; and for himself he would make his old clothes
serve for another year. The emigration scheme, which he
thought would have added to the comfort of his parents
and sisters, was not, however, carried into effect. The
advice to his family to emigrate proceeded from deep convictions. In a subsequent letter (4th December, 1850) he
writes: "If I could only be with you for a week, you would
soon be pushing on in the world. The world is ours. Our

Father made it to be inhabited, and many shall run to and fro, and knowledge shall be increased. *It will be increased more by emigration than by missionaries.*" He held it to be God's wish that the unoccupied parts of the earth should be possessed, and he believed in Christian colonization as a great means of spreading the gospel. We shall see afterward that to plant English and Scotch colonies in Africa became one of his master ideas and favorite schemes.

CHAPTER IV.

FIRST TWO STATIONS—MABOTSA AND CHONUANE.

A.D. 1843–1847.

Description of Mabotsa—A favorite hymn—Genera. reading—Mabotsa infested with lions—Livingstone's encounter—The native deacon who saved him—His Sunday-school—Marriage to Mary Moffat—Work at Mabotsa—Proposed institution for training native agents—Letter to his mother—Trouble at Mabotsa—Noble sacrifice of Livingstone—Goes to Sechéle and the Bakwains—New station at Chonuane—Interest shown by Sechéle—Journeys eastward—The Boers and the Transvaal—Their occupation of the country, and treat-ment of the natives—Work among the Bakwains—Livingstone's desire to move on—Theological conflict at home—His view of it—His scientific labors and miscellaneous employments.

DESCRIBING what was to be his new home to his friend Watt from Kuruman, 27th September, 1843, Livingstone says: " The Bakhatla have cheerfully offered to remove to a more favorable position than they at present occupy. We have fixed upon a most delightful valley, which we hope to make the centre of our sphere of operations in the interior. It is situated in what poetical gents like you would call almost an amphitheatre of mountains. The mountain range immediately in the rear of the spot where we have fixed our residence is called Mabotsa, or a mar-riage-feast. May the Lord lift upon us the light of his countenance, so that by our feeble instrumentality many may thence be admitted to the marriage-feast of the Lamb. The people are as raw as may well be imagined; they have not the least desire but for the things of the earth, and it must be a long time ere we can gain their attention to the things which are above."

Something led him in his letter to Mr. Watt to talk

of the old monks, and the spots they selected for their
establishments. He goes on to write lovingly of what was
good in some of the old fathers of the mediæval Church,
despite the strong feeling of many to the contrary; indicating thus early the working of that catholic spirit which
was constantly expanding in later years, which could
separate the good in any man from all its evil surroundings, and think of it thankfully and admiringly. In the
following extract we get a glimpse of a range of reading
much wider than most would probably have supposed
likely:

"Who can read the sermons of St. Bernard, the meditations of St.
Augustine, etc., without saying, whatever other faults they had: They
thirsted, and now they are filled. That hymn of St. Bernard, on the
name of Christ, although in what might be termed dog-Latin, pleases
me so; it rings in my ears as I wander across the wide, wide wilderness,
and makes me wish I was more like them—

"Jesu, dulcis memoria,   Jesu, spes pœnitentibus,
  Dans cordi vera gaudia;   Quam pius es petentibus!
  Sed super mel et omnia,   Quam bonus es quærentibus!
  Ejus dulcis præsentia.   Sed quid invenientibus!

 Nil canitur suavius,   Jesu, dulcedo cordium,
 Nil auditur jucundius,   Fons, rivus, lumen mentium,
 Nil cogitatur dulcius,   Excedens omne gaudium,
 Quam Jesus Dei filius.   Et omne desiderium."

Livingstone was in the habit of fastening inside the
boards of his journals, or writing on the fly-leaf, verses
that interested him specially. In one of these volumes
this hymn is copied at full length. In another we find a
very yellow newspaper clipping of the "Song of the Shirt."
In the same volume a clipping containing "The Bridge of
Sighs," beginning

"One more unfortunate,
Weary of breath,
Rashly importunate,
Gone to her death."

In another we have Coleridge's lines:

> " He prayeth well who loveth well
> Both man and bird and beast.
> He prayeth best who loveth best
> All things both great and small;
> For the dear God who loveth us,
> He made and loveth all."

In another, hardly legible on the marble paper, we find:

> "So runs my dream: but what am I?
> An infant crying in the night:
> An infant crying for the light:
> And with no language but a cry."

All Livingstone's personal friends testify that, considering the state of banishment in which he lived, his acquaintance with English literature was quite remarkable. When a controversy arose in America as to the genuineness of his letters to the *New York Herald*, the familiarity of the writer with the poems of Whittier was made an argument against him. But Livingstone knew a great part of the poetry of Longfellow, Whittier, and others by heart.

There was one drawback to the new locality: it was infested with lions. All the world knows the story of the encounter at Mabotsa, which was so near ending Livingstone's career, when the lion seized him by the shoulder, tore his flesh, and crushed his bone. Nothing in all Livingstone's history took more hold of the popular imagination, or was more frequently inquired about when he came home.[1] By a kind of miracle his life was saved, but the encounter left him lame for life of the arm which the lion

---

[1] He did not speak of it spontaneously, and sometimes he gave unexpected answers to questions put to him about it. To one person who asked very earnestly what were his thoughts when the lion was above him, he answered, "I was thinking what part of me he would eat first"—a grotesque thought, which some persons considered strange in so good a man, but which was quite in accordance with human experience in similar circumstances.

crunched.[1] But the world generally does not know that
Mebalwe, the native who was with him, and who saved his
life by diverting the lion when his paw was on his head,
was the teacher whom Mrs. M'Robert's twelve pounds had
enabled him to employ. Little did the good woman think
that this offering would indirectly be the means of preserving the life of Livingstone for the wonderful work
of the next thirty years! When, on being attacked by
Mebalwe, the lion left Livingstone, and sprang upon him,
he bit his thigh, then dashed toward another man, and
caught him by the shoulder, when in a moment, the previous shots taking effect, he fell down dead. Sir Bartle
Frere, in his obituary notice of Livingstone read to the
Royal Geographical Society, remarked: "For thirty years
afterward all his labors and adventures, entailing such
exertion and fatigue, were undertaken with a limb so
maimed that it was painful for him to raise a fowling-
piece, or in fact to place the left arm in any position above
the level of the shoulder."

In his *Missionary Travels* Livingstone says that but for
the importunities of his friends, he meant to have kept
this story in store to tell his children in his dotage. How
little he made of it at the time will be seen from the following allusion to it in a letter to his father, dated 27th
July, 1844. After telling how the attacks of the lions
drew the people of Mabotsa away from the irrigating
operations he was engaged in, he says:

"At last, one of the lions destroyed nine sheep in broad daylight on a
hill just opposite our house. All the people immediately ran over to it,
and, contrary to my custom, I imprudently went with them, in order to
see how they acted, and encourage them to destroy him. They surrounded him several times, but he managed to break through the circle.
I then got tired. In coming home I had to come near to the end of the
hill. They were then close upon the lion and had wounded him. He

---

[1] The false joint in the crushed arm was the mark by which the body of
Livingstone was identified when brought home by his followers in 1874.

rusnea out from the bushes which concealed him from view, and bit me on the arm so as to break the bone. It is now nearly well, however, feeling weak only from having been confined in one position so long; and I ought to praise Him who delivered me from so great a danger. I hope I shall never forget his mercy. You need not be sorry for me, for long before this reaches you it will be quite as strong as ever it was. Gratitude is the only feeling we ought to have in remembering the event. Do not mention this to any one. I do not like to be talked about."

In a letter to the Directors, Livingstone briefly adverts to Mebalwe's service on this occasion, but makes it a peg on which to hang some strong remarks on that favorite topic—the employment of native agency:

" Our native assistant Mebalwe has been of considerable value to the Mission. In endeavoring to save my life he nearly lost his own, for he was caught and wounded severely, but both before being laid aside, and since his recovery, he has shown great willingness to be useful. The cheerful manner in which he engages with us in manual labor in the station, and his affectionate addresses to his countrymen, are truly grati. fying. Mr. E. took him to some of the neighboring villages lately, in order to introduce him to his work; and I intend to depart to-morrow for the same purpose to several of the villages situated northeast of this. In all there may be a dozen considerable villages situated at con· venient distances around us, and we each purpose to visit them statedly. It would be an *immense advantage* to the cause had we many such agents."

Another proof that his pleas for native agency, published in some of the Missionary Magazines, were telling at home, was the receipt of a contribution for the employment of a native helper, amounting to £15, from a Sunday-school in Southampton. Touched with this proof of youthful sympathy, Livingstone addressed a long letter of thanks to the Southampton teachers and children, desiring to deepen their interest in the work. and concluding with an account of his Sunday-school:

"I yesterday commenced school for the first time at Mabotsa, and the poor little naked things came with fear and trembling. A native teacher assisted, and the chief collected as many of them as he could, or I believe we should have had none. The reason is, the women make us the hobgoblins of their children, telling them ' these white men bite

8

children, feed them with dead men's brains, and all manner of nonsense. We are just commencing our mission among them."

A new star now appeared in Livingstone's horizon,
destined to give a brighter complexion to his life, and a
new illustration to the name Mabotsa. Till this year
(1844) he had steadily repudiated all thoughts of marriage,
thinking it better to be independent. Nor indeed had he
met with any one to induce him to change his mind.
Writing in the end of 1843 to his friend Watt, he had said:
"There's no outlet for me when I begin to think of getting
married but that of sending home an advertisement to the
Evangelical Magazine, and if I get very old, it must be for
some decent sort of widow. In the meantime I am too
busy to think of any thing of the kind." But soon after
the Moffats came back from England to Kuruman, their
eldest daughter Mary rapidly effected a revolution in
Livingstone's ideas of matrimony. They became engaged.
In announcing his approaching marriage to the Directors,
he makes it plain that he had carefully considered the
bearing which this step might have on his usefulness as a
missionary. No doubt if he had foreseen the very extraordinary work to which he was afterwards to be called, he
might have come to a different conclusion. But now,
apparently, he was fixed and settled. Mabotsa would become a centre from which native missionary agents would
radiate over a large circumference. His own life-work
would resemble Mr. Moffat's. For influencing the women
and children of such a place, a Christian lady was indispensable, and who so likely to do it well as one born in
Africa, the daughter of an eminent and honored missionary, herself familiar with missionary life, and gifted with
the winning manner and the ready helping hand that were
so peculiarly adapted for this work? The case was as clear
as possible, and Livingstone was very happy.

On his way home from Kuruman, after the engagement,
he writes to her cheerily from Motito, on 1st August, 1844,

chiefly about the household they were soon to get up ; asking her to get her father to order some necessary articles, and to write to Colesberg about the marriage-license (and if he did not get it, they would license themselves!), and concluding thus:

"And now, my dearest, farewell. May God bless you! Let your affection be towards Him much more than towards me; and, kept by his mighty power and grace, I hope I shall never give you cause to regret that you have given me a part. Whatever friendship we feel towards each other, let us always look to Jesus as our common friend and guide, and may He shield you with his everlasting arms from every evil!"

Next month he writes from Mabotsa with full accounts of the progress of their house, of which he was both architect and builder:

"*Mabotsa*, 12*th September*, 1844. — I must tell you of the progress I have made in architecture. The walls are nearly finished, although the dimensions are 52 feet by 20 outside, or almost the same size as the house in which you now reside. I began with stone, but when it was breast-high, I was obliged to desist from my purpose to build it entirely of that material by an accident, which, slight as it was, put a stop to my operations in that line. A stone falling was stupidly, or rather instinctively, caught by me in its fall by the left hand, and it nearly broke my arm over again. It swelled up again, and I fevered so much I was glad of a fire, although the weather was quite warm. I expected bursting and discharge, but Baba bound it up nicely, and a few days' rest put all to rights. I then commenced my architecture, and six days have brought the walls up a little more than six feet.

"The walls will be finished long before you receive this, and I suppose the roof too, but I have still the wood of the roof to seek. It is not, however, far off; and as Mr. E. and I, with the Kurumanites, got on the roof of the school in a week, I hope this will not be more than a fortnight or three weeks. Baba has been most useful to me in making door and window frames; indeed, if he had not turned out I should not have been advanced so far as I am. Mr. E.'s finger is the cause in part of my having no aid from him, but all will come right at last. It is pretty hard work, and almost enough to drive love out of my head, but it is not situated there; it is in my heart, and won't come out unless you behave so as to quench it! . . .

"You must try and get a maid of some sort to come with you, although it is only old Moyimang; you can't go without some one, and a Makhatla can't be had for either love or money. . . .

"You must excuse soiled paper, my hands won't wash clean after dabbling mud all day. And although the above does not contain evidence of it, you are as dear to me as ever, and will be as long as our lives are spared.—I am still your most affectionate

"D. LIVINGSTON."

A few weeks later he writes:

"As I am favored with another opportunity to Kuruman, I gladly embrace it, and wish I could embrace you at the same time; but as I cannot, I must do the next best to it, and while I give you the good news that our work is making progress, and of course the time of our separation becoming beautifully less, I am happy in the hope that, by the messenger who now goes, I shall receive the good news that you are well and happy, and remembering me with some of that affection which we bear to each other. . . . All goes on pretty well here; the school is sometimes well, sometimes ill attended. I begin to like it, and I once believed I could never have any pleasure in such employment. I had a great objection to school-keeping, but I find in that as in almost every-thing else I set myself to as a matter of duty, I soon became enamored of it. A boy came three times last week, and on the third time could act as monitor to the rest through a great portion of the alphabet. He is a real Mokhatla, but I have lost sight of him again. If I get them on a little, I shall translate some of your infant-school hymns into Sichu-ana rhyme, and you may yet, if you have time, teach them the tunes to them. I, poor mortal, am as mute as a fish in regard to singing, and Mr. Englis says I have not a bit of imagination. Mebalwe teaches them the alphabet in the 'auld lang syne' tune sometimes, and I heard it sung by some youths in the gardens yesterday—a great improvement over their old see-saw tunes indeed. Sometimes we have twenty, some-times two, sometimes none at all.

"Give my love to A., and tell her to be sure to keep my lecture warm. She must not be vexed with herself, that she was not more frank to me. If she is now pleased, all is right. I have sisters, and know all of you have your failings, but I won't love you less for these. And to mother, too, give my kindest salutation. I suppose I shall get a lecture from her, too, about the largeness of the house. If there are too many win-dows, she can just let me know. I could build them all up in two days, and let the light come down the chimney, if that would please. I'll do anything for peace, except fighting for it. And now I must again, my dear, dear Mary, bid you good-bye. Accept my expressions as literally true when I say, I am your most affectionate and still confiding lover,

"D. LIVINGSTON."

In due time the marriage was solemnized, and Livingstone brought his wife to Mabotsa. Here they went vigorously to work, Mrs. Livingstone with her infant-school, and her husband with all the varied agencies, medical, educational, and pastoral, which his active spirit could bring to bear upon the people. They were a very superstitious race, and, among other things, had great faith in rain-making. Livingstone had a famous encounter with one of their rain-makers, the effect of which was that the pretender was wholly nonplused; but instead of being convinced of the absurdity of their belief, the people were rather disposed to think that the missionaries did not want them to get rain. Some of them were workers in iron, who carried their superstitious notions into that department of life, too, believing that the iron could be smelted only by the power of medicines, and that those who had not the proper medicine need not attempt the work. In the hope of breaking down these absurdities, Livingstone planned a course of popular lectures on the works of God in creation and providence, to be carried out in the following way:

" I intend to commence with the goodness of God in giving iron ore, by giving, if I can, a general knowledge of the simplicity of the substance, and endeavoring to disabuse their minds of the idea which prevents them, in general, from reaping the benefit of that mineral which abounds in their country. I intend, also, to pay more attention to the children of the few believers we have with us as a class, for whom, as baptized ones, we are bound especially to care. May the Lord enable me to fulfill my resolutions! I have now the happy prospect before me of real missionary work. All that has preceded has been preparatory."

All this time Livingstone had been cherishing his plan of a training seminary for native agents. He had written a paper and brought the matter before the missionaries, but without success. Some opposed the scheme fairly, as being premature, while some insinuated that his object was to stand well with the Directors, and get himself made

Professor. This last objection induced him to withdraw his proposal. He saw that in his mode of prosecuting the matter he had not been very knowing; it would have been better to get some of the older brethren to adopt it. He feared that his zeal had injured the cause he desired to benefit, and in writing to his friend Watt, he said that for months he felt bitter grief, and could never think of the subject without a pang.[1]

A second time he brought forward his proposal, but again without success. Was he then to be beaten? Far from it. He would change his tactics, however. He would first set himself to show what could be done by native efforts; he would travel about, wherever he found a road, and after inquiries, settle native agents far and wide. The plan had only to be tried, under God's blessing, to succeed. Here again we trace the Providence that shaped his career. Had his wishes been carried into effect, he might have spent his life training native agents, and doing undoubtedly a noble work: but he would not have traversed Africa; he would not have given its death-blow to African slavery; he would not have closed the open sore of the world, nor rolled away the great obstacle to the evangelization of the Continent.

Some glimpses of his Mabotsa life may be got from a letter to his mother (14th May, 1845). Usually his letters for home were meant for the whole family and addressed accordingly; but with a delicacy of feeling, which many will appreciate, he wrote separately to his mother after a little experience of married life:

"I often think of you, and perhaps more frequently since I got married than before. Only yesterday I said to my wife, when I thought of the nice clean bed I enjoy now, 'You put me in mind of my mother; she

---

[1] Dr. Moffat favored the scheme of a training seminary, and when he came home afterward, helped to raise a large sum of money for the purpose. He was strongly of opinion that the institution should be built at Sechéle's; but, contrary to his view, and that of Livingstone, it has been placed at Kuruman.

was always particular about our beds and linen. I had had rough times of it before.' . . .

" I cannot perceive that the attentions paid to my father-in-law at home have spoiled him. He is, of course, not the same man he formerly must have been, for he now knows the standing he has among the friends of Christ at home. But the plaudits he received have had a bad effect, and tho' not on *his* mind, yet on that of his fellow-laborers. You, perhaps, cannot understand this, but so it is. If one man is praised, others think this is more than is deserved, and that they, too ('others,' they say, while they mean themselves), ought to have a share. Perhaps you were gratified to see my letters quoted in the *Chronicle.* In some minds they produced bitter envy, and if it were in my power, I should prevent the publication of any in future. But all is in the Lord's hands ; on Him I cast my care. His testimony I receive as it stands—He careth for us. Yes, He does ; for He says it, who is every way worthy of credit. He will give what is good for me. He will see to it that all things work together for good. Do thou for me, O Lord God Almighty ! May his blessing rest on you, my dear mother. . . .

" I received the box from Mr. D. The clothes are all too wide by four inches at least. Does he think that aldermen grow in Africa ? Mr. N., too, fell into the same fault, but he will be pleased to know his boots will be worn by a much better man—Mr. Moffat. I am not an atom thicker than when you saw me. . . .

" Respecting the mission here, we can say nothing. The people have not the smallest love to the gospel of Jesus. They hate and fear it, as a revolutionary spirit is disliked by the old Tories. It appears to them as that which, if not carefully guarded against, will seduce them, and destroy their much-loved domestic institutions. No pro-slavery man in the Southern States dreads more the abolition principles than do the Bakhatla the innovations of the Word of God. Nothing but power Divine can work the mighty change."

Unhappily Mr. and Mrs. Livingstone's residence at Mabotsa was embittered by a painful collision with the missionary who had taken part in rearing the station. Livingstone was accused of acting unfairly by him, of assuming to himself more than his due, and attempts were made to discredit him, both among the missionaries and the Directors. It was a very painful ordeal, and Livingstone felt it keenly. He held the accusation to be unjust, as most people will hold it to have been who know that one of the charges against him was that he was a " nonentity"! A tone of indignation pervades his letters:—that after having borne the heat and burden of the day, he should be accused of claiming for himself the credit due to one who had done so little in comparison. But the noble spirit of Livingstone rose to the occasion. Rather than have any scandal before the heathen, he would give up his house and garden at Mabotsa, with all the toil and money they had cost him, go with his young bride to some other place, and begin anew the toil of house and school building, and gathering the people around him. His colleague was so struck with his generosity that he said had he known his intention he never would have spoken a word against him. Livingstone had spent all his money, and out of a salary of a hundred pounds it was not easy to build a house every other year. But he stuck to his resolution. Parting with his garden evidently cost him a pang, especially when he thought of the tasteless hands into which it was to fall. "I like a garden," he wrote, "but paradise will make amends for all our privations and sorrows here." Self-denial was a firmly established habit with him; and the passion of "moving on" was warm in his blood. Mabotsa did not thrive after Livingstone left it, but the brother with whom he had the difference lived to manifest a very different spirit.

In some of his journeys, Livingstone had come into close contact with the tribe of the Bakwains, which, on the murder of their chief, some time before, had been divided into two, one part under Bubi, already referred to, and the other under Sechéle, son of the murdered chief, also already introduced. Both of these chiefs had shown much regard for Livingstone, and on the death of Bubi, Sechéle and his people indicated a strong wish that a missionary should reside among them. On leaving Mabotsa, Livingstone transferred his services to this tribe. The name of the new station was Chonuane; it was situated some forty miles from Mabotsa, and in 1846 it became the centre of

Livingstone's operations among the Bakwains and their chief Sechéle.

Livingstone had been disappointed with the result of his work among the Bakhatlas. No doubt much good had been done; he had prevented several wars; but where were the conversions?[1] On leaving he found that he had made more impressions on them than he had supposed. They were most unwilling to lose him, offered to do anything in their power for his comfort, and even when his oxen were "inspanned" and he was on the point of moving, they offered to build a new house without expense to him in some other place, if only he would not leave them. In a financial point of view, the removal to Chonuane was a serious undertaking. He had to apply to the Directors at home for a building-grant—only thirty pounds, but there were not wanting objectors even to that small sum. It was only in self-vindication that he was constrained to tell of the hardships which his family had borne:—

"We endured for a long while, using a wretched infusion of native corn for coffee, but when our corn was done, we were fairly obliged to go to Kuruman for supplies. I can bear what other Europeans would consider hunger and thirst without any inconvenience, but when we arrived, to hear the old woman who had seen my wife depart about two years before, exclaiming before the door, 'Bless me! how lean she is! Has he starved her? Is there no food in the country to which she has been?' was more than I could well bear."

---

[1] When some of Livingstone's "new light" friends heard that there were so few conversions, they seem to have thought that he was too much of an old Calvinist, and wrote to him to preach that the remedy was as extensive as the disease—Christ loved *you*, and gave himself for *you*. "You may think me heretical," replied he, "but we don't need to make the extent of the atonement the main topic of our preaching. We preach to men who don't know but they are beasts, who have no idea of God as a personal agent, or of sin as evil, otherwise than as an offense against each other, which may or may not be punished by the party offended. . . . Their consciences are seared, and moral perceptions blunted. Their memories retain scarcely anything we teach them, and so low have they sunk that the plainest text in the whole Bible cannot be understood by them."

From the first, Sechéle showed an intelligent interest in Livingstone's preaching. He became a great reader especially of the Bible, and lamented very bitterly that he had got involved in heathen customs, and now did not know what to do with his wives. At one time he expressed himself quite willing to convert all his people to Christianity by the litupa, *i.e.* whips of rhinoceros hide; but when he came to understand better, he lamented that while he could make his people do anything else he liked, he could not get one of them to believe. He began family worship, and Livingstone was surprised to hear how well he conducted prayer in his own simple and beautiful style. When he was baptized, after a profession of three years, he sent away his superfluous wives in a kindly and generous way; but all their connections became active and bitter enemies of the gospel, and the conversion of Sechéle, instead of increasing the congregation, reduced it so much that sometimes the chief and his family were almost the only persons present. A bell-man of a somewhat peculiar order was once employed to collect the people for service—a tall gaunt fellow. "Up he jumped on a sort of platform, and shouted at the top of his voice, 'Knock that woman down over there. Strike her, she is putting on her pot! Do you see that one hiding herself? Give her a good blow. There she is—see, see, knock her down!' All the women ran to the place of meeting in no time, for each thought herself meant. But, though a most efficient bell-man, we did not like to employ him."

While residing at Chonuane, Livingstone performed two journeys eastward, in order to attempt the removal of certain obstacles to the establishment of at least one of his native teachers in that direction. This brought him into connection with the Dutch Boers of the Cashan mountains, otherwise called Magaliesberg. The Boers were emigrants from the Cape, who had been dissatisfied with the British rule, and especially with the emancipation of their Hottentot slaves, and had created for themselves a republic in the north (the Transvaal), in order that they might pursue, unmolested, the proper treatment of the blacks. "It is almost needless to add," says Livingstone, "that proper treatment has always contained in it the essential element of slavery, viz., compulsory unpaid labor." The Boers had effected the expulsion of Mosilikatse, a savage Zulu warrior, and in return for this service they considered themselves sole masters of the soil. While still engaged in the erection of his dwelling-house at Chonuane, Livingstone received notes from the Commandant and Council of the emigrants, requesting an explanation of his intentions, and an intimation that they had resolved to come and deprive Sechéle of his fire-arms. About the same time he received several very friendly messages and presents from Mokhatla, chief of a large section of the Bakhatla, who lived about four days eastward of his station, and had once, while Livingstone was absent, paid a visit to Chonuane, and expressed satisfaction with the idea of obtaining Paul, a native convert, as his teacher. As soon as his house was habitable, Livingstone proceeded to the eastward, to visit Mokhatla, and to confer with the Boers.

On his way to Mokhatla he was surprised at the unusual density of the population, giving him the opportunity of preaching the gospel at least once every day. The chief, Mokhatla, whose people were quiet and industrious, was eager to get a missionary, but said that an arrangement must be made with the Dutch commandant. This involved some delay.

Livingstone then returned to Chonuane, finished the erection of a school there, and setting systematic instruction fairly in operation under Paul and his son, Isaac, again went eastward, accompanied this time by Mrs. Livingstone and their infant son, Robert Moffat[1]—all the three being

---

[1] He wrote to his father that he would have called him Neil, if it had not been such an ugly name, and all the people would have called him Ra-Neeley!

in indifferent health.   Mebalwe, the catechist, was also with
them.   Taking a different route, they came on another Bakhatla tribe, whose country abounded in metallic ores, and
who, besides cultivating their fields, span cotton, smelted
iron, copper, and tin, made an alloy of tin and copper,
and manufactured ornaments.   Livingstone had constantly
an eye to the industries and commercial capabilities of the
countries he passed through.   Social reform was certainly
much needed here; for the chief, though not twenty years
of age, had already forty-eight wives and twenty children.
They heard of another tribe, said to excel all others in
manufacturing skill, and having the honorable distinction,
"they had never been known to kill any one."   This lily
among thorns they were unable to visit.   Three tribes of
Bakhalaka whom they did visit were at contiuual war.

Deriving his information from the Boers themselves,
Livingstone learned that they had taken possession of nearly
all the fountains, so that the natives lived in the country
only by sufferance.   The chiefs were compelled to furnish
the emigrants with as much free labor as they required.
This was in return for the privilege of living in the country
of the Boers!   The absence of law left the natives open to
innumerable wrongs which the better-disposed of the emigrants lamented, but could not prevent.   Livingstone found
that the forcible seizure of cattle was a common occurrence,
but another custom was even worse.   When at war, the
Dutch forced natives to assist them, and sent them before
them into battle, to encounter the battle-axes of their opponents, while the Dutch fired in safety at their enemies over
the heads of their native allies.   Of course all the disasters
of the war fell on the natives; the Dutch had only the
glory and the spoil.   Such treatment of the natives burned
into the very soul of Livingstone.   He was specially distressed at the purpose expressed to pick a quarrel with
Sechéle, for whatever the emigrants might say of other
tribes, they could not but admit that the Bechuanas had
been always an honest and peaceable people.

When Livingstone met the Dutch commandant he received favorably his proposal of a native missionary, but another obstacle arose. Near the proposed station lived a Dutch emigrant who had shown himself the inveterate enemy of missions. He had not scrupled to say that the proper way to treat any native missionary was to kill him. Livingstone was unwilling to plant Mebalwe beside so bloodthirsty a neighbor, and as he had not time to go to him, and try to bring him to a better mind, and there was plenty of work to be done at the station, they all returned to Chonuane.

" We have now," says Livingstone (March, 1847), " been a little more than a year with the Bakwains. No conversions have taken place, but real progress has been made." He adverts to the way in which the Sabbath was observed, no work being done by the natives in the gardens that day, and hunting being suspended. Their superstitious belief in rain-making had got a blow. There was a real desire for knowledge, though hindered by the prevailing famine caused by the want of rain. There was also a general impression among the people that the missionaries were their friends. But civilization apart from conversion would be but a poor recompense for their labor.

But, whatever success might attend their work among the Bakwains, Livingstone's soul was soaring beyond them:

" I am more and more convinced," he writes to the Directors, " that in order to the permanent settlement of the gospel in any part, the natives must be taught to relinquish their reliance on Europe. An onward movement ought to be made whether men will hear or whether they will forbear. I tell my Bakwains that if spared ten years, I shall move on to regions beyond them. If our missions would move onward now to those regions I have lately visited, they would in all probability prevent the natives settling into that state of determined hatred to all Europeans which I fear now characterizes most of the Caffres near the Colony. If natives are not elevated by contact with Europeans, they are sure to be deteriorated. It is with pain I have observed that all the tribes I have lately seen are undergoing the latter process. The

9

country is fine. It abounds in streams, and has many considerable rivers. The Boers hate missionaries, but by a kind and prudent course of conduct one can easily manage them. Medicines are eagerly received, and I intend to procure a supply of Dutch tracts for distribution among them. The natives who have been in subjection to Mosiiikatse place unbounded confidence in missionaries."

In his letters to friends at home, whatever topic Livingstone may touch, we see evidence of one over-mastering idea—the vastness of Africa, and the duty of beginning a new area of enterprise to reach its people. Among his friends the Scotch Congregationalists, there had been a keen controversy on some points of Calvinism. Livingstone did not like it; he was not a high Calvinist theoretically, yet he could not accept the new views, "from a secret feeling of being absolutely at the divine disposal as a sinner;" but these were theoretical questions, and with dark Africa around him, he did not see why the brethren at home should split on them. Missionary influence in South Africa was directed in a wrong channel. There were three times too many missionaries in the colony, and vast regions beyond lay untouched. He wrote to Mr. Watt: "If you meet me down in the colony before eight years are expired, you may shoot me."

Of his employments and studies he gives the following account: "I get the *Evangelical, Scottish Congregational, Eclectic, Lancet, British and Foreign Medical Review.* I can read in journeying, but little at home. Building, gardening, cobbling, doctoring, tinkering, carpentering, gunmending, farriering, wagon-mending, preaching, schooling, lecturing on physics according to my means, beside a chair in divinity to a class of three, fill up my time."

With all his other work, he was still enthusiastic in science. "I have written Professor Buckland," he says to Mr. Watt (May, 1845), "and send him specimens too, but have not received any answer. I have a great lot by me now. I don't know whether he received my letter or not.

Could you ascertain? I am trying to procure specimens of the entire geology of this region, and will try and make a sort of chart. I am taking double specimens now, so that if one part is lost, I can send another. The great difficulty is transmission. I sent a dissertation on the decrease of water in Africa. Call on Professor Owen and ask if he wants anything in the four jars I still possess, of either rhinoceros, camelopard, etc., etc. If he wants these, or anything else these jars will hold, he must send me more jars and spirits of wine."

He afterward heard of the fate of one of the boxes of specimens he had sent home—that which contained the fossils of Bootchap. It was lost on the railway after reaching England, in custody of a friend. "The thief thought the box contained bullion, no doubt. You may think of one of the faces in *Punch* as that of the scoundrel, when he found in the box a lot of 'chuckystanes.'" He had got many nocturnal-feeding animals, but the heat made it very difficult to preserve them. Many valuable seeds he had sent to Calcutta, with the nuts of the desert, but had heard nothing of them. He had lately got knowledge of a root to which the same virtues were attached as to ergot of rye. He tells his friend about the tsetse, the fever, the north wind, and other African notabilia. These and many other interesting points of information are followed up by the significant question—

"WHO WILL PENETRATE THROUGH AFRICA?"

CHAPTER V.

THIRD STATION—KOLOBENG.

A.D. 1847–1852.

Want of rain at Chonuane—Removal to Kolobeng—House-building and public works—Hopeful prospects—Letters to Mr. Watt, his sister, and Dr. Bennett —The church at Kolobeng—Pure communion—Conversion of Sechéle— Letter from his brother Charles—His history—Livingstone's relations with the Boers—He cannot get native teachers planted in the East—Resolves to explore northwards—Extracts from Journal—Scarcity of water—Wild animals and other risks—Custom-house robberies and annoyances—Visit from Secretary of London Missionary Society—Manifold employments of Livingstone—Studies in Sichuana—His reflection on this period of his life while detained at Manyuema in 1870.

THE residence of the Livingstones at Chonuane was of short continuance. The want of rain was fatal to agriculture, and about equally fatal to the mission. It was necessary to remove to a neighborhood where water could be obtained. The new locality chosen was on the banks of the river Kolobeng, about forty miles distant from Chonuane. In a letter to the Royal Geographical Society, his early and warm friend and fellow-traveler, Mr. Oswell, thus describes Kolobeng: "The town stands in naked deformity on the side of and under a ridge of red ironstone; the mission-house on a little rocky eminence over the river Kolobeng." Livingstone had pointed out to the chief that the only feasible way of watering the gardens was to select some good never-failing river, make a canal, and irrigate the adjacent lands. The wonderful influence which he had acquired was apparent from the fact that the very morning after he told them of his intention to move to the Kolobeng, the whole tribe was in motion for the "flitting." Livingstone had to set to work at his old business—building a house—the third which he had reared with his own hands. It was a mere hut—for a permanent house he had to wait a year. The natives, of course, had their huts to rear and their gardens to prepare; but, besides this, Livingstone set them to public works. For irrigating their gardens, a dam had to be dug and a water-course scooped out; sixty-five of the younger men dug the dam, and forty of the older made the water-course. The erection of the school was undertaken by the chief Sechéle: "I desire," he said, "to build a house for God, the defender of my town, and that you be at no expense for it whatever." Two hundred of his people were employed in this work.

Livingstone had hardly had time to forget his building troubles at Mabotsa and Chonuane, when he began this new enterprise. But he was in much better spirits, much more hopeful than he had been. Writing to Mr. Watt on 13th February, 1848, he says:—

"All our meetings are good compared to those we had at Mabotsa, and some of them admit of no comparison whatever. Ever since we moved, we have been incessantly engaged in manual labor. We have endeavored, as far as possible, to carry on systematic instruction at the same time, but have felt it very hard pressure on our energies. . . . Our daily labors are in the following sort of order:

"We get up as soon as we can, generally with the sun in summer, then have family worship, breakfast, and school; and as soon as these are over we begin the manual operations needed, sowing, ploughing, smithy work, and every other sort of work by turns as required. My better-half is employed all the morning in culinary or other work; and feeling pretty well tired by dinner-time, we take about two hours' rest then; but more frequently, without the respite I try to secure for myself, she goes off to hold infant-school, and this, I am happy to say, is very popular with the youngsters. She sometimes has eighty, but the average may be sixty. My manual labors are continued till about five o'clock. I then go into the town to give lessons and talk to any one who may be disposed for it. As soon as the cows are milked we have a meeting, and this is followed by a prayer-meeting in Sechéle's house, which brings me home about half-past eight, and generally tired enough, too fatigued to think of any mental exertion. I do not enumerate these duties by way of telling how much we do, but to let you know a cause of sorrow I have that so little of my time is devoted to real missionary work."

First there was a temporary house to be built, then a permanent one, and Livingstone was not exempted from the casualties of mechanics. Once he found himself dangling from a beam by his weak arm. Another time he had a fall from the roof. A third time he cut himself severely with an axe. Working on the roof in the sun, his lips got all scabbed and broken. If he mentions such things to Dr. Bennett or other friend, it is either in the way of illustrating some medical point or to explain how he had never found time to take the latitude of his station till he was stopped working by one of these accidents. At best it was weary work. "Two days ago," he writes to his sister Janet (5th July, 1848), "we entered our new house. What a mercy to be in a house again! A year in a little hut through which the wind blew our candles into glorious icicles (as a poet would say) by night, and in which crowds of flies continually settled on the eyes of our poor little brats by day, makes us value our present castle. Oh, Janet, know thou, if thou art given to building castles in the air, that that is easy work to erecting cottages on the ground." He could not quite forget that it was unfair treatment that had driven him from Mabotsa, and involved him in these labors. "I often think," he writes to Dr. Bennett, "I have forgiven, as I hope to be forgiven; but the remembrance of slander often comes boiling up, although I hate to think of it. You must remember me in your prayers, that more of the spirit of Christ may be imparted to me. All my plans of mental culture have been broken through by manual labor. I shall soon, however, be obliged to give my son and daughter a jog along the path to learning. . . . Your family increases very fast, and I fear we follow in your wake. I cannot realize the idea of your sitting with four around you, and I can scarcely believe myself to be so far advanced as to be the father of two."

Livingstone never expected the work of real Christianity to advance rapidly among the Bakwains. They were a slow people and took long to move. But it was not his desire to have a large church of nominal adherents. "Nothing," he writes, "will induce me to form an impure church. Fifty added to the church sounds fine at home, but if only five of these are genuine, what will it profit in the Great Day ? I have felt more than ever lately that the great object of our exertions ought to be conversion." There was no subject on which Livingstone had stronger feelings than on purity of communion. For two whole years he allowed no dispensation of the Lord's Supper, because he did not deem the professing Christians to be living consistently. Here was a crowning proof of his hatred of all sham and false pretense. and his intense love of solid, thorough, finished work.

Hardly were things begun to be settled at Kolobeng, when, by way of relaxation, Livingstone (January, 1848) again moved eastward. He would have gone sooner, but "a mad sort of Scotchman,"[1] having wandered past them shooting elephants, and lost all his cattle by the bite of the tsetse-fly, Livingstone had to go to his help; and moreover the dam, having burst, required to be repaired. Sechéle set out to accompany him, and intended to go with him the whole way; but some friends having come to visit his tribe, he had to return, or at least did return, leaving Livingstone four gallons of porridge, and two servants to act in his stead. "He is about the only individual," says Livingstone, "who possesses distinct, consistent views on the subject of our mission. He is bound by his wives: has a curious idea—would like to go to another country for three or four years in order to study, with the hope that probably his wives would have married others in the meantime. He would then return, and be admitted to the Lord's Supper, and teach his people the knowledge he has acquired. He seems incapable of putting them away. He feels so

---

[1] Mr. Gordon Cumming.

attached to them, and indeed we, too, feel much attached to most of them. They are our best scholars, our constant friends. We earnestly pray that they, too, may be enlightened by the Spirit of God."

The prayer regarding Sechéle was answered soon. Reviewing the year 1844 in a letter to the Directors, Livingstone says: "An event that excited more open enmity than any other was the profession of faith and subsequent reception of the chief into the church."

During the first years at Kolobeng he received a long letter from his younger brother Charles, then in the United States, requesting him to use his influence with the London Missionary Society that he might be sent as a missionary to China. In writing to the Directors about his brother, in reply to this request, Livingstone disclaimed all idea of influencing them except in so far as he might be able to tell them facts. His brother's history was very interesting. In 1839, when David Livingstone was in England, Charles became earnest about religion, influenced partly by the thought that as his brother, to whom he was most warmly attached, was going abroad, he might never see him again in this world, and therefore he would prepare to meet him in the next. A strong desire sprang up in his mind to obtain a liberal education. Not having the means to get this at home, he was advised by David to go to America, and endeavor to obtain admission to one of the colleges there where the students support themselves by manual labor. To help him in this, David sent him five pounds, which he had just received from the Society, being the whole of his quarter's allowance in London. On landing at New York, after selling his box and bed, Charles found his whole stock of cash to amount to £2, 13s. 6d. Purchasing a loaf and a piece of cheese as *viaticum*, he started for a college at Oberlin, seven hundred miles off, where Dr. Finney was President. He contrived to get to the college without having ever begged.

In the third year he entered on a theological course, with the view of becoming a missionary. He did not wish, and could never agree, as a missionary, to hold an appointment from an American Society, on account of the relation of the American Churches to slavery; therefore he applied to the London Missionary Society. David had suggested to his father that if Charles was to be a missionary, he ought to direct his attention to China. Livingstone's first missionary love had not become cold, and much though he might have wished to have his brother in Africa, he acted consistently on his old conviction that there were enough of English missionaries there, and that China had much more need.

The Directors declined to appoint Charles Livingstone without a personal visit, which he could not afford to make. This circumstance led him to accept a pastorate in New England, where he remained until 1857, when he came to this country and joined his brother in the Zambesi Expedition. Afterward he was appointed H. M. Consul at Fernando Po, but being always delicate, he succumbed to the climate of the country, and died a few months after his brother, on his way home, in October, 1873. Sir Bartle Frere, as President of the Royal Geographical Society, paid a deserved tribute to his affectionate and earnest nature, his consistent Christian life, and his valuable help to Christian missions and the African cause generally.[1]

Livingstone's relations with the Boers did not improve. He has gone so fully into this subject in his *Missionary Travels* that a very slight reference to it is all that is needed here. It was at first very difficult for him to comprehend how the most flagrant injustice and inhumanity to the black race could be combined, as he found it to be, with kindness and general respectability, and even with

---

[1] Journal of the Royal Geographical Society, 1874, p. cxxviii.

the profession of piety. He only came to comprehend this when, after more experience, he understood the demoralization which the slave-system produces. It was necessary for the Boers to possess themselves of children for servants, and believing or fancying that in some tribe an insurrection was plotting, they would fall on that tribe and bring off a number of the children. The most foul massacres were justified on the ground that they were necessary to subdue the troublesome tendencies of the people, and therefore essential to permanent peace. Livingstone felt keenly that the Boers who came to live among the Bakwains made no distinction between them and the Caffres, although the Bechuanas were noted for honesty, and never attacked eithers Boers or English. On the principle of elevating vague rumors into alarming facts, the Boers of the Cashan Mountains, having heard that Sechéle was possessed of fire-arms (the number of his muskets was five!) multiplied the number by a hundred, and threatened him with an invasion. Livingstone, who was accused of supplying these arms, went to the commandant Krieger, and prevailed upon him to defer the expedition, but refused point-blank to comply with Krieger's wish that he should act as a spy on the Bakwains. Threatening messages continued to be sent to Sechéle, ordering him to surrender himself, and to prevent English traders from passing through his country, or selling fire-arms to his people. On one occasion Livingstone was told by Mr. Potgeiter, a leading Dutchman, that he would attack any tribe that might receive a native teacher. Livingstone was so thoroughly identified with the natives that it became the desire of the colonists to get rid of him and all his belongings, and complaints were made of him to the Colonial Government as a dangerous person that ought not to be let alone.

All this made it very clear to Livingstone that his favorite plan of planting native teachers to the eastward could not be carried into effect, at least for the present. His disappointment in this was only another link in the chain of causes that gave to the latter part of his life so unlooked-for but glorious a destination. It set him to inquire whether in some other direction he might not find a sphere for planting native teachers which the jealousy of the Boers prevented in the east.

Before we set out with him on the northward journeys, to which he was led partly by the hostility of the Boers in the east, and partly by the very distressing failure of rain at Kolobeng, a few extracts may be given from a record of the period entitled "A portion of a Journal lost in the destruction of Kolobeng (September, 1853) by the Boers of Pretorius." Livingstone appears to have kept journals from an early period of his life with characteristic care and neatness; but that ruthless and most atrocious raid of the Boers, which we shall have to notice hereafter, deprived him of all them up to that date. The treatment of his books on that occasion was one of the most exasperating of his trials. Had they been burned or carried off he would have minded it less; but it was unspeakably provoking to hear of them lying about with handfuls of leaves torn out of them, or otherwise mutilated and destroyed. From the wreck of his journals the only part saved was a few pages containing notes of some occurrences in 1848–49:

" *May* 20, 1848.—Spoke to Sechéle of the evil of trusting in medicines instead of God. He felt afraid to dispute on the subject, and said he would give up all medicine if I only told him to do so. I was gratified to see symptoms of tender conscience. May God enlighten him !

" *July* 10*th*.—Entered new house on 4th curt. A great mercy. Hope it may be more a house of prayer than any we have yet inhabited.

" *Sunday, August* 6.—Sechéle remained as a spectator at the celebration of the Lord's Supper, and when we retired he asked me how he ought to act with reference to his superfluous wives, as he greatly desired to conform to the will of Christ, be baptized, and observe his ordinances. Advised him to do according to what he saw written in God's Book, but to treat them gently, for they had sinned in ignorance, and if driven away hastily might be lost eternally.

"*Sept.* 1.—Much opposition, but none manifested to us as individuals. Some, however, say it was a pity the lion did not kill me at Mabotsa. They curse the chief (Sechéle) with very bitter curses, and these come from the mouths of those whom Sechéle would formerly have destroyed for a single disrespectful word. The truth will, by the aid of the Spirit of God, ultimately prevail.

"*Oct.* 1.—Sechéle baptized; also Setefano.

"*Nov.*—Long for rains. Everything languishes during the intense heat; and successive droughts having only occurred since the Gospel came to the Bakwains, I fear the effect will be detrimental. There is abundance of rain all around us. And yet we, who have our chief at our head in attachment to the Gospel, receive not a drop. Has Satan power over the course of the winds and clouds? Feel afraid he will obtain an advantage over us, but must be resigned entirely to the Divine will.

"*Nov.* 27.—O Devil! Prince of the power of the air, art thou hindering us? Greater is He who is for us than all who can be against us. I intend to proceed with Paul to Mokhatla's. He feels much pleased with the prospect of forming a new station. May God Almighty bless the poor unworthy effort! Mebalwe's house finished. Preparing woodwork for Paul's house.

"*Dec.* 16.—Passed by invitation to Hendrick Potgeiter. Opposed to building a school. . . . Told him if he hindered the Gospel the blood of these people would be required at his hand. He became much excited at this.

"*Dec.* 17.—Met Dr. Robertson, of Swellendam. Very friendly. Boers very violently opposed. . . . Went to Pilanies. Had large attentive audiences at two villages when on the way home. Paul and I looked for a ford in a dry river. Found we had got a she black rhinoceros between us and the wagon, which was only twenty yards off. She had calved during the night—a little red beast like a dog. She charged the wagon, split a spoke and a felloe with her horn, and then left. Paul and I jumped into a rut, as the guns were in the wagon."

The black rhinoceros is one of the most dangerous of the wild beasts of Africa, and travelers stand in great awe of it. The courage of Dr. Livingstone in exposing himself to the risk of such animals on this missionary tour was none the less that he himself says not a word regarding it; but such courage was constantly shown by him. The following instances are given on the authority of Dr. Moffat as samples of what was habitual to Dr. Livingstone in the performance of his duty.

In going through a wood, a party of hunters were startled by the appearance of a black rhinoceros. The furious beast dashed at the wagon, and drove his horn into the bowels of the driver, inflicting a frightful wound. A messenger was despatched in the greatest haste for Dr. Livingstone, whose house was eight or ten miles distant. The messenger in his eagerness ran the whole way. Livingstone's friends were horror-struck at the idea of his riding through the wood at night, exposed to the rhinoceros and other deadly beasts. " No, no; you must not think of it, Livingstone; it is certain death." Livingstone believed it was a Christian duty to try to save the poor fellow's life, and he resolved to go, happen what might. Mounting his horse, he rode to the scene of the accident. The man had died, and the wagon had left, so that there was nothing for Livingstone but to return and run the risk of the forest anew, without even the hope that he might be useful in saving life.

Another time, when he and a brother missionary were on a tour a long way from home, a messenger came to tell his companion that one of his children was alarmingly ill. It was but natural for him to desire Livingstone to go back with him. The way lay over a road infested by lions. Livingstone's life would be in danger; moreover, as we have seen, he was intensely desirous to examine the fossil bones at the place. But when his friend expressed the desire for him to go, he went without hesitation. His firm belief in Providence sustained him in these as in so many other dangers.

Medical practice was certainly not made easier by what happened to some of his packages from England. Writing to his father-in-law, Mr. Moffat (18th January, 1849), he says:

" Most of our boxes which come to us from England are opened, and usually lightened of their contents. You will perhaps remember one in which Sechéle's cloak was. It contained, on leaving Glasgow, besides

10

the articles wnicn came here, a parcel of surgical instruments which I ordered, and of course paid for. One of these was a valuable cupping apparatus. The value at which the instruments were purchased for me was £4, 12s., their real value much more.

"The box which you kindly packed for us and despatched to Glasgow has, we hear, been gutted by the Custom-House thieves, and only a very few plain karosses left in it. When we see a box which has been opened we have not half the pleasure which we otherwise should in unpacking it. . . . Can you give me any information how these annoyances may be prevented? Or must we submit to it as one of the crooked things of this life, which Solomon says cannot be made straight?"

Not only in these scenes of active missionary labor, but everywhere else, Livingstone was in the habit of preaching to the natives, and conversing seriously with them on religion, his favorite topics being the love of Christ, the Fatherhood of God, the resurrection, and the last judgment. His preaching to them, in Dr. Moffat's judgment, was highly effective. It was simple, scriptural, conversational, went straight to the point, was well fitted to arrest the attention, and remarkably adapted to the capacity of the people. To his father he writes (5th July, 1848): "For a long time I felt much depressed after preaching the unsearchable riches of Christ to apparently insensible hearts; but now I like to dwell on the love of the great Mediator, for it always warms my own heart, and I know that the gospel is the power of God—the great means which He employs for the regeneration of our ruined world."

In the beginning of 1849 Livingstone made the first of a series of journeys to the north, in the hope of planting native missionaries among the people. Not to interrupt the continuous account of these journeys, we may advert here to a visit paid to him at Kolobeng, on his return from the first of them, in the end of the year, by Mr. Freeman of the London Missionary Society, who was at that time visiting the African stations. Mr. Freeman, to Livingstone's regret, was in favor of keeping up all Colonial stations, because the London Society alone paid attention to the black population. He was no⁺ much in sympathy with Livingstone.

"Mr. Freeman," he writes confidentially to Mr. Watt, "gave us no hope to expect any new field to be taken up. 'Expenditure to be reduced in Africa' was the word, when I proposed the new region beyond us, and there is nobody willing to go except Mr. Moffat and myself. Six hundred miles additional land-carriage, mosquitoes in myriads, sparrows by the million, an epidemic frequently fatal, don't look well in a picture. I am 270 miles from Kuruman; land-carriage for all that we use makes a fearful inroad into the £100 of salary, and then 600 miles beyond this makes one think unutterable things, for nobody likes to call for more salary. I think the Indian salary ought to be given to those who go into the tropics. I have a very strong desire to go and reduce the new language to writing, but I cannot perform impossibilities. I don't think it quite fair for the Churches to expect their messenger to live, as if he were the Prodigal Son, on the husks that the swine do eat, but I should be ashamed to say so to any one but yourself."

" I cannot perform impossibilities," said Livingstone; but few men could come so near doing it. His activity of mind and body at this outskirt of civilization was wonderful. A Jack-of-all-trades, he is building houses and schools, cultivating gardens, scheming in every manner of way how to get water, which in the remarkable drought of the season becomes scarcer and scarcer; as a missionary he is holding meetings every other night, preaching on Sundays, and taking such other opportunities as he can find to gain the people to Christ; as a medical man he is dealing with the more difficult cases of disease, those which baffle the native doctors; as a man of science he is taking observations, collecting specimens, thinking out geographical, geological, meteorological, and other problems bearing on the structure and condition of the continent; as a missionary statesman he is planning how the actual force might be disposed of to most advantage, and is looking round in this direction and in that, over hundreds of miles, for openings for native agents; and to promote these objects he is writing long letters to the Directors, to the *Missionary Chronicle,* to the

British Banner, to private friends, to any one likely to take an interest in his plans.

But this does not exhaust his labors. He is deeply interested in philological studies, and is writing on the Sichuana language:

"I have been hatching a grammar of the Sichuana language," he writes to Mr. Watt. "It is different in structure from any other language, except the ancient Egyptian. Most of the changes are effected by means of prefixes or affixes, the radical remaining unchanged. Attempts have been made to form grammars, but all have gone on the principle of establishing a resemblance between Sichuana, Latin, and Greek; mine is on the principle of analysing the language without reference to any others. Grammatical terms are only used when I cannot express my meaning in any other way. The analysis renders the whole language very simple, and I believe the principle elicited extends to most of the languages between this and Egypt. I wish to know whether I could get 20 or 30 copies printed for private distribution at an expense not beyond my means. It would be a mere tract, and about the size of this letter when folded, 40 or 50 pages perhaps.[1] Will you ascertain the cost, and tell me whether, in the event of my continuing hot on the subject half a year hence, you would be the corrector of the press? . . . Will you examine catalogues to find whether there is any dictionary of ancient Egyptian within my means, so that I might purchase and compare? I should not grudge two or three pounds for it. Professor Vater has written on it, but I do not know what dictionary he consulted. One Tattam has written a Coptic grammar; perhaps that has a vocabulary, and might serve my purpose. I see Tattam advertised by John Russell Smith, 4 Old Compton Street, Soho, London,—'Tattam (H.), *Lexicon Egyptiaco-Latinum e veteribus linguae Egyptiacae monumentis;* thick 8vo, bds., 10s., Oxf., 1835.' Will you purchase the above for me?"

At Mabotsa and Chonuane the Livingstones had spent but a little time; Kolobeng may be said to have been the only permanent home they ever had. During these years several of their children were born, and it was the only considerable period of their lives when both had their children about them. Looking back afterward on this period, and its manifold occupations, whilst detained in

---

[1] This gives a correct idea of the length of many of his letters.

Manyuema, in the year 1870, Dr. Livingstone wrote the following striking words:

"I often ponder over my missionary career among the Bakwains or Bakwaina, and though conscious of many imperfections, not a single pang of regret arises in the view of my conduct, except that I did not feel it to be my duty, while spending all my energy in teaching the heathen, to devote a special portion of my time to play with my children. But generally I was so much exhausted with the mental and manual labor of the day, that in the evening there was no fun left in me. I did not play with my little ones while I had them, and they soon sprung up in my absences, and left me conscious that I had none to play with."

The heart that felt this one regret in looking back to this busy time must have been true indeed to the instincts of a parent. But Livingstone's case was no exception to that mysterious law of our life in this world, by which, in so many things, we learn how to correct our errors only after the opportunity is gone. Of all the crooks in his lot, that which gave him so short an opportunity of securing the affections and moulding the character of his children seems to have been the hardest to bear. His long detention at Manyuema appears, as we shall see hereafter, to have been spent by him in learning more completely the lesson of submission to the will of God; and the hard trial of separation from his family, entailing on them what seemed irreparable loss, was among the last of his sorrows over which he was able to write the words with which he closes the account of his wife's death in the *Zambesi and its Tributaries,*—"FIAT, DOMINE, VOLUNTUS TUA!"

CHAPTER VI.

KOLOBENG *continued*—LAKE 'NGAMI.

A.D. 1849–1852.

Kolobeng failing through drought—Sebituane's country and the Lake 'Ngami —Livingstone sets out with Messrs. Oswell and Murray—Rivers Zouga and Tamanak'le—Old ideas of the interior revolutionized—Enthusiasm of Livingstone—Discovers Lake 'Ngami—Obliged to return—Prize from Royal Geographical Society—Second expedition to the lake, with wife and children— Children attacked by fever—Again obliged to return—Conviction as to healthier spot beyond—Idea of finding passage to sea either west or east— Birth and death of a child—Family visits Kuruman—Third expedition, again with family—He hopes to find a new locality—Perils of the journey—He reaches Sebituane—The chief's illness and death—Distress of Livingstone— Mr. Oswell and he go on the Linyanti—Discovery of the Upper Zambesi—No locality found for settlement—More extended journey necessary—He returns —Birth of Oswald Livingstone—Crisis in Livingstone's life—His guiding principles—New plans—The Makololo begin to practice slave-trade—New thoughts about commerce—Letters to Directors—The Bakwains—*Pros* and *cons* of his new plan—His unabated missionary zeal—He goes with his family to the Cape—His literary activity.

WHEN Sechéle turned back after going so far with Livingstone eastward, it appeared that his courage had failed him. "Will you go with me northward?" Livingstone once asked him, and it turned out that he was desirous to do so. He wished to see Sebituane, a great chief living to the north of Lake 'Ngami, who had saved his life in his infancy, and otherwise done him much service. Sebituane was a man of great ability, who had brought a vast number of tribes into subjection, and now ruled over a very extensive territory, being one of the greatest magnates of Africa. Livingstone, too, had naturally a strong desire to become acquainted with so influential a man. The fact of his living near the lake revived the project that had slumbered for years in his mind—to be the first of the missionaries who should look on its waters. At Kolobeng, too, the settlement was in such straits, owing to the excessive drought which dried up the very river, that the people would be compelled to leave it and settle elsewhere. The want of water, and consequently of food, in the gardens, obliged the men to be absent collecting locusts, so that there was hardly any one to come either to church or school. Even the observance of the Sabbath broke down. If Kolobeng should have to be abandoned, where would Livingstone go next? It was certainly worth his while to look if a suitable locality could not be found in Sebituane's territory. He had resolved that he would not stay with the Bakwains always. If the new region were not suitable for himself, he might find openings for native teachers; at all events, he would go northward and see. Just before he started, messengers came to him from Lechulatebe, chief of the people of the lake, asking him to visit his country, and giving such an account of the quantity of ivory that the cupidity of the Bakwain guides was roused, and they became quite egar to be there.

On 1st June, 1849, Livingstone accordingly set out from Kolobeng. Sechéle was not of the party, but two English hunting friends accompanied him, Mr. Oswell and Mr. Murray—Mr. Oswell generously defraying the cost of the guides. Sekomi, a neighboring chief who secretly wished the expedition to fail, lest his monopoly of the ivory should be broken up, remonstrated with them for rushing on to certain death—they must be killed by the sun and thirst, and if he did not stop them, people would blame him for the issue. "No fear," said Livingstone, "people will only blame our own stupidity."

The great Kalahari desert, of which Livingstone has given so full an account, lay between them and the lake. They passed along its northeast border, and had traversed

about half of the distance, when one day it seemed most
unexpectedly that they had got to their journey's end.
Mr. Oswell was a little in advance, and having cleared an
intervening thick belt of trees, beheld in the soft light of
the setting sun what seemed a magnificent lake twenty
miles in circumference; and at the sight threw his hat
in the air, and raised a shout which made the Bakwains
think him mad. He fancied it was 'Ngami, and, indeed,
it was a wonderful deception, caused by a large salt-pan
gleaming in the light of the sun; in fact, the old, but ever
new phenomenon of the mirage. The real 'Ngami was
yet 300 miles farther on.

Livingstone has given ample details of his progress in
the *Missionary Travels*, dwelling especially on his joy when
he reached the beautiful river Zouga, whose waters flowed
from 'Ngami. Providence frustrated an attempt to rouse
ill-feeling against him on the part of two men who had
been sent by Sekomi, apparently to help him, but who now
went before him and circulated a report that the object of
the travelers was to plunder all the tribes living on the
river and the lake. Half-way up, the principal man was
attacked by fever, and died; the natives thought it a
judgment, and seeing through Sekomi's reason for wishing
the expedition not to succeed, they by and by became quite
friendly, under Livingstone's fair and kind treatment.

A matter of great significance in his future history
occurred at the junction of the rivers Tamanak'le and
Zouga:

"I inquired," he says, "whence the Tamanak'le came. 'Oh! from
a country full of rivers,—so many, no one can tell their number, and
full of large trees.' This was the first confirmation of statements I
had heard from the Bakwains who had been with Sebituane, that the
country beyond was not the 'large sandy plateau' of the philosophers.
The prospect of a highway, capable of being traversed by boats to an
entirely unexplored and very populous region, grew from that time forward stronger and stronger in my mind; so much so, that when we
actually came to the lake, this idea occupied such a large portion of

my mental vision, that the actual discovery seemed of but little importance. I find I wrote, when the emotions caused by the magnificent prospects of the new country were first awakened in my breast, that they 'might subject me to the charge of enthusiasm, a charge which I deserved, as nothing good or great had ever been accomplished in the world without it.' "[1]

Twelve days after, the travelers came to the northeast end of Lake 'Ngami, and it was on 1st August, 1849, that this fine sheet of water was beheld for the first time by Europeans. It was of such magnitude that they could not see the farther shore, and they could only guess its size from the reports of the natives that it took three days to go round it.

Lechulatebe, the chief who had sent him the invitation, was quite a young man, and his reception by no means corresponded to what the invitation implied. He had no idea of Livingstone going on to Sebituane, who lived two hundred miles farther north, and perhaps supplying him with fire-arms which would make him a more dangerous neighbor. He therefore refused Livingstone guides to Sebituane, and sent men to prevent him from crossing the river. Livingstone was not to be baulked, and worked many hours in the river trying to make a raft out of some rotten wood,—at the imminent risk of his life, as he afterward found, for the Zouga abounds with alligators. The season was now far advanced, and as Mr. Oswell volunteered to go down to the Cape and bring up a boat next year, the expedition was abandoned for the time.

Returning home by the Zouga, they had better opportunity to mark the extraordinary richness of the country, and the abundance and luxuriance of its products, both animal and vegetable. Elephants existed in crowds, and ivory was so abundant that a trader was purchasing it at the rate of ten tusks for a musket worth fifteen shillings.

---

[1] *Missionary Travels*, p. 65.

Two years later, after effect had been given to Livingstone's discovery, the price had risen very greatly.

Writing to his friend Watt, he dwells with delight on the river Zouga:

"It is a glorious river; you never saw anything so grand. The banks are extremely beautiful, lined with gigantic trees, many quite new. One bore a fruit a foot in length and three inches in diameter. Another measured seventy feet in circumference. Apart from the branches it looked like a mass of granite; and then the Bakoba in their canoes— did I not enjoy sailing in them? Remember how long I have been in a parched-up land, and answer. The Bakoba are a fine frank race of men, and seem to understand the message better than any people to whom I have spoken on Divine subjects for the first time. What think you of a navigable highway into a large section of the interior? yet that the Tamanak'le is. . . . Who will go into that goodly land? Who? Is it not the Niger of this part of Africa? . . . I greatly enjoyed sailing in their canoes, rude enough things, hollowed out of the trunks of single trees, and visiting the villages along the Zouga. I felt but little when I looked on the lake; but the Zouga and Tamanak'le awakened emotions not to be described. I hope to go up the latter next year."

The discovery of the lake and the river was communicated to the Royal Geographical Society in extracts from Livingstone's letters to the London Missionary Society, and to his friend and former fellow-traveler, Captain Steele. In 1849 the Society voted him a sum of twenty-five guineas "for his successful journey, in company with Messrs. Oswell and Murray, across the South African desert, for the discovery of an interesting country, a fine river, and an extensive inland lake." In addressing Dr. Tidman and Alderman Challis, who represented the London Missionary Society, the President (the late Captain, afterward Rear-Admiral, W. Smyth, R.N., who distinguished himself in early life by his journey across the Andes to Lima, and thence to the Atlantic) adverted to the value of the discoveries in themselves, and in the influence they would have on the regions beyond. He spoke also of the help which Livingstone had derived as an explorer from his influence as a missionary. The journey he had performed successfully had hitherto baffled the best-furnished travelers. In 1834, an expedition under Dr. Andrew Smith, the largest and best-appointed that ever left Cape Town, had gone as far as 23° south latitude; but that proved to be the utmost distance they could reach, and they were compelled to return. Captain Sir James E. Alexander, the only scientific traveler subsequently sent out from England by the Geographical Society, in despair of the lake, and of discovery by the oft-tried eastern route, explored the neighborhood of the western coast instead.[1] The President frankly ascribed Livingstone's success to the influence he had acquired as a missionary among the natives, and Livingstone thoroughly believed this. "The lake," he wrote to his friend Watt, "belongs to missionary enterprise." "Only last year," he subsequently wrote to the Geographical Society, "a party of engineers, in about thirty wagons, made many and persevering efforts to cross the desert at different points, but though inured to the climate, and stimulated by the prospect of gain from the ivory they expected to procure, they were compelled, for want of water, to give up the undertaking." The year after Livingstone's first visit, Mr. Francis Galton tried, but failed, to reach the lake, though he was so successful in other directions as to obtain the Society's gold medal in 1852.

Livingstone was evidently gratified at the honor paid him, and the reception of the twenty-five guineas from the Queen. But the gift had also a comical side. It carried him back to the days of his Radical youth, when he and his friends used to criticise pretty sharply the destination of the nation's money. "The Royal Geographical Society," he writes to his parents (4th December, 1850), "have awarded twenty-five guineas for the discovery of the lake.

---

[1] Journal of the Royal Geographical Society, vol. xx. p. xxviii.

It is from the Queen. You must be very loyal, all of you. Next time she comes your way, shout till you are hoarse. Oh, you Radicals, don't be thinking it came out of your pockets! Long live Victoria!"[1]

Defeated in his endeavor to reach Sebituane in 1849, Livingstone, the following season, put in practice his favorite maxim, " Try again." He left Kolobeng in April, 1850, and this time he was accompanied by Sechéle, Mebalwe, twenty Bakwains, Mrs. Livingstone, and their whole troop of infantry, which now amounted to three. Traveling in the charming climate of South Africa in the roomy wagon, at the pace of two miles and a half an hour, is not like traveling at home; but it was a proof of Livingstone's great unwillingness to be separated from his family, that he took them with him, notwithstanding the risk of mosquitoes, fever, and want of water. The people of Kolobeng were so engrossed at the time with their employments, that till harvest was over, little missionary work could be done.

The journey was difficult, and on the northern branch of the Zouga many trees had to be cut down to allow the wagons to pass. The presence of a formidable enemy was reported on the banks of the Tamanak'le,—the tsetsefly, whose bite is so fatal to oxen. To avoid it, another route had to be chosen. When they got near the lake, it was found that fever had recently attacked a party of Englishmen, one of whom had died, while the rest recovered under the care of Dr. and Mrs. Livingstone. Livingstone took his family to have a peep at the lake; " the children," he wrote, " took to playing in it as ducklings do. Paidling in it was great fun." Great fun to them, who had seen little

---

[1] In a more serious vein he wrote in a previous letter: "I wonder you do not go to see the Queen. I was as disloyal as others when in England, for though I might have seen her in London, I never went. Do you ever pray for her?" This letter is dated 5th February, 1850, and must have been written before he heard of the prize.

enough water for a while; and in a quiet way, great fun to their father too,—his own children " paidling" in his own lake! He was beginning to find that in a missionary point of view, the presence of his wife and children was a considerable advantage; it inspired the natives with confidence, and promoted tender feelings and kind relations. The chief, Lechulatebe, was at last propitiated at a considerable sacrifice, having taken a fancy to a valuable rifle of Livingstone's, the gift of a friend, which could not be replaced. The chief vowed that if he got it he would give Livingstone everything he wished, and protect and feed his wife and children into the bargain, while he went on to Sebituane. Livingstone at once handed him the gun. " It is of great consequence," he said, " to gain the confidence of these fellows at the beginning." It was his intention that Mrs. Livingstone and the children should remain at Lechulatebe's until he should have returned. But the scheme was upset by an outburst of fever. Among others, two of the children were attacked. There was no help but to go home. The gun was left behind in the hope that ere long Livingstone would get back to claim the fulfillment of the chief's promise. It was plain that the neighborhood of the lake was not habitable by Europeans. Hence a fresh confirmation of his views as to the need of native agency, if intertropical Africa was ever to be Christianized.

But Livingstone was convinced that there must be a healthier spot to the north. Writing to Mr. Watt (18th August, 1850), he not only expresses this conviction, but gives the ground on which it rested. The extract which we subjoin gives a glimpse of the sagacity that from apparently little things drew great conclusions; but more than that, it indicates the birth of the great idea that dominated the next period of Livingstone's life—the desire and determination to find a passage to the sea, either on the east or the west coast:

"A more salubrious climate must exist farther up to the north, and that the country is higher, seems evident from the fact mentioned by the Bakoba, that the water of the Teoge, the river that falls into the 'Ngami at the northwest point of it, flows with great rapidity. Canoes ascending, punt all the way, and the men must hold on by reeds in order to prevent their being carried down by the current. Large trees, spring-bucks and other antelopes are sometimes brought down by it. Do you wonder at my pressing on in the way we have done? The Bechuana mission has been carried on in a *cul-de-sac.* I tried to break through by going among the Eastern tribes, but the Boers shut up that field. A French missionary, Mr. Fredoux, of Motito, tried to follow on my trail to the Bamangwato, but was turned back by a party of armed Boers. When we burst through the barrier on the north, it appeared very plain that no mission could be successful there, unless we could get a well-watered country leaving a passage to the sea on either the east or west coast. This project I am almost afraid to meet, but nothing else will do. I intend (D. V.) to go in next year and remain a twelvemonth. My wife, poor soul—I pity her!—proposed to let me go for that time while she remained at Kolobeng. You will pray for us both during that period."

A week later (August 24, 1850) he writes to the Directors that no convenient access to the region can be obtained from the south, the lake being 870 miles from Kuruman:

"We must have a passage to the sea on either the eastern or western coast. I have hitherto been afraid to broach the subject on which my perhaps dreamy imagination dwells. You at home are accustomed to look on a project as half finished when you have received the co-operation of the ladies. My better half has promised me a twelvemonth's leave of absence for mine. Without promising anything, I mean to follow a useful motto in many circumstances, and *Try again.*"

On returning to Kolobeng, Mrs. Livingstone was delivered of a daughter—her fourth child. An epidemic was raging at the time, and the child was seized and cut off, at the age of six weeks. The loss, or rather the removal, of the child affected Livingstone greatly. "It was the first death in our family," he says in his Journal, "but was just as likely to have happened had we remained at home, and we have now one of our number in heaven."

To his parents he writes (4th December, 1850):

"Our last child, a sweet little girl with blue eyes, was taken from us to join the company of the redeemed, through the merits of Him of whom she never heard. It is wonderful how soon the affections twine round a little stranger. We felt her loss keenly. She was attacked by the prevailing sickness, which attacked many native children, and bore up under it for a fortnight. We could not apply remedies to one so young, except the simplest. She uttered a piercing cry previous to expiring, and then went away to see the King in his beauty, and the land—the glorious land, and its inhabitants. Hers is the first grave in all that country marked as the resting-place of one of whom it is believed and confessed that she shall live again."

Mrs. Livingstone had an attack of serious illness, accompanied by paralysis of the right side of the face, and rest being essential for her, the family went, for a time, to Kuruman. Dr. Livingstone had a strong desire to go to the Cape for the excision of his uvula, which had long been troublesome. But, with characteristic self-denial, he put his own case out of view, staying with his wife, that she might have the rest and attention she needed. He tried to persuade his father-in-law to perform the operation, and, under his direction, Dr. Moffat went so far as to make a pair of scissors for the purpose; but his courage, so well tried in other fields, was not equal to the performance of such a surgical operation.

Some glimpses of Livingstone's musings at this time, showing, among other things, how much more he thought of his spiritual than his Highland ancestry, occur in a letter to his parents, written immediately after his return from his second visit to the lake (28th July, 1850). If they should carry out their project of emigration to America, they would have an interesting family gathering:

"One, however, will be 'over the hills and far away' from your happy meeting. The meeting which we hope will take place in Heaven will be unlike a happy one, in so far as earthly relationships are concerned. One will be so much taken up in looking at Jesus, I don't know when we shall be disposed to sit down and talk about the days of lang syne.

And then there will be so many notables whom we should like to notice
and shake hands with—Luke, for instance, the beloved physician, and
Jeremiah, and old Job, and Noah, and Enoch, that if you are wise, you
will make the most of your union while you are together, and not fail
to write me fully, while you have the opportunity here. . . .

"Charles thinks we are not the descendants of the Puritans. I don't
know what you are, but I am. And if you dispute it, I shall stick to
the answer of a poor little boy before a magistrate. *M.*—'Who were
your parents?' *Boy* (rubbing his eyes with his jacket-sleeve)—"Never
had none, sir.' Dr. Wardlaw says that the Scotch Independents are the
descendants of the Puritans, and I suppose the pedigree is through
Rowland Hill and Whitefield. But I was a member of the very church
in which John Howe, the chaplain of Oliver Cromwell, preached, and
exercised the pastorate. I was ordained, too, by English Independents.
Moreover, I am a Doctor too. Agnes and Janet, get up this moment
and curtsy to his Reverence! John and Charles, remember the dream
of the sheaves! *I* descended from kilts and Donald Dhus? Na, na, I
won't believe it.

"We have a difficult, difficult field to cultivate here. All I can say
is, that I think knowledge is increasing. But for the belief that the
Holy Spirit works, and will work for us, I should give up in despair.
Remember us in your. prayers, that we grow not weary in well-doing.
It is hard to work for years with pure motives, and all the time be
looked on by most of those to whom our lives are devoted, as having
some sinister object in view. Disinterested labor—benevolence—is so
out of their line of thought, that many look upon us as having some
ulterior object in view. But He who died for us, and whom we ought
to copy, did more for us than we can do for any one else. He endured
the contradiction of sinners. May we have grace to follow in his
steps!'

The third, and at last successful, effort to reach Sebituane
was made in April, 1851. Livingstone was again accompanied by his family, and by Mr. Oswell. He left Kolobeng
with the intention not to return, at least not immediately,
but to settle with his family in such a spot as might be
found advantageous, in the hilly region, of whose existence
he was assured. They found the desert drier than ever,
no rain having fallen throughout an immense extent of
territory. To the kindness of Mr. Oswell the party was
indebted for most valuable assistance in procuring water,
wells having been dug or cleared by his people beforehand

at various places, and at one place at the hazard of Mr. Oswell's life, under an attack from an infuriated lioness. In his private Journal, and in his letters to home, Livingstone again and again acknowledges with deepest gratitude the numberless acts of kindness done by Mr. Oswell to him and his family, and often adds the prayer that God would reward him, and of His grace give him the highest of all blessings. "Though I cannot repay, I may record with gratitude his kindness, so that, if spared to look upon these, my private memoranda, in future years, proper emotions may ascend to Him who inclined his heart to show so much friendship."

The party followed the old route, around the bed of the Zouga, then crossed a piece of the driest desert they had ever seen, with not an insect or a bird to break the stillness. On the third day a bird chirped in a bush, when the dog began to bark! Shobo, their guide, a Bushman, lost his way, and for four days they were absolutely without water. In his *Missionary Travels*, Livingstone records quietly, as was his wont. his terrible anxiety about his children·

"The supply of water in the wagons had been wasted by one of our servants, and by the afternoon only a small portion remained for the children. This was a bitterly anxious night; and next morning, the less there was of water, the more thirsty the little rogues became. The idea of their perishing before our eyes was terrible; it would almost have been a relief to me to have been reproached with being the entire cause of the catastrophe, but not one syllable of upbraiding was uttered by their mother, though the tearful eye told the agony within. In the afternoon of the fifth day, to our inexpressible relief, some of the men returned with a supply of that fluid of which we had never before felt the true value."

"No one," he remarks in his Journal, "knows the value of water till he is deprived of it. We never need any spirits to qualify it, or prevent an immense draught of it from doing us harm. I have drunk water swarming with insects, thick with mud, putrid from other mixtures, and no stinted draughts of it either, yet never felt any inconvenience from it."

"My opinion is," he said on another occasion, "that the most severe labors and privations may be undergone without alcoholic stimulus, because those who have endured the most had nothing else but water, and not always enough of that."

One of the great charms of Livingstone's character, and one of the secrets of his power—his personal interest in each individual, however humble—appeared in connection with Shobo, the Bushman guide, who misled them and took the blunder so coolly. "What a wonderful people," he says in his Journal, "the Bushmen are! always merry and laughing, and never telling lies wantonly like the Bechuana. They have more of the appearance of worship than any of the Bechuana. When will these dwellers in the wilderness bow down before their Lord? No man seems to care for the Bushman's soul. I often wished I knew their language, but never more than when we traveled with our Bushman guide, Shobo."

Livingstone had given a fair trial to the experiment of traveling along with his family. In one of his letters at this time he speaks of the extraordinary pain caused by the mosquitoes of those parts, and of his children being so covered with their bites, that not a square inch of whole skin was to be found on their bodies. It is no wonder that he gave up the idea of carrying them with him in the more extended journey he was now contemplating. He could not leave them at Kolobeng, exposed to the raids of the Boers; to Kuruman there were also invincible objections; the only possible plan was to send them to England, though he hoped that when he got settled in some suitable part of Sebituane's dominions, with a free road to the sea, they would return to him, and help him to bring the people to Christ.

In the *Missionary Travels* Livingstone has given a full account of Sebituane, chief of the Makololo, "unquestionably the greatest man in all that country"—his remarkable career, his wonderful warlike exploits (for which he could always bring forward justifying reasons), his interesting and attractive character, and wide and powerful influence. In one thing Sebituane was very like Livingstone himself; he had the art of gaining the affections both of his own people and of strangers. When a party of poor men came to his town to sell hoes or skins, he would sit down among them, talk freely and pleasantly to them, and probably cause some lordly dish to be brought, and give them a feast on it, perhaps the first they had ever shared. Delighted beyond measure with his affability and liberality, they felt their hearts warm toward him; and as he never allowed a party of strangers to go away without giving every one of them—servants and all—a present, his praises were sounded far and wide. "He has a heart! he is wise!" were the usual expressions Livingstone heard before he saw him.

Sebituane received Livingstone with great kindness, for it had been one of the dreams of his life to have intercourse with the white man. He placed full confidence in him from the beginning, and was ready to give him everything he might need. On the first Sunday when the usual service was held he was present, and Livingstone was very thankful that he was there, for it turned out to be the only proclamation of the gospel he ever heard. For just after realizing what he had so long and ardently desired, he was seized with severe inflammation of the lungs, and died after a fortnight's illness. Livingstone, being a stranger, feared to prescribe, lest, in the event of his death, he should be accused of having caused it. On visiting him, and seeing that he was dying, he spoke a few words respecting hope after death. But being checked by the attendants for introducing the subject, he could only commend his soul to God. The last words of Sebituane were words of kindness to Livingstone's son: "Take him to Maunku (one of his wives) and tell her to give him some milk." Livingstone was deeply affected by his death. A deeper sense of brotherhood, a warmer glow of affection had been kindled in his heart toward Sebituane than had seemed possible. With his very tender conscience and deep sense of spiritual realities, Livingstone was afraid, as in the case of Sehamy eight years before, that he had not spoken to him so pointedly as he might have done. It is awfully affecting to follow him into the unseen world, of which he had heard for the first time just before he was called away. In his Journal, Livingstone gives way to his feelings as he very seldom allowed himself to do. His words bring to mind David's lament for Jonathan or for Absalom, although he had known Sebituane less than a month, and he was one of the race whom many Boers and slave-stealers regarded as having no souls:

"Poor Sebituane, my heart bleeds for thee; and what would I not do for thee now? I will weep for thee till the day of my death. Little didst thou think when, in the visit of the white man, thou sawest the long cherished desires of years accomplished, that the sentence of death had gone forth! Thou thoughtest that thou shouldest procure a weapon from the white man which would be a shield from the attacks of the fierce Matebele; but a more deadly dart than theirs was aimed at thee; and though thou couldest well ward off a dart—none ever better—thou didst not see that of the king of terrors. I will weep for thee, my brother, and I will cast forth my sorrows in despair for thy condition! But I know that thou wilt receive no injustice whither thou art gone; 'Shall not the Judge of all the earth do right?' I leave thee to Him. Alas! alas! Sebituane. I might have said more to him. God forgive me. Free me from blood-guiltiness. If I had said more of death I might have been suspected as having foreseen the event, and as guilty of bewitching him. I might have recommended Jesus and his great atonement more. It is, however, very difficult to break through the thick crust of ignorance which envelops their minds."

The death of Sebituane was a great blow in another sense. The region over which his influence extended was immense, and he had promised to show it to Livingstone and to select a suitable locality for his residence. This heathen chief would have given to Christ's servant what the Boers refused him! Livingstone would have h~d his wish—an entirely new country to work upon, where the name of Christ had never yet been spoken. So at least he thought. Sebituane's successor in the chiefdom was his daughter, Ma-mochisane. From her he received liberty to visit any part of the country he chose. While waiting for a reply (she was residing at a distance), he one day fell into a great danger from an elephant which had come on him unexpectedly. "We were startled by his coming a little way in the direction in which we were standing, but he did not give us chase. I have had many escapes. We seem immortal till our work is done."

Mr. Oswell and he then proceeded in a northeasterly direction, passing through the town of Linyanti, and on the 3d of August they came on the beautiful river at Seshéke :

"We thanked God for permitting us to see this glorious river. All we said to each other was ' How glorious ! how magnificent ! how beau-tiful !' . . . In crossing, the waves lifted up the canoe and made it roll beautifully. The scenery of the Firths of Forth and Clyde was brought vividly to my view, and had I been fond of indulging in sen-timental effusions, my lachrymal apparatus seemed fully charged. But then the old man who was conducting us across might have said, 'What on earth are you blubbering for ? Afraid of these crocodiles, eh ?' The little sentimentality which exceeded was forced to take its course down the inside of the nose. We have other work in this world than indulging in sentimentality of the 'Sonnet to the Moon' variety."

The river, which went here by the name of Seshéke, was found to be the Zambesi, which had not previously been known to exist in that region. In writing about it to his brother Charles, he says, "It was the first *river* I ever saw." Its discovery in this locality constituted one of the great geographical feats with which the name of Livingstone is connected. He heard of rapids above, and of great water-falls below; but it was reserved for him on a future visit to behold the great Victoria Falls, which in the popular imagination have filled a higher place than many of his more useful discoveries.

The travelers were still a good many days' distance from Ma-mochisane, without whose presence nothing could be settled; but besides, the reedy banks of the rivers were found to be unsuitable for a settlement, and the higher regions were too much exposed to the attacks of Mosilikatse. Livingstone saw no prospect of obtaining a suitable station, and with great reluctance he made up his mind to retrace the weary road, and return to Kolobeng. The people were very anxious for him to stay, and offered to make a garden for him, and to fulfill Sebituane's promise to give him oxen in return for those killed by the tsetse.

Setting out with the wagons on 13th August, 1851, the party proceeded slowly homeward. On 15th September, 1851, Livingstone's Journal has this unexpected and simple entry: "A son, William Oswell Livingstone,[1] born at a place we always call Bellevue." On the 18th: "Thomas attacked by fever; removed to a high part on his account. Thomas was seized with fever three times at about an interval of a fortnight." Not a word about Mrs. Livingstone, but three pages of observations about medical treatment of fever, thunderstorms, constitutions of Indian and African people, leanness of the game, letter received from Directors approving generally of his course, a gold watch sent by Captain Steele, and Gordon Cumming's book, "a miserably poor thing." Amazed, we ask, Had Livingstone any heart? But ere long we come upon a copy of a letter, and some remarks connected with it, that give us an impression of the depth and strength of his nature, unsurpassed by anything that has yet occurred.

"The following extracts," he says, "show in what light our efforts are regarded by those who, as much as we do, desire that the 'gospel may be preached to all nations.'" Then follows a copy of a letter which had been addressed

---

[1] He had intended to call him Charles, and announced this to his father; but, finding that Mr. Oswell, to whom he was so much indebted, would be pleased with the compliment, he changed his purpose and the name accordingly.

to him before they set out by Mrs. Moffat, his mother-in-law, remonstrating in the strongest terms against his plan of taking his wife with him; reminding him of the death of the child, and other sad occurrences of last year; and in the name of everything that was just, kind, and even decent, beseeching him to abandon an arrangement which all the world would condemn. Another letter from the same writer informed him that much prayer had been offered that, if the arrangements were not in accordance with Christian propriety, he might in great mercy be prevented by some dispensation of Providence from carrying them out. Mrs. Moffat was a woman of the highest gifts and character, and full of admiration for Livingstone. The insertion of these letters in his Journal shows that, in carrying out his plan, the objections to which it was liable were before his mind in the strongest conceivable form. No man who knows what Livingstone was will imagine for a moment that he had not the most tender regard for the health, the comfort, and the feelings of his wife; in matters of delicacy he had the most scrupulous regard to propriety; his resolution to take her with him must, therefore, have sprung from something far stronger than even his affection for her. What was this stronger force?

It was his inviolable sense of duty, and his indefeasible conviction that his Father in heaven would not forsake him whilst pursuing a course in obedience to his will, and designed to advance the welfare of his children. As this furnishes the key to Livingstone's future life, and the answer to one of the most serious objections ever brought against it, it is right to spend a little time in elucidating the principles by which he was guided.

There was a saying of the late Sir Herbert Edwardes which he highly valued: "He who has to act on his own responsibility is a slave if he does not act on his own judgment." Acting on this maxim, he must set aside the views of others as to his duty, provided his own judgment was clear regarding it. He must even set aside the feelings and apparent interest of those dearest to him, because duty was above everything else. His faith in God convinced him that, in the long run, it could never be the worse for him and his that he had firmly done his duty. All true faith has in it an element of venture, and in Livingstone's faith this element was strong. Trusting God, he could expose to venture even the health, comfort, and welfare of his wife and children. He was convinced that it was his duty to go forth with them and seek a new station for the Gospel in Sebituane's country. If this was true, God would take care of them, and it was "better to trust in the Lord than to put confidence in man." People thoughtlessly accused him of making light of the interests of his family. No man suffered keener pangs from the course he had to follow concerning them, and no man pondered more deeply what duty to them required.

But to do all this, Livingstone must have had a very clear perception of the course of duty. This is true. But how did he get this? First, his singleness of heart, so to speak, attracted the light: "If thine eye be single, thy whole body shall be full of light." Then, he was very clear and very minute in his prayers. Further, he was most careful to scan all the providential indications that might throw light on the Divine will. And when he had been carried so far on in the line of duty, he had a strong presumption that the line would be continued, and that he would not be called to turn back. It was in front, not in rear, that he expected to find the pillar of cloud and the pillar of fire. In course of time, this hardened into a strong instinctive habit, which almost dispensed with the process of reasoning.

In Dean Stanley's *Sinai and Palestine* allusion is made to a kindred experience,—that which bore Abraham from Chaldea, Moses from Egypt, and the greater part of the tribes from the comfortable pastures of Gilead and Bashan to the rugged hill-country of Judah and Ephraim. Notwithstanding all the attractions of the richer countries, they were borne onward and forward, not knowing whither they went, instinctively feeling that they were fulfilling the high purposes to which they were called. In the later part of Livingstone's life, the necessity of going forward to the close of the career that had opened for him seemed to settle the whole question of duty.

But at this earlier stage, he had been conscientiously scrutinizing all that had any bearing on that question; and now that he finds himself close to his home, and can thank God for the safe confinement of his wife, and the health of the new-born child, he gathers together all the providences that showed that in this journey, which excited such horror even among his best friends, he had after all been following the guidance of his Father. First, in the matter of guides, he had been wonderfully helped, notwithstanding a deep plot to deprive him of any. Then there was the sickness of Sekómi, whose interest had been secured through his going to see him, and prescribing for him; this had propitiated one of the tribes. The services of Shobo, too, and the selection of the northern route, proposed by Kamati, had been of great use. Their going to Seshéke, and their detention for two months, thus allowing them time to collect information respecting the whole country; the river Chobe not rising at its usual time; the saving of Livingstone's oxen from the tsetse, notwithstanding their detention on the Zouga; his not going with Mr. Oswell to a place where the tsetse destroyed many of the oxen; the better health of Mrs. Livingstone during her confinement than in any previous one; a very opportune present they had got, just before her confinement, of two bottles of wine;[1] the approbation of the Directors, the presentation of a gold watch

---

[1] In writing to his father, Livingstone mentions that the wine was a gift from Mrs. Bysshe Shelley, in acknowledgment of his aid in repairing a wheel of her wagon.

by Captain Steele, the kind attentions of Mr. Oswell, and
the cookery of one of their native servants named George;
the recovery of Thomas, whereas at Kuruman a child had
been cut off; the commencement of the rains, just as they
were leaving the river, and the request of Mr. Oswell that
they should draw upon him for as much money as they
should need, were all among the indications that a faithful
and protecting Father in heaven had been ordering their
path, and would order it in like manner in all time to come.

Writing at this time to his father-in-law, Mr. Moffat, he
said, after announcing the birth of Oswell: "What you
say about difference of opinion is true. In my past life,
I have always managed to think for myself, and act
accordingly. I have occasionally met with people who
took it on themselves to act for me, and they have offered
their thoughts with an emphatic 'I think'; but I have
excused them on the score of being a little soft-headed in
believing they could think both for me and themselves."

While Kolobeng was Livingstone's headquarters, a new
trouble rose upon the mission horizon. The Makololo (as
Sebituane's people were called) began to practice the slave-
trade. It arose simply from their desire to possess guns.
For eight old muskets they had given to a neighboring
tribe eight boys, that had been taken from their enemies
in war, being the only article for which the guns could be
got. Soon after, in a fray against another tribe, two
hundred captives were taken, and, on returning, the Makololo met some Arab traders from Zanzibar, who for three
muskets received about thirty of their captives.

Another of the master ideas of his life now began to
take hold upon Livingstone. Africa was exposed to a
terrible evil through the desire of the natives to possess
articles of European manufacture, and their readiness for
this purpose to engage in the slave-trade. Though no
African had ever been known to sell his own children into
captivity, the tribes were ready enough to sell other chil-

dren that had fallen into their hands by war or otherwise. But if a legitimate traffic were established through which they might obtain whatever European goods they desired in exchange for ivory and other articles of native produce, would not this frightful slave-trade be brought to an end? The idea was destined to receive many a confirmation before Livingstone drew his last breath of African air. It naturally gave a great impulse to the purpose which had already struck its roots into his soul—to find a road to the sea either on the eastern or western coast. Interests wider and grander than even the planting of mission stations on the territories of Sebituane now rose to his view. The welfare of the whole continent, both spiritual and temporal, was concerned in the success of this plan of opening new channels to the enterprise of British and other merchants, always eager to hear of new markets for their goods. By driving away the slave-trade, much would be done to prepare the way for Christian missions which could not thrive in an atmosphere of war and commotion. An idea involving issues so vast was fitted to take a right powerful hold on Livingstone's heart, and make him feel that no sacrifice could be too great to be encountered, cheerfully and patiently, for such an end.

Writing to the Directors (October, 1851), he says:

"You will see by the accompanying sketch-map what an immense region God in his grace has opened up. If we can enter in and form a settlement, we shall be able in the course of a very few years to put a stop to the slave-trade in that quarter. It is probable that the mere supply of English manufacturers on Sebituane's part will effect this, for they did not like the slave-trade, and promised to abstain. I think it will be impossible to make a fair commencement unless I can secure two years devoid of family cares. I shall be obliged to go southward, perhaps to the Cape, to have my uvula excised and my arm mended (the latter, if it can be done, only). It has occurred to me that, as we must send our children to England, it would be no great additional expense to send them now along with their mother. This arrangement would enable me to proceed, and devote about two or perhaps three years to this new region; but I must beg your sanction, and if you please let it be given or withheld as soon as you can conveniently, so that it might meet me at the Cape. To orphanize my children will be like tearing out my bowels, but when I can find time to write you fully you will perceive it is the only way, except giving up that region altogether.

"Kuruman will not answer as a residence, nor yet the Colony. If I were to follow my own inclinations, they would lead me to settle down quietly with the Bakwains, or some other small tribe, and devote some of my time to my children; but *Providence seems to call me to the regions beyond,* and if I leave them anywhere in this country, it will be to let them become heathens. If you think it right to support them, I believe my parents in Scotland would attend to them otherwise."

Continuing the subject in a more leisurely way a few weeks later, he refers to the very great increase of traffic that had taken place since the discovery of Lake 'Ngami two years before; the fondness of the people for European articles; the numerous kinds of native produce besides ivory, such as beeswax, ostrich feathers, etc., of which the natives made little or no use, but which they would take care of if regular trade were established among them. He thought that if traders were to come up the Zambesi and make purchases from the producers they would both benefit themselves and drive the slave-dealer from the market. It might be useful to establish a sanatorium, to which missionaries might come from less healthy districts to recruit. This would diminish the reluctance of missionaries to settle in the interior. For himself, though he had reared three stations with much bodily labor and fatigue, he would cheerfully undergo much more if a new station would answer such objects. In referring to the countries drained by the Zambesi, he believed he was speaking of a large section of the slave-producing region of Africa. He then went on to say that to a certain extent their hopes had been disappointed; Mr. Oswell had not been able to find a passage to the sea, and he had not been able to find a station for missionary work. They had therefore returned together. "He assisted me," adds

Livingstone, "in every possible way. May God reward him !"

In regard to mission work for the future an important question arose, What should be done for the Bakwains? They could not remain at Kolobeng—hunger and the Boers decided that point. Was it not, then, his duty to find and found a new station for them? Dr. Livingstone thought not. He had always told them that he would remain with them only for a few years. One of his great ideas on missions in Africa was that a fair trial should be given to as many places as possible, and if the trial did not succeed the missionaries should pass on to other tribes. He had a great aversion to the common impression that the less success one had the stronger was one's duty to remain. Missionaries were only too ready to settle down and make themselves as comfortable as possible, whereas the great need was for men to move on, to strike out into the regions beyond, to go into all the world. He had far more sympathy for tribes that had never heard the gospel than for those who had had it for years. He used to refer to certain tribes near Griqualand that had got a little instruction, but had no stated missionaries; they used to send some of their people to the Griquas to learn what they could, and afterward some others; and these persons, returning, communicated what they knew, till a wonderful measure of knowledge was acquired, and a numerous church was formed. If the seed had once been sown in any place it would not remain dormant, but would excite the desire for further knowledge; and on the whole it would be better for the people to be thrown somewhat on their own resources than to have everything done for them by missionaries from Europe. In regard to the Bakwains, though they had promised well at first, they had not been a very teachable people. He was not inclined to blame them; they had been so pinched by hunger and badgered by the Boers that they could not attend to instruction; or rather, they had too good an excuse for not doing so. "I have much affection for them," he says in his Journal, "and though I pass from them I do not relinquish the hope that they will yet turn to Him to whose mercy and love they have often been invited. The seed of the living Word will not perish."

The finger of Providence clearly pointed to a region farther north in the country of the Barotse or beyond it. He admitted that there were *pros* and *cons* in the case. Against his plan,—some of his brethren did not hesitate to charge him with being actuated by worldly ambition. This was the more trying, for sometimes he suspected his own motives. Others dwelt on what was due to his family. Moreover, his own predilections were all for a quiet life. And there was also the consideration, that as the Directors could not well realize the distances he would have to travel before he reached the field, he might appear more as an explorer than a missionary. On the other hand:

"I am conscious," he says, "that though there is much impurity in my motives, they are in the main for the glory of Him to whom I have devoted myself. I never anticipated fame from the discovery of the Lake. I cared very little about it, but the sight of the Tamanak'le, and the report of other large rivers beyond, all densely populated, awakened many and enthusiastic feelings. . . . Then, again, consider the multitude that in the Providence of God have been brought to light in the country of Sebituane; the probability that in our efforts to evangelize we shall put a stop to the slave-trade in a large region, and by means of the highway into the North which we have discovered bring unknown nations into the sympathies of the Christian world. If I were to choose my work, it would be to reduce this new language, translate the Bible into it, and be the means of forming a small church. Let this be accomplished, I think I could then lie down and die contented. Two years' absence will be necessary. . . . Nothing but a strong conviction that the step will lead to the glory of Christ would make me orphanize my children. Even now my bowels yearn over them. They will forget me; but I hope when the day of trial comes, I shall not be found a more sorry soldier than those who serve an earthly sovereign. Should you not feel yourselves justified in incurring the expense of their support in England, I shall feel called upon to renounce the hope of carrying the gospel into that country, and labor among those who live in a more healthy country, viz., the Bakwains. But, stay, I am not sure; so powerfully convinced am I that it is the will of the Lord I should, *I will go, no matter who opposes;* but from you I expect nothing but encouragement. I know you wish as ardently as I can that all the world may be filled with the glory of the Lord. I feel relieved when I lay the whole case before you."

He proposed that a brother missionary, Mr. Ashton, should be placed among the Bamangwato, a people who were in the habit of spreading themselves through the Bakalahari, and should thus form a link between himself and the brethren in the south.

In a postscript, dated Bamangwato, 14th November, he gratefully acknowledges a letter from the Directors, in which his plans are approved of generally. They had recommended him to complete a dictionary of the Sichuana language. This he would have been delighted to do when his mind was full of the subject, but with the new projects now before him, and the probability of having to deal with a new language for the Zambesi district, he could not undertake such a work at present.

In a subsequent letter to the Directors (Cape Town, 17th March, 1852), Livingstone finds it necessary to go into full details with regard to his finances. Though he writes with perfect calmness, it is evident that his exchequer was sadly embarrassed. In fact, he had already not only spent all the salary (£100) of 1852, but fifty-seven pounds of 1853, and the balance would be absorbed by expenses in Cape Town. He had been as economical as possible; in personal expenditure most careful—he had been a teetotaler for twenty years. He did not hesitate to express his conviction that the salary was inadequate, and to urge the Directors to defray the extra expenditure which was now inevitable; but with characteristic generosity he urged Mr. Moffat's claims much more warmly than his own.

From expressions in Livingstone's letter to the Directors,

it is evident that he was fully aware of the risk he ran, in his new line of work, of appearing to sink the missionary in the explorer. There is no doubt that next to the charge of forgetting the claims of his family, to which we have already adverted, this was the most plausible of the objections taken to his susequent career. But any one who has candidly followed his course of thought and feeling from the moment when the sense of unseen realities burst on him at Blantyre, to the time at which we have now arrived, must see that this view is althogether destitute of support. The impulse of divine love that had urged him first to become a missionary had now become with him the settled habit of his life. No new ambition had flitted across his path, for though he had become known as a geographical discoverer, he says he thought very little of the fact, and his life shows this to have been true. Twelve years of missionary life had given birth to no sense of weariness, no abatement of interest in these poor black savages, no reluctance to make common cause with them in the affairs of life, no despair of being able to do them good. On the contrary, he was confirmed in his opinion of the efficacy of his favorite plan of native agency, and if he could but get a suitable base of operations, he was eager to set it going, and on every side he was assured of native welcome. Shortly before (5th February, 1850), when writing to his father with reference to a proposal of his brother Charles that he should go and settle in America, he had said: "I am a missionary, heart and soul. God had an only Son, and He was a missionary and a physician. A poor, poor imitation of Him I am, or wish to be. In this service I hope to live, in it I wish to die." The spectre of the slave-trade had enlarged his horizon, and shown him the necessity of a commercial revolution for the whole of Africa, before effectual and permanent good could be done in any part of it. The plan which he had now in view multiplied the risks he ran, and compelled him to think anew whether he was ready to sacrifice himself, and if so, for what. All that Livingstone did was thus done with open eyes and well-considered resolution. Adverting to the prevalence of fever in some parts of the country, while other parts were comparatively healthy, he says in his Journal: "I offer myself as a forlorn hope in order to ascertain whether there is a place fit to be a sanatorium for more unhealthy spots. May God accept my service, and use me for his glory. A great honor it is to' be a fellow-worker with God." "It is a great venture," he writes to his sister (28th April, 1851). "Fever may cut us all off. I feel much when I think of the children dying. But who will go if we don't? Not one. I would venture everything for Christ. Pity I have so little to give. But He will accept us, for He is a good master. Never one like Him. He can sympathize. May He forgive, and purify, and bless us."

If in his spirit of high consecration he was thus unchanged, equally far was he from having a fanatical disregard of life, and the rules of provident living.

"Jesus," he says, "came not to judge,—κρίνω,—condemn judicially, or execute vengeance on any one. His was a message of peace and love. He shall not strive nor cry, neither shall his voice be heard in the streets. Missionaries ought to follow his example. Neither insist on our rights, nor appear as if we could allow our goods to be destroyed without regret: for if we are righteous overmuch, or stand up for our rights with too much vehemence, we beget dislikes, and the people see no difference between ourselves and them. And if we appear to care nothing for the things of this world, they conclude we are rich, and when they beg, our refusal is ascribed to niggardliness, and our property, too, is wantonly destroyed. 'Ga ba tloke'=they are not in need, is the phrase employed when our goods are allowed to go to destruction by the neglect of servants. . . . In coming among savage people, we ought to make them feel we are of them, 'we seek not yours, but you'; but while very careful not to make a gain of them, we ought to be as careful to appear thankful, and appreciate any effort they may make for our comfort or subsistence."

On reaching Kolobeng from 'Ngami they found the station deserted. The Bakwains had removed to Limaüe. Sechéle came down the day after, and presented them with an ox —a valuable gift in his circumstances. Sechéle had much yet to bear from the Boers; and after being, without provocation, attacked, pillaged, and wasted, and robbed of his children, he was bent on going to the Queen of England to state his wrongs. This, however, he could not accomplish, though he went as far as the Cape. Coming back afterward to his own people, he gathered large numbers about him from other tribes, to whose improvement he devoted himself with much success. He still survives, with the one wife whom he retained; and, though not without some drawbacks (which Livingstone ascribed to the bad example set him by some), he maintains his Christian profession. His people are settled at some miles' distance from Kolobeng, and have a missionary station, supported by a Hanoverian Society. His regard for the memory of Livingstone is very great, and he reads with eagerness all that he can find about him. He has ever been a warm friend of missions has a wonderful knowledge of the Bible, and can preach well. The influence of Livingstone in his early days was doubtless a real power in mission-work. Mebalwe, too, we are informed by Dr. Moffat, still survives; a useful man, an able preacher, and one who has done much to bring his people to Christ.

It was painful to Livingstone to say good-bye to the Bakwains, and (as Mrs. Moffat afterward reminded him) his friends were not all in favor of his doing so; but he regarded his departure as inevitable. After a short stay at Kuruman, he and his family went on to Cape Town, where they arrived on the 16th of March, 1852, and had new proofs of Mr. Oswell's kindness. After eleven years' absence, Livingstone's dress-coat had fallen a little out of fashion, and the whole costume of the party was somewhat in the style of Robinson Crusoe. The generosity of "the best friend we have in Africa" made all comfortable, Mr. Oswell remarking that Livingstone had as good a right as he to the money drawn from the " preserves on his estate "—the elephants. Mentally, Livingstone traces to its source the kindness of his friend, thinking of One to whom he owed all—" O divine Love, I have not loved Thee strongly, deeply, warmly enough." The retrospect of his eleven years of African labor, unexampled though they had been, only awakened in him the sense of unprofitable service.

Before closing the record of this period, we must take a glance at the remarkable literary activity which it witnessed. We have had occasion to refer to Livingstone's first letters to Captain Steele, for the Geographical Society; additional letters were contributed from time to time. His philological researches have also been noticed. In addition to these, we find him writing two articles on African Missions for the *British Quarterly Review,* only one of which was published. He likewise wrote two papers for the *British Banner* on the Boers. While crossing the desert, after leaving the Cape on his first great journey, he wrote a remarkable paper on " Missionary Sacrifices," and another of great vigor on the Boers. Still another paper on Lake 'Ngami was written for a Missionary Journal contemplated, but never started, under the editorship of the late Mr. Isaac Taylor; and he had one in his mind on the religion of the Bechuanas, presenting a view which differed somewhat from that of Mr. Moffat. Writing to Mr. Watt from Linyanti (3d October, 1853), on printing one of his papers, he says:

"But the expense, my dear man. What a mess I am in, writing papers which cannot pay their own way ! Pauper papers, in fact, which must go to the workhouse for support. Ugh ! Has the Caffre War paper shared the same fate? and the Language paper too ? Here I have two by me, which I will keep in their native obscurity. One is on the South African Boers and slavery, in which I show that their church is, and always has been, the great bulwark of slavery, cattle-lifting, and Caffre-marauding; and I correct the mistaken views of some writers who describe the Boers as all that is good, and of others who describe them as all that is bad, by showing who are the good and who are the bad. The other, which I rather admire,—what father doesn't his own progeny ?—is on the missionary work, and designed to aid young men of piety to form a more correct idea of it than is to be had from much of the missionary biography of 'sacrifices.' I magnify the enterprise, exult in the future, etc., etc. It was written in coming across the desert, and if it never does aught else, it imparted comfort and encouragement to myself.[1] . . . I feel almost inclined to send it. . . . If the Caffre War one is rejected, then farewell to spouting in Reviews."

If he had met with more encouragement from editors he would have written more. But the editorial cold shoulder was beyond even his power of endurance. He laid aside his pen in a kind of disgust, and this doubtless was one of the reasons that made him unwilling to resume it on his return to England. Editors were wiser then; and the offer from one London Magazine of £400 for four articles, and from *Good Words* of £1000 for a number of papers to be fixed afterward,—offers which, however, were not accepted finally,—showed how the tide had turned.

---

[1] For extracts from the paper on " Missionary Sacrifices," see Appendix No. I. For part of the paper on the Boers, see *Catholic Presbyterian,* December, 1879 (London, Nisbet and Co.).

CHAPTER VII.

FROM THE CAPE TO LINYANTI.

A.D. 1852–1853.

Unfavorable feeling at Cape Town—Departure of Mrs. Livingstone and children —Livingstone's detention and difficulties—Letter to his wife—To Agnes— Occupations at Cape Town—The Astronomer-Royal—Livingstone leaves the Cape and reaches Kuruman—Destruction of Kolobeng by the Boers—Letters to his wife and Rev. J. Moore—His resolution to open up Africa *or perish*— Arrival at Linyanti—Unhealthiness of the country—Thoughts on setting out for coast—Sekelétu's kindness—Livingstone's missionary activity—Death of Mpepe, and of his father—Meeting with Ma-mochisane—Barotse country— Determines to go to Loanda—Heathenism unadulterated—Taste for the beautiful—Letter to his children—to his father—Last Sunday at Linyanti— Prospect of his falling.

WHEN Livingstone arrived at the Cape, he found the authorities in a state of excitement over the Caffre War, and very far from friendly toward the London Missionary Society, some of whose missionaries—himself among the number—were regarded as "unpatriotic." He had a very poor opinion of the officials, and their treatment of the natives scandalized him. He describes the trial of an old soldier, Botha, as "the most horrid exhibition I ever witnessed." The noble conduct of Botha in prison was a beautiful contrast to the scene in court. This whole Caffre War had exemplified the blundering of the British authorities, and was teaching the natives developments, the issue of which could not be foreseen. As for himself, he writes to Mr. Moffat, that he was cordially hated, and perhaps he might be pulled up; but he knew that some of his letters had been read by the Duke of Wellington and Lord Brougham with pleasure, and, possibly, he might get

13

justice. He bids his father-in-law not to be surprised if he saw him abused in the newspapers.

On the 23d April, 1852, Mrs. Livingstone and the four children sailed from Cape Town for England. The sending of his children to be brought up by others was a very great trial, and Dr. Livingstone seized the opportunity to impress on the Directors that those by whom missionaries were sent out had a great duty to the children whom their parents were compelled to send away. Referring to the filthy conversation and ways of the heathen, he says:

"Missionaries expose their children to a contamination which they have had no hand in producing. We expose them and ourselves for a time in order to elevate those sad captives of sin and Satan, who are the victims of the degradation of ages. None of those who complain about missionaries sending their children home ever descend to this. And again, as Mr. James in his *Young Man from Home* forcibly shows, a greater misfortune cannot befall a youth than to be cast into the world without a home. In regard to even the vestige of a home, my children are absolutely vagabonds. When shall we return to Kolobeng? When to Kuruman? *Never.* The mark of Cain is on your foreheads, your father is a missionary. Our children ought to have both the sympathies and prayers of those at whose bidding we become strangers for life."

Was there ever a plea more powerful or more just? It is sad to think that the coldness of Christians at home should have led a man like Livingstone to fancy that, because his children were the children of a missionary, they would bear the mark of Cain, and be homeless vagabonds. Why are we at home so forgetful of the privilege of refreshing the bowels of those who take their lives in their hands for the love of Christ, by making a home for their offspring? In a higher state of Christianity there will be hundreds of the best families at home delighted, for the love of their Master, to welcome and bring up the missionary's children. And when the Great Day comes, none will more surely receive that best of all forms of repayment, "Inasmuch as ye did it unto the least of these my brethren, ye did it unto Me."

Livingstone, who had now got the troublesome uvula cut out, was detained at the Cape nearly two months after his family left. He was so distrusted by the authorities that they would hardly sell powder and shot to him, and he had to fight a battle that demanded all his courage and perseverance for a few boxes of percussion-caps. At the last moment, a troublesome country postmaster, to whom he had complained of an overcharge of postage, threatened an action against him for defamation of character, and, rather than be further detained, deep in debt though he was, Livingstone had to pay him a considerable sum. His family were much in his thoughts; he found some relief in writing by every mail. His letters to his wife are too sacred to be spread before the public; we confine ourselves to a single extract, to show over what a host of suppressed emotions he had to march in this expedition:

" *Cape Town, 5th May,* 1852.—My DEAREST MARY,—How I miss you now, and the children! My heart yearns incessantly over you. How many thoughts of the past crowd into my mind! I feel as if I would treat you all much more tenderly and lovingly than ever. You have been a great blessing to me. You attended to my comfort in many, many ways. May God bless you for all your kindnesses! I see no face now to be compared with that sunburnt one which has so often greeted me with its kind looks. Let us do our duty to our Saviour, and we shall meet again. I wish that time were now. You may read the letters over again which I wrote at Mabotsa, the sweet time you know. As I told you before, I tell you again, they are true, true; there is not a bit of hypocrisy in them. I never show all my feelings; but I can say truly, my dearest, that I loved you when I married you, and the longer I lived with you, I loved you the better. . . . Let us do our duty to Christ, and He will bring us through the world with honor and usefulness. He is our refuge and high tower; let us trust in Him at all times, and in all circumstances. Love Him more and more, and diffuse his love among the children. Take them all round you, and kiss them for me. Tell them I have left them for the love of Jesus, and they must love Him too, and avoid sin, for that displeases Jesus. I shall be delighted to hear of you all safe in England. . . ."

A few days later, he writes to his eldest daughter, then in her fifth year:

" *Cape Town,* 18*th May,* 1852.—My DEAR AGNES,—This is your own little letter. Mamma will read it to you, and you will hear her just as if I were speaking to you, for the words which I write are those which she will read. I am still at Cape Town. You know you left me there when you all went into the big ship and sailed away. Well, I shall leave Cape Town soon. Malatsi has gone for the oxen, and then I shall go away back to Sebituane's country, and see Seipone and Meriye, who gave you the beads and fed you with milk and honey. I shall not see you again for a long time, and I am very sorry. I have no Nannie now. I have given you back to Jesus, your Friend—your Papa who is in heaven. He is above you, but He is always near you. When we ask things from Him, that is praying to Him; and if you do or say a naughty thing ask Him to pardon you, and bless you, and make you one of his children. Love Jesus much, for He loves you, and He came and died for you. Oh, how good Jesus is! I love Him, and I shall love Him as long as I live. You must love Him too, and you must love your brothers and mamma, and never tease them or be naughty, for Jesus does not like to see naughtiness.—Good-bye, my dear Nannie,

D. LIVINGSTON."

Among his other occupations at Cape Town, Livingstone put himself under the instructions of the Astronomer-Royal, Mr. (afterward Sir Thomas) Maclear, who became one of his best and most esteemed friends. His object was to qualify himself more thoroughly for taking observations that would give perfect accuracy to his geographical explorations. He tried English preaching too, but his throat was still tender, and he felt very nervous, as he had done at Ongar. "What a little thing," he writes to Mr. Moffat, " is sufficient to bring down to old-wifeishness such a rough tyke as I consider myself! Poor, proud human nature is a great fool after all." A second effort was more successful. " I preached," he writes to his wife, " on the text, 'Why will ye die?' I had it written out and only referred to it twice, which is an improvement in English. I hope good was done. The people were very attentive indeed. I felt less at a loss than in Union Chapel."[1] He arranged with

---

[1] The manuscript of this sermon still exists. The sermon is very simple, scriptural, and earnest, in the style of Bishop Ryle, or of Mr. Moody.

a mercantile friend, Mr. Rutherfoord, to direct the operations of a native trader, George Fleming, whom that gentleman was to employ for the purpose of introducing lawful traffic in order to supplant the slave-trade.

It was not till the 8th of June that he left the Cape. His wagon was loaded to double the usual weight from his good nature in taking everybody's packages. His oxen were lean, and he was too poor to provide better. He reached Griqua Town on the 15th August, and Kuruman a fortnight later. Many things had occasioned unexpected delay, and the last crowning detention was caused by the breaking down of a wheel. It turned out, however, that these delays were probably the means of saving his life. Had they not occurred he would have reached Kolobeng in August. But this was the very time when the commando of the Boers, numbering 600 colonists and many natives besides, were busy with the work of death and destruction. Had he been at Kolobeng, Pretorius would probably have executed his threat of killing him; at the least he would have deen deprived of all the property that he carried with him, and his projected enterprise would have been brought to an end.

In a letter to his wife, Livingstone gives full details of the horrible outrage perpetrated shortly before by the Boers at Kolobeng:

" *Kuruman, 20th September,* 1852.—Along with this I send you a long letter; this I write in order to give you the latest news. The Boers gutted our house at Kolobeng; they brought four wagons down and took away sofa, table, bed, all the crockery, your desk (I hope it had nothing in it—Have you the letters?), smashed the wooden chairs, took away the iron ones, tore out the leaves of all the books, and scattered them in front of the house, smashed the bottles containing medicines, windows, oven-door, took away the smith-bellows, anvil, all the tools,—in fact everything worth taking; three corn-mills, a bag of coffee, for which I paid six pounds, and lots of coffee, tea, and sugar, which the gentlemen who went to the north left; took all our cattle and Paul's and Mebalwe's. They then went up to Limaüe, went to church morning and afternoon, and heard Mebalwe preach! After the second service they told Sechéle that they had come to fight, because he allowed Englishmen to proceed to the North, though they had repeatedly ordered him not to do so. He replied that he was a man of peace, that he could not molest Englishmen, because they had never done him any harm, and always treated him well. In the morning they commenced firing on the town with swivels, and set fire to it. The heat forced some of the women to flee, the men to huddle together on the small hill in the middle of the town; the smoke prevented them seeing the Boers, and the cannon killed many, sixty (60) Bakwains. The Boers then came near to kill and destroy them all, but the Bakwains killed thirty-five (35), and many horses. They fought the whole day, but the Boers could not dislodge them. They stopped firing in the evening, and then the Bakwains retired on account of having no water. The above sixty are not all men; women and children are among the slain. The Boers were 600, and they had 700 natives with them. All the corn is burned. Parties went out and burned Bangwaketse town, and swept off all the cattle. Sebubi's cattle are all gone. All the Bakhatla cattle gone. Neither Bangwaketse nor Bakhatla fired a shot. All the corn burned of the whole three tribes. Everything edible is taken from them. How will they live? They told Sechéle that the Queen had given off the land to them, and henceforth they were the masters,—had abolished chieftainship. Sir Harry Smith tried the same, and England has paid two millions of money to catch one chief, and he is still as free as the winds of heaven. How will it end? I don't know, but I will tell you the beginning. There are two parties of Boers gone to the Lake. These will to a dead certainty be cut off. They amount to thirty-six men. Parties are sent now in pursuit of them. The Bakwains will plunder and murder the Boers without mercy, and by and by the Boers will ask the English Government to assist them to put down rebellion, and of this rebellion I shall have, of course, to bear the blame. They often expressed a wish to get hold of me. I wait here a little in order to get information when the path is clear. Kind Providence detained me from falling into the very thick of it. God will preserve me still. He has work for me or He would have allowed me to go in just when the Boers were there. We shall remove more easily now that we are lightened of our furniture. They have taken away our sofa. I never had a good rest on it. We had only got it ready when we left. Well, they can't have taken away all the stones. We shall have a seat in spite of them, and that, too, with a merry heart which doeth good like a medicine. I wonder what the Peace Society would do with these worthies. They are Christians. The Dutch predicants baptize all their children, and admit them to the Lord's Supper. . . ."

Dr. Livingstone was not disposed to restrain his indignation and grief over his losses. For one so patient and good, he had a very large vial of indignation, and on occasion poured it out right heartily over all injustice and cruelty. On no heads was it ever discharged more freely than on these Transvaal Boers. He made a formal representation of his losses both to the Cape and Home authorities, but never received a farthing of compensation. The subsequent history of the Transvaal Republic will convince many that Livingstone was not far from the truth in his estimate of the character of the free and independent Boers.

But while perfectly sincere in his indignation over the treatment of the natives and his own losses, his playful fancy could find a ludicrous side for what concerned himself, and grim enjoyment in showing it to his friends. "Think," he writes to his friend Watt, "think of a big fat Boeress drinking coffee out of my kettle, and then throwing her tallowy corporeity on my sofa, or keeping her needles in my wife's writing-desk! Ugh! and then think of foolish John Bull paying so many thousands a year for the suppression of the slave-trade, and allowing Commissioner Aven to make treaties with Boers who carry on the slave-trade. . . . The Boers are mad with rage against me because my people fought bravely. It was I, they think, who taught them to shoot Boers. Fancy your reverend friend teaching the young idea how to shoot Boers, and praying for a blessing on the work of his hands!"

In the same spirit he writes to his friend Moore:

"I never knew I was so rich until I recounted up the different articles that were taken away. They cannot be replaced in this country under £300. Many things brought to our establishment by my better-half were of considerable value. Of all I am now lightened, and they want to ease me of my head. . . . The Boers kill the blacks without compunction, and without provocation, because they believe they have no souls. . . . Viewing the dispensation apart from the extreme wickedness of the Boers, it seemed a judgment on the blacks for their rejection of the gospel. They have verily done despite unto the Spirit of grace. . . . Their enmity was not manifested to us, but to the gospel. I am grieved for them, and still hope that the good seed will yet vegetate."[1]

But while he could relax playfully at the thought of the desolation at Kolobeng, he knew how to make it the occasion likewise of high resolves. The Boers, as he wrote the Directors, were resolved to shut up the interior. He was determined, with God's help, to open the country. Time would show which would be most successful in resolution, —they or he. To his brother-in-law he wrote that he would open a path through the country, *or perish.*

As for the contest with the Boers, we may smile at their impotent wrath. It is a singular fact, that while Sechéle still retains the position of an independent chief, the republic of the Boers has passed away. It is now part of the British Empire.

The country was so unsettled that for a long time Dr. Livingstone could not get guides at Kuruman to go with him to Sebituane's. At length, however, he succeeded, and leaving Kuruman finally about the end of December, 1852, in company with George Fleming, Mr. Rutherfoord's trader, he set out in a new direction, to the west of the old,

---

[1] This letter to Mr. Moore contains a trait of Livingstone, very trifling in the occasion out of which it arose, but showing vividly the nature of the man. He had promised to send Mr. Moore's little son some curiosities, but had forgotten when his family went to England. Being reminded of his promise in a postscript the little fellow had added to a letter from his father, Livingstone is "overwhelmed with shame and confusion of face." He feels he has disappointed the boy and forgotten his promise. Again and again Livingstone returns to the subject, and feels assured that his young friend would forgive him if he knew how much he suffered for his fault. That in the midst of his own overwhelming troubles he should feel so much for the disappointment of a little heart in England, shows how terrible a thing it was to him to cause needless pain, and how profoundly it distressed him to seem forgetful of a promise. Years afterward he wrote that he had brought an elephant's tail for Henry, but one of the men stole all the hairs and sold them. He had still a tusk of a hippopotamus for him, and a tooth for his brother, but he had brought no curiosities, for he could scarcely get along himself.

in order to give a wide berth to the Boers. Traveling rapidly he passed through Sebituane's country, and in June, 1858, arrived at Linyanti, the capital of the Makololo. He wrote to his wife that he had been very anxious to go to Kolobeng and see with his own eyes the destruction wrought by the savages. He had a great longing, too, to visit once more the grave of Elizabeth, their infant daughter, but he heard that the Boers were in the neighborhood, and were anxious to catch him, and he thought it best not to go. Two years before, he had been at Linyanti with Mr. Oswell. Many details of the new journey are given in the *Missionary Travels,* which it is unnecessary to repeat. It may be enough to state that he found the country flooded, and that on the way it was no unusual thing for him to be wet all day, and to walk through swamps, and water three or four feet deep. Trees, thorns, and reeds offered tremendous resistance, and he and his people must have presented a pitiable sight when forcing their way through reeds with cutting edges. "With our own hands all raw and bloody, and knees through our trousers, we at length emerged." It was a happy thought to tear his pocket-handkerchief into two parts and tie them over his knees. "I remember," he says in his Journal, referring to last year's journey, "the toil which our friend Oswell endured on our account. He never spared himself." It is not to be supposed that his guides were happy in such a march; it required his tact stretched to its very utmost to prevent them from turning back. "At the Malopo," he writes to his wife, "there were other dangers besides. When walking before the wagon in the morning twilight, I observed a lioness about fifty yards from me, in the squatting way they walk when going to spring. She was followed by a very large lion, but seeing the wagon, she turned back." Though he escaped fever at first, he had repeated attacks afterward, and had to be constantly using remedies against it. The unhealthiness of the region to Europeans forced itself painfully on his attention, and made him wonder in what way God would bring the light of the gospel to the poor inhabitants. As a physician his mind was much occupied with the nature of the disease, and the way to cure it. If only he could discover a remedy for that scourge of Africa, what an invaluable boon would he confer on its much-afflicted people!

"I would like," he says in his Journal, "to devote a portion of my life to the discovery of a remedy for that terrible disease, the African fever.[1] I would go into the parts where it prevails most, and try to dis-cover if the natives have a remedy for it. I must make many inquiries of the river people in this quarter. What an unspeakable mercy it is to be permitted to engage in this most holy and honorable work! What ▂n infinity of lots in the world are poor, miserable, and degraded com-pared with mine! I might have been a common soldier, a day-laborer, a factory operative, a mechanic, instead of a missionary. If my faculties had been left to run riot or to waste as those of so many young men, I should now have been used up, a dotard, as many of my school-fellows are. I am respected by the natives, their kind expressions often make me ashamed, and they are sincere. So much deference and favor manifested without any effort on my part to secure it comes from the Author of every good gift. I acknowledge the mercies of the great God with devout and reverential gratitude."

Dr. Livingstone had declined a considerate proposal that another missionary should accompany him, and deliber-ately resolved to go this great journey alone. He knew, in fact, that except Mr. Moffat, who was busy with his translation of the Bible, no other missionary would go with him.[2] But in the absence of all to whom he could unburden his spirit, we find him more freely than usual pouring out his feelings in his Journal, and it is but an act of justice to himself that it should be made known how

---

[1] Livingstone's Remedy for African fever. See Appendix No. II.

[2] Dr. Moffat informs us that Livingstone's desire for his company was most intense, and that he pressed him in such a way as would have been irresistible, had his going been possible. But for his employment in translating, Dr. Moffat would have gone with all his heart.

his thoughts were running, with so bold and difficult an undertaking before him:

"*28th September,* 1852.—Am I on my way to die in Sebituane's country? Have I seen the end of my wife and children? The breaking up of all my connections with earth, leaving this fair and beautiful world, and knowing so little of it? I am only learning the alphabet of it yet, and entering on an untried state of existence. Following Him who has entered in before me into the cloud, the veil, the Hades, is a serious prospect. Do we begin again in our new existence to learn much by experience, or have we full powers? My soul, whither wilt thou emigrate? Where wilt thou lodge the first night after leaving this body? Will an angel soothe thy flutterings, for sadly flurried wilt thou be in entering upon eternity? Oh! if Jesus speak one word of peace, that will establish in thy breast an everlasting calm! O Jesus, fill me with Thy love now, and I beseech Thee, accept me, and use me a little for Thy glory. I have done nothing for Thee yet, and I would like to do something. O do, do, I beseech Thee, accept me and my service, and take Thou all the glory. . . ."

"*23d January,* 1853.—I think much of my poor children. . . ."

"*4th February,* 1853.—I am spared in health, while all the company have been attacked by the fever. If God has accepted my service, then my life is charmed till my work is done. And though I pass through many dangers unscathed while working the work given me to do, when that is finished, some simple thing will give me my quietus. Death is a glorious event to one going to Jesus. Whither does the soul wing its way? What does it see first? There is something sublime in passing into the second stage of our immortal lives if washed from our sins. But, oh! to be consigned to ponder over all our sins with memories excited, every scene of our lives held up as in a mirror before our eyes, and we looking at them and waiting for the day of judgment!"

"*17th February.*—It is not the encountering of difficulties and dangers in obedience to the promptings of the inward spiritual life, which constitutes tempting of God and Providence; but the acting without faith, proceeding on our own errands with no previous convictions of duty, and no prayer for aid and direction."

"*22d May.*—I will place no value on anything I have or may possess, except in relation to the kingdom of Christ. If anything will advance the interests of that kingdom, it shall be given away or kept, only as by giving or keeping of it I shall most promote the glory of Him to whom I owe all my hopes in time and eternity. May grace and strength sufficient to enable me to adhere faithfully to this resolution be imparted to me, so that in truth, not in name only, all my interests and those of my

children may be identified with his cause. . . . I will try and remember always to approach God in secret with as much reverence in
speech, posture, and behavior as in public. Help me, Thou who knowest
my frame and pitiest as a father his children."

When Livingstone reached the Makololo, a change had
taken place in the government of the tribe. Ma-mochisane,
the daughter of Sebituane, had not been happy in her
chiefdom, and had found it difficult to get along with the
number of husbands whom her dignity as chief required
her to maintain. She had given over the government to
her brother Sekelétu, a youth of eighteen, who was generally recognized, though not without some reluctance, by
his brother, Mpepe. Livingstone could not have foreseen
how Sekelétu would receive him, but to his great relief
and satisfaction he found him actuated by the most kindly
feelings. He found him, boy as he was, full of vague
expectations of benefits, marvelous and miraculous, which
the missionaries were to bring. It was Livingstone's first
work to disabuse his mind of these expectations, and let
him understand that his supreme object was to teach them
the way of salvation through Jesus Christ. To a certain
extent Sekelétu was interested in this:

"He asked many sensible questions about the system of Christianity
in connection with the putting away of wives. They are always furnished with objections sooner than with the information. I commended
him for asking me, and will begin a course of instruction to-morrow.
He fears that learning to read will change his heart, and make him put
away his wives. Much depends on his decision. May God influence
his heart to decide aright!"

Two days after Livingstone says in his Journal:

"*1st June.*—The chief presented eight large and three small tusks this
morning. I told him and his people I would rather see them trading
than giving them to me. They replied that they would get trade with
George Fleming, and that, too, as soon as he was well; but these they
gave to their father, and they were just as any other present. They
asked after the gun-medicine, believing that now my heart would be
warm enough to tell them anything, but I could not tell them a lie. I

offered to show Sekelétu how to shoot, and that was all the medicine I knew. I felt as if I should have been more pleased had George been amassing ivory than I. Yet this may be an indispensable step in the progress toward opening the west. I must have funds; and here they come pouring in. It would be impossible to overlook his providence who has touched their hearts. I have used no undue influence. Indeed I have used none directly for the purpose Kindness shown has been appreciated here, while much greater kindness shown to tribes in the south has resulted in a belief we missionaries must be fools. I do thank my God sincerely for his favor, and my hearty prayer is that He may continue it, and make whatever use He pleases of me, and may He have mercy on this people!"

Dr. Livingstone was careful to guard against the supposition that he allowed Sekelétu to enrich him without recompense, and in his Journal he sets down a list of the various articles presented by himself to the chief, including three goats, some fowls, powder, wire, flints, percussion-caps, an umbrella and a hat, the value of the whole being £31, 16s. When Sekelétu knew Dr. Livingstone's plans, he undertook that he should be provided with all requisites for his journey. But he was most anxious to retain him, and for some time would not let him go. Livingstone had fascinated him. Sekelétu said that he had found a new father. And Livingstone pondered the possibility of establishing a station here. But the fever, the fever! could he bring his family? He must pass on and look for a healthier spot. His desire was to proceed to the country of the Barotse. At length, on the 16th June, Sekelétu gives his answer:

"The chief has acceded to my request to proceed to Barotse and see the country. I told him my heart was sore, because having left my family to explore his land, and, if possible, find a suitable location for a mission, I could not succeed, because detained by him here. He says he will take me with him. He does not like to part with me at all. He is obliged to consult with those who gave their opinion against my leaving. But it is certain I am permitted to go. Thanks be to God for influencing their hearts!"

Before we set out with the chief on this journey, it will

14

be well to give a few extracts from Livingstone's Journal, showing how unwearied were his efforts to teach the people:

"*Banks of Chobe, Sunday, May 15th.*—Preached twice to about sixty people. Very attentive. It is only divine power which can enlighten dark minds as these. . . . The people seem to receive ideas on divine subjects slowly. They listen, but never suppose that the truths must become embodied in actual life. They will wait until the chief becomes a Christian, and if he believes, then they refuse to follow,—as was the case among the Bakwains. Procrastination seems as powerful an instrument of deception here as elsewhere."

"*Sunday, 12th June.*—A good and very attentive audience. We introduce entirely new motives, and were these not perfectly adapted for the human mind and heart by their divine Author, we should have no success."

"*Sunday, 19th June.*—A good and attentive audience, but immediately after the service I went to see a sick man, and when I returned toward the Kotla, I found the chief had retired into a hut to drink beer; and, as the custom is, about forty men were standing singing to him, or, in other words, begging beer by that means. A minister who had not seen so much pioneer service as I have done would have been shocked to see so little effect produced by an earnest discourse concerning the future judgment, but time must be given to allow the truth to sink into the dark mind, and produce its effect. The earth shall be filled with the knowledge of the glory of the Lord—that is enough. We can afford to work in faith, for Omnipotence is pledged to fulfill the promise. The great mountains become a plain before the Almighty arm. The poor Bushman, the most degraded of all Adam's family, shall see his glory, and the dwellers in the wilderness shall bow before Him. The obstacles to the coming of the Kingdom are mighty, but come it will for all that:

> "'Then let us pray that come it may,
> As come it will for a' that,
> That man to man the world o'er
> Shall brothers be for a' that.'

"The hard and cold unbelief which distinguished the last century, and which is still aped by would-be philosophers in the present, would sneer at our faith, and call it superstition, enthusiasm, etc. But were we believers in human progress and no more, there must be a glorious future for our world. Our dreams must come true, even though they are no more than dreams. The world is rolling on to the golden age.

. . . Discoveries and inventions are cumulative. Another century must present a totally different aspect from the present. And when we view the state of the world and its advancing energies, in the light afforded by childlike, or call it childish, faith, we see the earth filling with the knowledge of the glory of God,—ay, all nations seeing his glory and bowing before Him whose right it is to reign. Our work and its fruits are cumulative. We work toward another state of things. Future missionaries will be rewarded by conversions for every sermon. We are their pioneers and helpers. Let them not forget the watchmen of the night—us, who worked when all was gloom, and no evidence of success in the way of conversion cheered our paths. They will doubtless have more light than we, but we served our Master earnestly, and proclaimed the same gospel as they will do."

Of the services which Livingstone held with the people, we have the following picture:

"When I stand up, all the women and children draw near, and, having ordered silence, I explain the plan of salvation, the goodness of God in sending his Son to die, the confirmation of his mission by miracles, the last judgment or future state, the evil of sin, God's commands respecting it, etc.; always choosing one subject only for an address, and taking care to make it short and plain, and applicable to them. This address is listened to with great attention by most of the audience. A short prayer concludes the service, all kneeling down, and remaining so till told to rise. At first we have to enjoin on the women who have children to remain sitting, for when they kneel, they squeeze their children, and a simultaneous skirl is set up by the whole troop of youngsters, who make the prayer inaudible."

When Livingstone and Sekelétu had gone about sixty miles on the way to the Barotse, they encountered Mpepe, Sekelétu's half-brother and secret rival. It turned out that Mpepe had a secret plan for killing Sekelétu, and that three times on the day of their meeting that plan was frustrated by apparently accidental causes. On one of these occasions, Livingstone, by covering Sekelétu, prevented him from being speared. Mpepe's treachery becoming known, he was arrested by Sekelétu's people, and promptly put to death. The episode was not agreeable, but it illustrated savage life. It turned out that Mpepe favored the

slave-trade, and was closely engaged with certain Portuguese traders in intrigues for establishing and extending it.
Had Sekelétu been killed, Livingstone's enterprise would
certainly have been put an end to, and very probably likewise Livingstone himself.

The party, numbering about one hundred and sixty,
proceeded up the beautiful river which on his former visit
Livingstone had first known as the Sesheke, but which was
called by the Barotse the Liambai or Leeambye. The
term means "the large river," and Luambeji, Luambesi,
Ambezi, Yimbezi, and Zambezi are names applied to it at
different parts of its course. In the progress of their journey
they came to the town of the father of Mpepe, where, most
unexpectedly, Livingstone encountered a horrible scene.
Mpepe's father and another headman were known to have
favored the plan for the murder of Sekelétu, and were
therefore objects of fear to the latter. When all were met,
and Mpepe's father was questioned why he did not stop his
son's proceedings, Sekelétu suddenly sprang to his feet and
gave the two men into custody. All had been planned
beforehand. Forthwith they were led away, surrounded
by Sekelétu's warriors, all dream of opposition on their
part being as useless as interference would have been on
Livingstone's. Before his eyes he saw them hewn to pieces
with axes, and cast into the river to be devoured by the
alligators. Within two hours of their arrival the whole
party had left the scene of this shocking tragedy, Livingstone being so horrified that he could not remain. He
did his best to show the sin of blood-guiltiness, and bring
before the people the scene of the Last Judgment, which
was the only thing that seemed to make any impression.

Farther on his way he had an interview with Ma-mochisane, the daughter of Sebituane who had resigned in favor
of Sekelétu. He was the first white man she had ever seen.
The interview was pleasing and not without touches of
womanly character; the poor woman had felt an *embarras*

de richesses in the matter of husbands, and was very uncomfortable when married women complained of her
taking their spouses from them. Her soul recoiled from
the business; she wished to have a husband of her own
and to be like other women.

So anxious was Livingstone to find a healthy locality,
that, leaving Sekelétu, he proceeded to the farthest limit
of the Barotse country, but no healthy place could be found.
It is plain, however, that in spite of all risk, and much as
he suffered from the fever, he was planning, if no better
place could be found, to return himself to Linyanti and be
the Makololo missionary. Not just immediately, however.
Having failed in the first object of his journey—to find a
healthy locality—he was resolved to follow out the second,
and endeavor to discover a highway to the sea. First he
would try the west coast, and the point for which he would
make was St. Paul de Loanda. He might have found a
nearer way, but a Portuguese trader whom he had met,
and from whom he had received kindness, was going by
that route to St. Philip de Benguela. The trader was im-
plicated in the slave-trade, and Livingstone knew what a
disadvantage it would be either to accompany or to follow
him. He therefore returned to Linyanti; and there began
preparations for the journey to Loanda on the coast.

During the time thus spent in the Barotse country,
Livingstone saw heathenism in its most unadulterated
form. It was a painful, loathsome, and horrible spectacle.
His views of the Fall and of the corruption of human
nature were certainly not lightened by the sight. In his
Journal he is constantly letting fall expressions of weariness at the noise, the excitement, the wild savage dancing,
the heartless cruelty, the utter disregard of feelings, the
destruction of children, the drudgery of the old people, the
atrocious murders with which he was in contact. Occasionally he would think of other scenes of travel; if a friend,
for example, were going to Palestine, he would say how

gladly he would kiss the dust that had been trod by the Man of Sorrows. One day a poor girl comes hungry and naked to the wagons, and is relieved from time to time; then disappears to die in the woods of starvation or be torn in pieces by the hyenas. Another day, as he is preaching, a boy, walking along with his mother, is suddenly seized by a man, utters a shriek as if his heart had burst, and becomes, as Livingstone finds, a hopeless slave. Another time, the sickening sight is a line of slaves attached by a chain. That chain haunts and harrows him.

Amid all his difficulties he patiently pursued his work as missionary. Twice every Sunday he preached, usually to good audiences, the number rising on occasions so high as a thousand. It was a great work to sow the good seed so widely, where no Christian man had ever been, proclaiming every Lord's Day to fresh ears the message of Divine love. Sometimes he was in great hopes that a true impression had been made. But usually, whenever the service was over, the wild savage dance with all its demon noises succeeded, and the missionary could but look on and sigh. So ready was he for labor that when he could get any willing to learn, he commenced teaching them the alphabet. But he was continually met by the notion that his religion was a religion of medicines, and that all the good it could do was by charms. Intellectual culture seemed indispensable to dissipate this inveterate superstition regarding Christian influence.

A few extracts from his Journal in the Barotse country will more vividly exhibit his state of mind:

"*27th August*, 1853.—The more intimately I become acquainted with barbarians, the more disgusting does heathenism become. It is inconceivably vile. They are always boasting of their fierceness, yet dare not visit another tribe for fear of being killed. They never visit anywhere but for the purpose of plunder and oppression. They never go anywhere but with a club or spear in hand. It is lamentable to see those who might be children of God, dwelling in peace and love, so utterly the children of the devil, dwelling in fear and continual irritation. They bestow honors and flattering titles on me in confusing profusion. All from the least to the greatest call me Father, Lord, etc., and bestow food without recompense, out of pure kindness. They need a healer. May God enable me to be such to them. . . .

"*31st August.*—The slave-trade seems pushed into the very centre of the continent from both sides. It must be profitable. . . .

"*September* 25, *Sunday.*—A quiet audience to-day. The seed being sown, the least of all seeds now, but it will grow a mighty tree. It is as it were a small stone cut out of a mountain, but it will fill the whole earth. He that believeth shall not make haste. Surely if God can bear with hardened impenitent sinners for thirty, forty, or fifty years, waiting to be gracious, we may take it for granted that his is the best way. He could destroy his enemies, but He waits to be gracious. To become irritated with their stubbornness and hardness of heart is ungodlike. . . .

"*13th October.*—Missionaries ought to cultivate a taste for the beautiful. We are necessarily compelled to contemplate much moral impurity and degradation. We are so often doomed to disappointment. We are apt to become either callous or melancholy, or, if preserved from these, the constant strain on the sensibilities is likely to injure the bodily health. On this account it seems necessary to cultivate that faculty for the gratification of which God has made such universal provision. See the green earth and blue sky, the lofty mountain and the verdant valley, the glorious orbs of day and night, and the starry canopy with all their celestial splendor, the graceful flowers so chaste in form and perfect in coloring. The various forms of animated life present to him whose heart is at peace with God through the blood of his Son an indescribable charm. He sees in the calm beauties of nature such abundant provision for the welfare of humanity and animate existence. There appears on the quiet repose of earth's scenery the benignant smile of a Father's love. The sciences exhibit such wonderful intelligence and design in all their various ramifications, some time ought to be devoted to them before engaging in missionary work. The heart may often be cheered by observing the operation of an ever-present intelligence, and we may feel that we are leaning on his bosom while living in a world clothed in beauty, and robed with the glorious perfections of its maker and preserver. We must feel that there is a Governor among the nations who will bring all his plans with respect to our human family to a glorious consummation. He who stays his mind on his ever-present, ever-energetic God, will not fret himself because of evil-doers. He that believeth shall not make haste."

"*26th October.*—I have not yet met with a beautiful woman among the black people, and I have seen many thousands in a great variety of tribes. I have seen a few who might be called passable, but none at all to be compared to what one may meet among English servant-girls. Some beauties are said to be found among the Caffres, but among the people I have seen I cannot conceive of any European being captivated with them. The whole of my experience goes toward proving that civilization alone produces beauty, and exposure to the weather and other vicissitudes tend to the production of deformation and ugliness. . . .

"*28th October.*—The conduct of the people whom we have brought from Kuruman shows that no amount of preaching or instruction will insure real piety. . . . The old superstitions cannot be driven out of their minds by faith implanted by preaching. They have not vanished in either England or Scotland yet, after the lapse of centuries of preaching. Kuruman, the entire population of which amounted in 1853 to 638 souls, enjoys and has enjoyed the labors of at least two missionaries,—four sermons, two prayer-meetings, infant schools, adult schools, sewing schools, classes, books, etc., and the amount of visible success is very gratifying, a remarkable change indeed from the former state of these people. Yet the dregs of heathenism still cleave fast to the minds of the majority. They have settled deep down into their souls, and one century will not be sufficient to elevate them to the rank of Christians in Britain. The double influence of the spirit of commerce and the gospel of Christ has given an impulse to the civilization of men. The circulation of ideas and commodities over the face of the earth, and the discovery of the gold regions, have given enhanced rapidity to commerce in other countries, and the diffusion of knowledge. But what for Africa? God will do something else for it; something just as wonderful and unexpected as the discovery of gold."

It needs not to be said that his thoughts were very often with his wife and children. A tender letter to the four little ones shows that though some of them might be beginning to forget him, their names were written imperishably on his heart:

"*Sekelétu's Town, Linyanti, 2d October.*—MY DEAR ROBERT, AGNES, AND THOMAS AND OSWELL,—Here is another little letter for you all. I should like to see you much more than write to you, and speak with my tongue rather than with my pen; but we are far from each other—very, very far. Here are Seipone, and Meriye and others who saw you as the first white children they ever looked at. Meriye came the other day and brought a round basket for Nannie. She made it of the leaves of the palmyra. Others put me in mind of you all by calling me Rananee, and Rarobert, and there is a little Thomas in the town, and when I think of you I remember, though I am far off, Jesus, our good and gracious Jesus, is ever near both you and me, and then I pray to Him to bless you and make you good.

"He is ever near. Remember this if you feel angry or naughty. Jesus is near you, and sees you, and He is so good and kind. When He was among men, those who heard Him speak said, ' Never man spake like this man,' and we now say, ' Never did man love like Him.' You see little Zouga is carried on mamma's bosom. You are taken care of by Jesus with as much care as mamma takes of Zouga. He is always watching you and keeping you in safety. It is very bad to sin, to do any naughty things, or speak angry or naughty words before Him.

"My dear children, take Him as your Guide, your Helper, your Friend, and Saviour through life. Whatever you are troubled about ask Him to keep you. Our God is good. We thank Him that we have such a Saviour and Friend as He is. Now you are little, but you will not always be so, hence you must learn to read and write and work. All clever men can both read and write, and Jesus needs clever men to do his work. Would you not like to work for Him among men? Jesus is wishing to send his gospel to all nations, and He needs clever men to do this. Would you like to serve Him? Well, you must learn now, and not get tired learning. After some time you will like learning better than playing, but you must play, too, in order to make your bodies strong and be able to serve Jesus.

"I am glad to hear that you go to the academy. I hope you are learning fast. Don't speak Scotch. It is not so pretty as English. Is the Tau learning to read with mamma? I hope you are all kind to mamma. I saw a poor woman in a chain with many others, up at the Barotse. She had a little child, and both she and her child were very thin. See how kind Jesus was to you. No one can put you in chains unless you become bad. If, however, you learn bad ways, beginning only by saying bad words or doing little bad things, Satan will have you in the chains of sin, and you will be hurried on in his bad ways till you are put into the dreadful place which God hath prepared for him and all who are like him. Pray to Jesus to deliver you from sin, give you new hearts, and make you his children. Kiss Zouga, mamma, and each other for me.—Your ever affectionate father,

"D. LIVINGSTON."

A letter to his father and other relations at Hamilton, 30th September, 1853, is of a somewhat apologetic and explanatory cast. Some of the friends had the notion that he should have settled somewhere, "preaching the simple gospel," and converting people by every sermon:

"You see what they make of the gospel, and my conversation on it, in which my inmost heart yearned for their conversion. Many now think Jesus and Sebituane very much the same sort of person. I was prevented by fever and other matters from at once following up the glorious object of this journey: viz., while preaching the gospel beyond every other man's line of things made ready to our hands, to discover a healthy location for a mission, and I determined to improve the time by teaching to read. This produced profound deliberation and lengthened palavers, and at length the chief told me that he feared learning to read would change his heart and make him content with one wife like Sechéle. He has four. It was in vain I urged that the change contemplated made the affair as voluntary as if he would now change his mind from four to thirty, as his father had. He could not realize the change that would give relish to any other system than the present. He felt as the man who is mentioned by Serles as saying he would not like to go to heaven to be employed for ever singing and praising on a bare cloud without anything to eat or drink. . . .

"The conversion of a few, however valuable their souls may be, cannot be put into the scale against the knowledge of the truth spread over the whole country. In this I do and will exult. As in India, we are doomed to perpetual disappointment; but the knowledge of Christ spreads over the masses. We are like voices crying in the wilderness. We prepare the way for a glorious future in which missionaries telling the same tale of love will convert by every sermon. I am trying now to establish the Lord's kingdom in a region wider by far than Scotland. Fever seems to forbid; but I shall work for the glory of Christ's kingdom —fever or no fever. All the intelligent men who direct our society and understand the nature of my movements support me warmly. A few, I understand, in Africa, in writing home, have styled my efforts as 'wanderings.' The very word contains a lie coiled like a serpent in its bosom. It means traveling without an object, or uselessly. I am now performing the duty of writing you. If this were termed 'dawdling,' it would be as true as the other. . . . I have actually seen letters to the Directors in which I am gravely charged with holding the views of the Plymouth Brethren. So very sure am I that I am in the path which God's Providence has pointed out, as that by which Christ's kingdom is to be promoted, that if the Society should object, I would consider it my duty to withdraw from it. . .

"*P.S.*—My throat became well during the long silence of traveling across the desert. It plagues again now that I am preaching in a moist climate."

Dr. Livingstone now began his preparations for the journey from Linyanti to Loanda. Sekelétu was kind and generous. The road was impracticable for wagons, and the native trader, George Fleming, returned to Kuruman. The Kuruman guides had not done well, so that Livingstone resolved to send them back, and to get Makololo men instead. Here is the record of his last Sunday at Linyanti:

"*6th Nov.*, 1853.—Large audience. Kuruman people don't attend. If it is a fashion to be church-going, many are drawn into its observance. But placed in other circumstances, the true character comes out. This is the case with many Scotchmen. May God so imbue my mind with the spirit of Christianity that in all circumstances I may show my Christian character! Had a long conversation with Motlube, chiefly on a charm for defending the town or for gun medicine. They think I know it but will not impart the secret to them. I used every form of expression to undeceive him, but to little purpose. Their belief in medicine which will enable them to shoot well is very strong, and simple trust in an unseen Saviour to defend them against such enemies as the Matebele is too simple for them. I asked if a little charcoal sewed up in a bag were a more feasible protector than He who made all things, and told them that one day they would laugh heartily at their own follies in bothering me so much for gun medicine. A man who has never had to do with a raw heathen tribe has yet to learn the Missionary A B C."

On the 8th he writes:

"Our intentions are to go up the Leeba till we reach the falls, then send back the canoe and proceed in the country beyond as best we can. Matiamvo is far beyond, but the Cassantse (probably Cassange) live on the west of the river. May God in mercy permit me to do something for the cause of Christ in these dark places of the earth! May He accept my children for his service, and sanctify them for it! My blessing on my wife. May God comfort her! If my watch comes back after I am cut off, it belongs to Agnes. If my sextant, it is Robert's. The Paris medal to Thomas. Double-barreled gun to Zouga. Be a Father to the fatherless, and a Husband to the widow, for Jesus' sake."

The probability of his falling was full in his view. But the thought was ever in his mind, and ever finding expression in letters both to the Missionary and the Geographical

Societies, and to all his friends,—"Can the love of Christ not carry the missionary where the slave-trade carries the trader?" His wagon and goods were left with Sekelétu, and also the Journal from which these extracts are taken.[1] It was well for him that his conviction of duty was clear as noonday. A year after, he wrote to his father-in-law:

"I had fully made up my mind as to the path of duty before starting. I wrote to my brother-in-law, Robert Moffat: 'I shall open up a path into the interior, or perish.' I never have had the shadow of a shade of doubt as to the propriety of my course, and wish only that my exertions may be honored so far that the gospel may be preached and believed in all this dark region."

---

[1] This Journal is mentioned in the *Missionary Travels* as having been lost (p. 229). It was afterward recovered. It contains, among other things, some important notes on Natural History.

CHAPTER VIII.

FROM LINYANTI TO LOANDA.

A.D. 1853–1854.

Difficulties and hardships of journey—His traveling kit—Four books—His Journal—Mode of traveling—Beauty of country—Repulsiveness of the people—Their religious belief—The negro—Preaching—The magic-lantern—Loneliness of feeling—Slave-trade—Management of the natives—Danger from Chiboque—from another chief—Livingstone ill of fever—At the Quango—Attachment of followers—" The good time coming "—Portuguese settlements—Great kindness of the Portuguese—Arrives at Loanda—Received by Mr. Gabriel—His great friendship—No letters—News through Mr. Gabriel—Livingstone becomes aquainted with naval officers—Resolves to go back to Linyanti and make for East Coast—Letter to his wife—Correspondence with Mr. Maclear—Accuracy of his observations—Sir John Herschel—Geographical Society award their gold metal—Remarks of Lord Ellesmere.

THE journey from Linyanti to Loanda occupied from the 11th November, 1853, to 31st May, 1854. It was in many ways the most difficult and dangerous that Livingstone had yet performed, and it drew out in a very wonderful manner the rare combination of qualities that fitted him for his work. The route had never been traversed, so far as any trustworthy tradition went, by any European. With the exception of a few of Sekelétu's tusks, the oxen needed for carrying, and a trifling amount of coffee, cloth, beads, etc., Livingstone had neither stores of food for his party, nor presents with which to propitiate the countless tribes of rapacious and suspicious savages that lined his path. The Barotse men who accompanied him, usually called the "Makololo," though on the whole faithful and patient, "the best that ever accompanied me," were a burden in one sense, as much as a help in another; chicken-hearted, ready to succumb to every trouble, and to be cowed by any chief that wore a threatening face. Worse if possible, Livingstone himself was in wretched health. During this part of the journey he had constant attacks of intermittent fever,[1] accompanied in the latter stages of the road with dysentery of the most distressing kind. In the intervals of fever he was often depressed alike in body and in mind. Often the party were destitute of food of any sort, and never had they food suitable for a fever-stricken invalid. The vexations he encountered were of no common kind: at starting, the greater part of his medicines was stolen, much though he needed them; in the course of the journey, his pontoon was left behind; at one time, while he was under the influence of fever, his riding-ox threw him, and he fell heavily on his head; at another, while crossing a river, the ox tossed him into the water; the heavy rains, and the necessity of wading through streams three or four times a day, kept him almost constantly wet; and occasionally, to vary the annoyance, mosquitos would assail him as fiercely as if they had been waging a war of extermination. The most critical moments of peril, demanding the utmost coolness and most dauntless courage, would sometimes occur during the stage of depression after fever; it was then he had to extricate himself from savage warriors, who vowed that he must go back, unless he gave them an ox, a gun, or a man. The ox he could ill spare, the gun not at all, and as for giving the last—a man—to make a slave of, he would sooner die. At the best, he was a poor ragged skeleton when he reached those who had hearts to feel for him and hands to help him. Had he not been a prodigy of patience, faith, and courage, had he not known where to find help in all time of his tribulation, he would never have reached the haunts of civilized men.

His traveling-kit was reduced to the smallest possible bulk; that he minded little, but he was vexed to be able to

---

[1] The number of attacks was thirty-one.

take so few books. A few days after setting out, he writes
in his private Journal:

"I feel the want of books in this journey more than anything else.
A Sichuana Pentateuch, a lined journal, Thomson's Tables, a Nautical
Almanac, and a Bible, constitute my stock. The last constitutes my
chief resource; but the want of other mental pabulum is felt severely.
There is little to interest in the conversation of the people. Loud disputes often about the women, and angry altercations in which the same
string of abuse is used, are more frequent than anything else."

The "lined journal," of which mention is made here, was
probably the most wonderful thing of the kind ever taken
on such a journey. It is a strongly bound quarto volume
of more then 800 pages, with a lock and key. The writing
is so neat and clear that it might almost be taken for litho-
graph. Occasionally there is a page with letters beginning
to sprawl, as if one of those times had come when he tells
us that he could neither think nor speak, nor tell any one's
name—possibly not even his own, if he had been asked it.
He used to jot his observations on little note-books, and
extend them when detained by rain or other causes.

The journal differs in some material respects from the
printed record of this journey. It is much more explicit in
setting forth the bad treatment he often received. When
he spoke of these things to the public, he made constant
use of the mantle of charity, and the record of many a bad
deed and many a bad character is toned down. Naturally,
too, the journal is more explicit on the subject of his own
troubles, and more free in recording the play of his feelings.
It does not hide the communings of his heart with his
heavenly Father. It is built up in a random-rubble style;
here a solemn prayer, in the next line a note of lunar
observations; then a dissertation on the habits of the hippopotamus. Notes bearing on the character, the superstitions,
and the feelings of the natives are of frequent occurrence.
The explanation is, that Livingstone put down everything
as it came, reserving the arranging and digesting of the

whole to a future time. The extremely hurried manner in which he was obliged to write his *Missionary Travels* prevented him from fulfilling all his plan, and compelled him to content himself with giving to the public then what could be put most readily together. There are indications that he contemplated in the end a much more thorough use of his materials. It is not to be supposed that his published volumes contained all that he deemed worthy of publication, or that a censure is due to those who reproduce some portions which he passed over. As to the neat and finished form in which the Journal exists, it was one of the many fruits of a strong habit of orderliness and self-respect which he had begun to learn at the hand of his mother, and which he practiced all his life. Even in the matter of personal cleanliness and dress he was uniformly most attentive in his wanderings among savages. "I feel certain," he said, "that the lessons of cleanliness rigidly instilled by my mother in childhood helped to maintain that respect which these people entertain for European ways."

The course of the journey was first along the river Zambesi, as he had gone before with Sekelétu, to its junction with the Leeba, then along the Leeba to the country of Lobale on the left and Londa on the right. Then, leaving the canoes, he traveled on oxback first N.N.W. and then W. till he reached St. Paul de Loanda on the coast. His Journal, like the published volume, is full of observations on the beauty and wonderful capacity and productiveness of the country through which he passed after leaving the river. Instinctively he would compare it with Scotland. A beautiful valley reminds him of his native vale of Clyde, seen from the spot where Mary Queen of Scots saw the battle of Langside; only the Scottish scene is but a miniature of the much greater and richer landscape before him. At the sight of the mountains he would feel his Highland blood rushing through him, banishing all thoughts of fever and fatigue. If only the blessings of the gospel could be spread among the people, what a glorious land it would become! But alas for the people! In most cases they were outwardly very repulsive. Never seen without a spear or a club in their hands, the men seemed only to delight in plunder and slaughter, and yet they were utter cowards. Their mouths were full of cursing and bitterness. The execrations they poured on each other were incredible. In very wantonness, when they met they would pelt each other with curses, and then perhaps burst into a fit of laughter. The women, like the men, went about in almost total nudity, and seemed to know no shame. So reckless were the chiefs of human life, that a man might be put to death for a single distasteful word; yet sometimes there were exhibitions of very tender feeling. The headman of a village once showed him, with much apparent feeling, the burnt house of a child of his, adding,—"She perished in it, and we have all removed from our own huts and built here round her, in order to weep over her grave." From some of the people he received great kindness; others were quite different. Their character, in short, was a riddle, and would need to be studied more. But the prevalent aspect of things was both distressing and depressing. If he had thought of it continually, he would have become the victim of melancholy. It was a characteristic of his large and buoyant nature, that, besides having the resource of spiritual thought, he was able to make use of another divine corrective to such a tendency, to find delightful recreation in science, and especially in natural history, and by this means turn the mind away for a time from the dark scenes of man's depravity.

The people all seemed to recognize a Supreme Being; but it was only occasionally, in times of distress, that they paid Him homage. They had no love for Him like that of Christians for Jesus—only terror. Some of them, who were true negroes, had images, simple but grotesque. Their strongest belief was in the power of medicines acting as charms. They fully recognized the existence of the soul after death. Some of them believed in the metamorphosis of certain persons into alligators or hippopotamuses, or into lions. This belief could not be shaken by any arguments—at least on the part of man. The negroes proper interested him greatly; they were numerous, prolific, and could not be extirpated. He almost regretted that Mr. Moffat had translated the Bible into Sichuana. That language might die out; but the negro might sing, "Men may come and men may go, but I go on for ever."

The incessant attacks of fever from which Livingstone suffered in this journey, the continual rain occurring at that season of the year, the return of the affection of the throat for which he had got his uvula excised, and the difficulty of speaking to tribes using different dialects, prevented him from holding his Sunday services as regularly as before. Such entries in his Journal as the following are but too frequent:

"*Sunday*, 19*th*.—Sick all Sunday and unable to move. Several of the people were ill too, so that I could do nothing but roll from side to side in my miserable little tent, in which, with all the shade we could give it, the thermometer stood upward of 90°."

But though little able to preach, Livingstone made the most of an apparatus which in some degree compensated his lack of speech—a magic-lantern which his friend, a former fellow-traveler, Mr. Murray, had given him. The pictures of Abraham offering up Isaac, and other Bible scenes, enabled him to convey important truths in a way that attracted the people. It was, he says, the only service he was ever asked to repeat. The only uncomfortable feeling it raised was on the part of those who stood on the side where the slides were drawn out. They were terrified lest the figures, as they passed along, should take possession of them, entering like spirits into their bodies!

The loneliness of feeling engendered by the absence of all human sympathy was trying. "Amidst all the beauty and loveliness with which I am surrounded, there is still a feeling of want in the soul,—as if something more were needed to bathe the soul in bliss than the sight of the perfection in working and goodness in planning of the great Father of our spirits. I need to be purified—fitted for the eternal, to which my soul stretches away, in ever returning longings. I need to be made more like my blessed Saviour, to serve my God with all my powers. Look upon me, Spirit of the living God, and supply all Thou seest lacking."

It was Livingstone's great joy to begin this long journey with a blessed act of humanity, boldly summoning a trader to release a body of captives, so that no fewer than eighteen souls were restored to freedom. As he proceeded he obtained but too plain evidence of the extent to which the slave traffic prevailed, uniformly finding that wherever slavers had been, the natives were more difficult to deal with and more exorbitant in their demands. Slaves in chains were sometimes met with—a sight which some of his men had never beheld before.

Livingstone's successful management of the natives constituted the crowning wonder of this journey. Usually the hearts of the chiefs were wonderfully turned to him, so that they not only allowed him to pass on, but supplied him with provisions. But there were some memorable occasions on which he and his company appeared to be doomed. When he passed through the Chiboque country, the provisions were absolutely spent; there was no resource but to kill a riding-ox, a part of which, according to custom, was sent to the chief. Next day was Sunday. After service the chief sent an impudent message demanding much more valuable presents. His people collected round Livingstone, brandishing their weapons, and one young man all but brought down his sword on his head.

It seemed impossible to avoid a fight; yet Livingstone's management prevailed—the threatened storm passed away.

Some days after, in passing through a forest in the dominions of another chief, he and his people were in momentary expectation of an attack. They went to the chief's village and spoke to the man himself; and here, on a Sunday, while ill of fever, Livingstone was able to effect a temporary settlement. The chief sent them some food; then yams, a goat, fowl, and meat. Livingstone gave him a shawl, and two bunches of beads, and he seemed pleased. During these exciting scenes he felt no fever; but when they were over the constant wettings made him experience a sore sense of sinking, and this Sunday was a day " of perfect uselessness." Monday came, and while Livingstone was as low as possible, the inexorable chief renewed his demands. " It was," he says, " a day of torture."

" After talking nearly the whole day we gave the old chief an ox, but he would not take it, but another. I was grieved exceedingly to find that our people had become quite disheartened, and all resolved to return home. All I can say has no effect. I can only look up to God to influence their minds, that the enterprise fail not, now that we have reached the very threshold of the Portuguese settlements. I am greatly distressed at this change, for what else can be done for this miserable land I do not see. It is shut. O Almighty God, help, help! and leave not this wretched people to the slave-dealer and Satan. The people have done well hitherto, I see God's good influence in it. Hope He has left only for a little season. No land needs the gospel more than this miserable portion. I hope I am not to be left to fail in introducing it."

On Wednesday morning, however, final arrangements were made, and the party passed on in peace. Ten days later, again on a Sunday, they were once more pestered by a great man demanding dues. Livingstone replied by simply defying him. He might kill him, but God would judge. And on the Monday they left peaceably, thankful for their deliverance, some of the men remarking, in view of it, that they were " children of Jesus," and Livingstone thanking God devoutly for his great mercy. Next day they were again stopped at the river Quango. The poor Makololo had parted in vain with their copper ornaments, and Livingstone with his razors, shirts, etc.; yet he had made up his mind (as he wrote to the Geographical Society afterward) to part with his blanket and coat to get a passage, when a young Portuguese sergeant, Cypriano de Abrao, made his appearance, and the party were allowed to pass.

There were many proofs that, though a poor set of fellows, Livingstone's own followers were animated with extraordinary regard for him. No wonder! They had seen how sincere he was in saying that he would die rather than give any of them up to captivity. And all his intercourse with them had been marked by similar proofs of his generosity and kindness. When the ox flung him into the river, about twenty of them made a simultaneous rush for his rescue, and their joy at his safety was very great.

Amid all that was discouraging in the present aspect of things, Livingstone could always look forward and rejoice in the good time coming:

" *Sunday 22d.*—This age presents one great fact in the Providence of God ; missions are sent forth to all quarters of the world,—missions not of one section of the Church, but of all sections, and from nearly all Christian nations. It seems very unfair to judge of the success of these by the number of conversions which have followed. These are rather proofs of the missions being of the right sort. They show the direction of the stream which is set in motion by Him who rules the nations, and is destined to overflow the world. The fact which ought to stimulate us above all others is, not that we have contributed to the conversion of a few souls, however valuable these may be, but that we are diffusing a knowledge of Christianity throughout the world. The number of conversions in India is but a poor criterion of the success which has followed the missionaries there. The general knowledge is the criterion; and there, as well as in other lands where missionaries in the midst of masses of heathenism seem like voices crying in the wilderness—Reformers before the Reformation, future missionaries will see conversions follow every sermon. We prepare the way for them. May they not forget the pioneers who worked in the thick gloom with few rays to cheer, except such as flow from faith in God's promises! We work for a glorious future which we are not destined to see—the golden age which has not been, but will yet be. We are only morning-stars shining in the dark, but the glorious morn will break, the good time coming yet. The present mission-stations will all be broken up. No matter how great the outcry against the instrumentality which God employs for his purposes, whether by French soldiery as in Tahiti, or tawny Boers as in South Africa, our duty is onward, onward, proclaiming God's Word whether men will hear or whether they will forbear. A few conversions show whether God's Spirit is in a mission or not. No mission which has his approbation is entirely unsuccessful. His purposes have been fulfilled, if we have been faithful. 'The nation or kingdom that will not serve Thee shall utterly be destroyed'—this has often been preceded by free offers of friendship and mercy, and many missions which He has sent in the olden time seemed bad failures. Noah's preaching was a failure, Isaiah thought his so too. Poor Jeremiah is sitting weeping tears over his people, everybody cursing the honest man, and he ill-pleased with his mother for having borne him among such a set. And Ezekiel's stiff-necked, rebellious crew were no better. Paul said, 'All seek their own, not the things of Jesus Christ,' and he knew that after his departure grievous wolves would enter in, not sparing the flock. Yet the cause of God is still carried on to more enlightened developments of his will and character, and the dominion is being given by the power of commerce and population unto the people of the saints of the Most High. And this is an everlasting kingdom, a little stone cut out of a mountain without hands which shall cover the whole earth. For this time we work; may God accept our imperfect service!"

At length Livingstone began to get near the coast, reaching the outlying Portuguese stations. He was received by the Portuguese gentlemen with great kindness, and his wants were generously provided for. One of them gave him the first glass of wine he had taken in Africa. Another provided him with a suit of clothing. Livingstone invoked the blessing of Him who said, "I was naked and ye clothed me." His Journal is profuse in its admiration of some of the Portuguese traders, who did not like the slave-trade— not they, but had most enlightened views for the welfare of Africa. But opposite some of these eulogistical passages of the Journal there were afterward added an expressive series of marks of interrogation. At a later date he saw reason to doubt the sincerity of some of the professions of these gentlemen. Ingenuous and trustful, he could at first think nothing but good of those who had shown him such marked attention. Afterward, the inexorable logic of facts proved too strong, even for his unsuspecting soul. But the kindness of the Portuguese was most genuine, and Livingstone never ceased to be grateful for a single kind act. It is important to note that whatever he came to think of their policy afterward, he was always ready to make this acknowledgment.

Arrived at Loanda, 31st May, 1854, with his twenty-seven followers, he was most kindly received by Mr. Edmund Gabriel, the British Commissioner for the suppression of the slave-trade there, and everything was done by him for his comfort. The sensation of lying on an English bed, after six months lying on the ground, was indescribably delightful. Mr. Gabriel was equally attentive to him during a long and distressing attack of fever and dysentery that prostrated him soon after his arrival at Loanda. In his Journal the warmest benedictions are poured on Mr. Gabriel, and blessings everlasting besought for his soul. One great disappointment he suffered at Loanda—not a single letter was awaiting him. His friends must have thought he could never reach it. This want of letters was a very frequent trial, especially to one who wrote so many, and of such length. The cordial friendship of Mr. Gabriel, however, was a great solace. He gave him much information, not only on all that concerned the slave-trade—now more than ever attracting his attention—but also on the natural history of the district, and he entered *con amore* into the highest objects of his mission. Afterward, in acknowledging to the Directors of the London Missionary Society receipt of a letter for Dr. Livingstone, intrusted to his care, Mr. Gabriel wrote as follows (20th March, 1856):

" Dr. Livingstone, after the noble objects he has achieved, most assuredly wants no testimony from me. I consult, therefore, the impulse of my own mind alone, when I declare that in no respect was my intercourse more gratifying to me than in the opportunities afforded to me of observing his *earnest, active, and unwearied solicitude for the advancement of Christianity.* Few, perhaps, have had better opportunities than myself of estimating *the benefit the Christian cause in this country has derived from Dr. Livingstone's exertions.* It is indeed fortunate for that sacred cause, and highly honorable to the London Missionary Society, *when qualities and dispositions like his are employed in propagating its blessings among men.* Irrespective, moreover, of his *laudable and single-minded conduct as a minister of the Gospel,* and his attainments in making observations which have determined the true geography of the interior, the Directors, I am sure, will not have failed to perceive how interesting and valuable are all the communications they receive from him—as sketches of the social condition of the people, and the material, fabrics, and produce of these lands. I most fervently pray that the kind Providence, which has hitherto carried him through so many perils and hardships, may guide him safely to his present journey's end."

The friendship of Mr. Gabriel was honorable both to himself and to Dr. Livingstone. At a very early period he learned to appreciate Livingstone thoroughly, he saw how great as well as how good a man he was, and felt that to be the friend of such a man was one of the highest distinctions he could have. After Livingstone left Loanda, and while he was detained within reach of letters, a brisk correspondence passed between them; Mr. Gabriel tells him about birds, helps him in his schemes for promoting lawful commerce, goes into ecstasies over a watch-chain which he had got from him, tells him the news of the battle of the Alma in the Crimea, in which his friend, Colonel Steele, had distinguished himself, and of the success of the Rae Expedition in finding the remains of the party under Sir John Franklin. In an official communication to Lord Clarendon, after Livingstone had left, Mr. Gabriel says, 5th August, 1855: "I am grieved to say that this excellent man's health has suffered a good deal [on the return journey]. He nevertheless wrote in cheerful spirits, sanguine

of success in doing his duty under the guidance and protection of that kind Providence who had always carried
him through so many perils and hardships. He assures
me that since he knew the value of Christianity, he has
ever wished to spend his life in propagating its blessings
among men, and adds that the same desire remains still
as strong as ever."

While Livingstone was at Loanda, he made several
acquaintances among the officers of Her Majesty's navy,
engaged in the suppression of the slave-trade. For many
of these gentlemen he was led to entertain a high regard.
Their humanity charmed him, and so did their attention
to their duties. In his early days, sharing the feeling then
so prevalent in his class, he had been used to think of
epauleted gentlemen as idlers, or worse—"*fruges consumere
nati.*" Personal acquaintance, as in so many other cases,
rubbed off the prejudice. In many ways Livingstone's
mind was broadening. His intensely sympathetic nature
drew powerfully to all who were interested in what was
rapidly becoming his own master-idea—the suppression of
the slave-trade. We shall see proofs not a few, how this
sympathetic affection modified some of his early opinions,
and greatly widened the sphere of his charity.

After all the illness and dangers he had encountered,
Livingstone might quite honorably have accepted a berth
in one of Her Majesty's cruisers, and returned to England.
But the men who had come with him from the Barotse
country to Loanda had to return, and Livingstone knew
that they were quite unable to perform the journey without
him. That consideration determined his course. All the
risks and dangers of that terrible road—the attacks of
fever and dysentery, the protracted absence of those for
whom he pined, were not to be thought of when he had a
duty to these poor men. Besides, he had not yet accomplished his object. He had, indeed, discovered a way by
which his friend Sekelétu might sell his tusks to far greater

16

advantage, and which would thus help to introduce a legitimate traffic among the Makololo, and expel the slave-trade; but he had discovered no healthy locality for a mission, nor any unexceptional highway to the sea for the purpose of general traffic. The east coast seemed to promise better than the west. That great river, the Zambesi, might be found to be a navigable highway to the sea. He would return to Linyanti, and set out from it to find a way to the eastern shore. Loaded with kindness from many quarters, and furnished with presents for Sekelétu, and for the chiefs along the way, Livingstone bade farewell to Loanda on 20th September, 1854.

The following letter to Mrs. Livingstone, written a month afterward, gives his impressions of Loanda and the neighborhood:

" *Golungo Alto*, 25*th October*, 1854.—It occurs to me, my dearest Mary, that if I send you a note from different parts on the way through this colony, some of them will surely reach you; and if they carry any of the affection I bear to you in their composition, they will not fail to comfort you. I got everything in Loanda I could desire; and were there only a wagon-path for us, this would be as good an opening into the interior as we could wish. I remained rather a long time in the city in consequence of a very severe attack of fever and dysentery which reduced me very much; and I remained a short time longer than that actually required to set me on my legs, in longing expectation of a letter from you. None came, but should any come up to the beginning of November, it will come after me by post to Cassangé.

" The [Roman Catholic] Bishop, who was then acting-governor, gave a horse, saddle, and bridle, a colonel's suit of clothes, etc., for Sekelétu, and a dress of blue and red cloth, with a white cotton blanket and cap to each of my companions, who are the best set of men I ever traveled with except Malatzi and Mebalwe. The merchants of Loanda gave Sekelétu a large present of cloth, beads, etc., and one of them, a Dutchman, gave me an order for ten oxen as provisions on the way home to the Zambesi. This is all to encourage the natives to trade freely with the coast, and will have a good effect in increasing our influence for that which excels everything earthly. Everything has, by God's gracious blessing, proved more auspicious than I anticipated. We have a most warm-hearted friend in Mr. Gabriel. He acted a brother's part, and now writes me in the most affectionate manner. I thank God for his goodness in influencing the hearts of so many to show kindness, to whom I was a total stranger. The Portuguese have all been extremely kind. In coming through the coffee plantations I was offered more coffee than I could take or needed, and the best in the world. One spoonful makes it stronger than three did of that we used. It is found wild on the mountains.

"Mr. Gabriel came about 30 miles with me, and ever since, though I spoke freely about the slave-trade, the very gentlemen who have been engaged in it, and have been prevented by our ships from following it, and often lost much, treated me most kindly in their houses, and often accompanied me to the next place beyond them, bringing food for all in the way. The common people are extremely civil, and a very large proportion of the inhabitants in one district called Ambaca can read and write well. They were first taught by the Roman Catholic missionaries, and now teach each other so well, it is considered a shame in an Amba-cista not to be able to write his own name at least. But they have no Bibles. They are building a church at Ambaca, and another is in course of erection here, though they cannot get any priests. May God grant that we may be useful in some degree in this field also. . . . Give my love to all the children, they will reap the advantage of your remaining longer at home than we anticipated. I hope Robert, Agnes, and Tom are each learning as fast as they can. When will they be able to write a letter to me? How happy I shall be to meet them and you again! I hope a letter from you may be waiting for me at Zambesi. Love to all the children. How tall is Zouga? Accept the assurance of unabated love.

"DAVID LIVINGSTON."

It must not be forgotten that all this time Dr. Livingstone was making very careful astronomical observations, in order to determine his exact positions, and transmitting elaborate letters to the Geographical Society. His astronomical observations were regularly forwarded to his friend the Astronomer-Royal at the Cape, Mr. Maclear, for verification and correction.

Writing to Livingstone on 27th March, 1854, with reference to some of his earlier observations, after noticing a few trifling mistakes, Mr. Maclear says: "It is both interesting and amusing to trace your improvement as an observer. Some of your early observations, as you remark, are rough, and the angles ascribed to objects misplaced in transcribing. But upon the whole I do not hesitate to assert that no explorer on record has determined his path with the precision you have accomplished." A year afterward, 11th August, 1855, but with reference to papers received from Sekelétu's place, Mr. Maclear details what he had done in reducing his observations, preparing abstracts of them, sending them to the authorities, and publishing them in the Cape papers. He informs him that Sir John Herschel placed them before the Geographical Society, and that a warm eulogium on his labors and discoveries, and particularly on the excellent series of observations which fixed his track so exactly, appeared in the President's Address.

Then, referring to his wonderful journey to Loanda, and remarkable escapes, he says: "Nor is your escape with life from so many attacks of fever other than miraculous. Perhaps there is nothing on record of the kind, and it can only be explained by Divine interference for a good purpose. O may life be continued to you, my dear friend! You have accomplished more for the happiness of mankind than has been done by all the African travelers hitherto put together."

Mr. Maclear's reference to Livingstone's work, in writing to Sir John Herschel, was in these terms: "Such a man deserves every encouragement in the power of his country to give. He has done that which few other travelers in Africa can boast of—he has fixed his geographical points with very great accuracy, and yet he is only a poor missionary."

Nor did Dr. Livingstone pass unrewarded in other quarters. In the Geographical Society, his journey to Loanda, of which he sent them an account, excited the liveliest interest. In May, 1855, on the motion of Sir Roderick Murchison, the Society testified its appreciation by awarding him their gold medal—the highest honor they had to bestow. The occasion was one of great interest. From the chair, Lord Ellesmere spoke of Livingstone's work in science as but subordinate to those higher ends which he had ever prosecuted in the true spirit of a missionary. The simplicity of his arrangements gave additional wonder to the results. There had just appeared an account of a Portuguese expedition of African exploration from the East Coast:

" I advert to it," said his Lordship, " to point out the contrast between the two. Colonel Monteiro was the leader of a small army—some twenty Portuguese soldiers, and a hundred and twenty Caffres. The contrast is as great between such military array and the solitary grandeur of the missionary's progress, as it is between the actual achievements of the two—between the rough knowledge obtained by the Portuguese of some three hundred leagues of new country, and the scientific precision with which the unarmed and unassisted Englishman has left his mark on so many important stations of regions hitherto a blank."

About the time when these words were spoken, Dr. Livingstone was at Cabango on his return journey, recovering from a very severe attack of rheumatic fever which had left him nearly deaf; besides, he was almost blind in consequence of a blow received on the eye from a branch of a tree in riding through the forest. Notwithstanding, he was engaged in writing a despatch to the Geographical Society, through Sir Roderick Murchison, of which more anon, reporting progress, and explaining his views of the structure of Africa. But we must return to Loanda, and set out with him and his Makololo in proper form, on their homeward tour.

CHAPTER IX.

FROM LOANDA TO QUILIMANE.

A.D. 1854–1856.

Livingstone sets out from Loanda—Journey back—Effects of slavery—Letter to his wife—Severe attack of fever—He reaches the Barotse country—Day of thanksgiving—His efforts for the good of his men—Anxieties of the Moffats—Mr. Moffat's journey to Mosilikatse—Box at Linyanti—Letter from Mrs. Moffat—Letters to Mrs. Livingstone, Mr. Moffat, and Mrs. Moffat— Kindness of Sekelétu—New escort—He sets out for the East Coast—Discovers the Victoria Falls—The healthy longitudinal ridges—Pedestrianism— Great dangers—Narrow escapes—Triumph of the spirit of trust in God— Favorite texts—Reference to Captain Maclure's experience—Chief subjects of thought—Structure of the continent—Sir Roderick Murchison antici- pates his discovery—Letters to Geographical Society—First letter from Sir Roderick Murchison—Missionary labor—Monasteries—Protestant mission- stations wanting in self-support—Letter to Directors—Fever not so serious an obstruction as it seemed—His own hardships—Theories of mission-work —Expansion *v.* Concentration—Views of a missionary statesman—He reaches Tette—Letter to King of Portugal—To Sir Roderick Murchison—Reaches Senna—Quilimane—Retrospect—Letter from Directors—Goes to Mauritius —Voyage home—Narrow escape from shipwreck in Bay of Tunis—He reaches England, Dec., 1856—News of his father's death.

DR. LIVINGSTONE left St. Paul de Loanda on 24th Sep- tember, 1854, arrived at his old quarters at Linyanti on 11th September, 1855, set out eastward on 3d November, 1855, and reached Quilimane on the eastern coast on 20th May, 1856. His journey thus occupied a year and eight months, and the whole time from his leaving the Cape on 8th June, 1852, was within a few days of four years. The return journey from Loanda to Linyanti took longer than the journey outward. This arose from detention of various kinds:[1] the sicknesses of Livingstone and his men, the

---

[1] Dr. Livingstone observed that traders generally traveled ten days in the month, and rested twenty, making seven geographical miles a day, or seventy

heavy rains, and in one case, at Pungo Andongo, the necessity of reproducing a large packet of letters, journals, maps, and despatches, which he had sent off from Loanda. These were despatched by the mail-packet " Forerunner," which unhappily went down off Madeira, all the passengers but one being lost. But for his promise to the Makololo to return with them to their country, Dr. Livingstone would have been himself a passenger in the ship. Hearing of the disaster while paying a visit to a very kind and hospitable Portuguese gentleman at Pungo Andongo, on his way back, Livingstone remained there some time to reproduce his lost papers. The labor thus entailed must have been very great, for his ordinary letters covered sheets almost as large as a newspaper, and his maps and despatches were produced with extraordinary care.

He found renewed occasion to acknowledge in the warmest terms the kindness he received from the Portuguese; and his prayers that God would reward and bless them were not the less sincere that in many important matters he could not approve of their ways.

In traversing the road backward along which he had already come, not many things happened that demand special notice in this brief sketch. We find him both in his published book and still more in his private Journal repeating his admiration of the country and its glorious scenery. This revelation of the marvelous beauty of a country hitherto deemed a sandy desert was one of the most astounding effects of Livingstone's travels on the public mind. But the more he sees of the people the more profound does their degradation appear, although the many instances of remarkable kindness to himself, and occasional cases of genuine feeling one toward another,

per month. In his case in this journey the proportion was generally reversed— twenty days of traveling and ten of rest, and his rate per day was about ten geographical miles, or two hundred per month. As he often zigzagged, the geographical mile represented considerably more. See letter to Royal Geographical Society, October 16, 1855.

convinced him that there was a something in them not quite barbarised. On one point he was very clear—the Portuguese settlements among them had not improved them. Not that he undervalued the influences which the Portuguese had brought to bear on them; he had a much more favorable opinion of the Jesuit missions than Protestants have usually allowed themselves to entertain, and felt both kindly and respectfully toward the padres, who in the earlier days of these settlements had done, he believed, a useful work. But the great bane of the Portuguese settlements was slavery. Slavery prevented a good example, it hindered justice, it kept down improvement. If a settler took a fancy to a good-looking girl, he had only to buy her, and make her his concubine. Instead of correcting the polygamous habits of the chiefs and others, the Portuguese adopted like habits themselves. In one thing indeed they were far superior to the Boers—in their treatment of the children born to them by native mothers. But the whole system of slavery gendered a blight which nothing could counteract; to make Africa a prosperous land, liberty must be proclaimed to the captive, and the slave system, with all its accursed surroundings, brought conclusively to an end. Writing to Mrs. Livingstone from Bashinge, 20th March, 1855, he gives some painful particulars of the slave-trade. Referring to a slave-agent with whom he had been, he says:

"This agent is about the same in appearance as Mebalwe, and speaks Portuguese as the Griquas do Dutch. He has two chainsful of women going to be sold for the ivory. Formerly the trade went from the interior into the Portuguese territory; now it goes the opposite way. This is the effect of the Portuguese love of the trade: they cannot send them abroad on account of our ships of war on the coast, yet will sell them to the best advantage. These women are decent-looking, as much so as the general run of Kuruman ladies, and were caught lately in a skirmish the Portuguese had with their tribe; and they will be sold for about three tusks each. Each has an iron ring round the wrist, and that is attached to the chain, which she carries in the hand to prevent it jerking and hurting the wrist. How would Nannie like to be thus treated? and yet it is only by the goodness of God in appointing our lot in different circumstances that we are not similarly degraded, for we have the same evil nature, which is so degraded in them as to allow of men treating them as beasts.

"I long for the time when I shall see you again. I hope in God's mercy for that pleasure. How are my dear ones? I have not seen any equal to them since I put them on board ship. My brave little dears! I only hope God will show us mercy, and make them good too. . . .

"I work at the interior languages when I have a little time, and also at Portuguese, which I like from being so much like Latin. Indeed, when I came I understood much that was said from its similarity to that tongue, and when I interlarded my attempts at Portuguese with Latin, or spoke it entirely, they understood me very well. The Negro language is not so easy, but I take a spell at it every day I can. It is of the same family of languages as the Sichuana. . . .

"We have passed two chiefs who plagued us much when going down, but now were quite friendly. At that time one of them ordered his people not to sell us anything, and we had at last to force our way past him. Now he came running to meet us, saluting us, etc., with great urbanity. He informed us that he would come in the evening to receive a present, but I said unless he brought one he should receive nothing. He came in the usual way. The Balonda show the exalted position they occupy among men, viz., riding on the shoulders of a spokesman in the way little boys do in England. The chief brought two cocks and some eggs. I then gave a little present too. The alteration in this gentleman's conduct—the Peace Society would not credit it—is attributable solely to my people possessing guns. When we passed before, we were defenseless. May every needed blessing be granted to you and the dear children, is the earnest prayer of your ever most affectionate

"D. LIVINGSTON."

It was soon after the date of this letter that Livingstone was struck down by that severe attack of rheumatic fever, accompanied by great loss of blood, to which reference has already been made. "I got it," he writes to Mr. Maclear, "by sleeping in the wet. There was no help for it. Every part of a plain was flooded ankle-deep. We got soaked by going on, and sodden if we stood still." In his former journey he had been very desirous to visit Matiamvo, paramount chief of the native tribes of Londa, whose friendship would have helped him greatly in his journey; but at

that time he found himself too poor to attempt the enterprise. The loss of time and consumption of goods caused
by his illness on the way back prevented him from accomplishing his purpose now.

Not only was the party now better armed than before,
but the good name of Livingstone had also become better
known along the line, and during his return journey he
did not encounter so much opposition. We connot fail to
be struck with his extraordinary care for his men. It was
his earnest desire to bring them all back to their homes,
and in point of fact the whole twenty-seven returned in
good health. How carefully he must have nursed them
in their attacks of fever, and kept them from unnecessary
exposure, it is hardly possible for strangers adequately to
understand.

On reaching the country of the Barotse, the home of
most of them, a day of thanksgiving was observed (23d
July, 1855). The men had made little fortunes in Loanda,
earning sixpence a day for weeks together by helping to
discharge a cargo of coals or, as they called them, "stones
that burned." But, like Livingstone, they had to part
with everything on the way home, and now they were in
rags; yet they were quite as cheerful and as fond of their
leader as ever, and felt that they had not traveled in vain.
They quite understood the benefit the new route would
bring in the shape of higher prices for tusks and the other
merchandise of home. On the thanksgiving day—

"The men decked themselves out in their best, for all had managed
to preserve their suits of European clothing, which, with their white
and red caps, gave them a rather dashing appearance. They tried to
walk like soldiers, and called themselves 'my braves.' Having been
again saluted with salvos from the women, we met the whole population,
and having given an address on divine things, I told them we had come
that day to thank God before them all for his mercy in preserving us
from dangers, from strange tribes and sicknesses. We had another
service in the afternoon. They gave us two fine oxen to slaughter, and
the women have supplied us abundantly with milk and meal. This is

all gratuitous, and I feel ashamed that I can make no return. My men explain the whole expenditure on the way hither, and they remark gratefully : "It does not matter, you have opened a path for us, and we shall have sleep.' Strangers from a distance come flocking to see me, and seldom come empty-handed. I distribute all presents among my men."

Several of the poor fellows on reaching home found domestic trouble—a wife had proved inconstant and married another man. As the men had generally more wives than one, Livingstone comforted them by saying that they still had as many as he.

Amid the anxieties and sicknesses of the journey, and multiplied subjects of thought and inquiry, Livingstone was as earnest as ever for the spiritual benefit of the people. Some extracts from his Journal will illustrate his efforts in this cause, and the flickerings of hope that would spring out of them, dimmed, however, by many fears :

"*August* 5, 1855.—A large audience listened attentively to my address this morning, but it is impossible to indulge any hopes of such feeble efforts. God is merciful, and will deal with them in justice and kindness. This constitutes a ground of hope. Poor degraded Africa! A permanent station among them might effect something in time, but a considerable time is necessary. Surely some will pray to their merciful Father in their extremity, who never would have thought of Him but for our visit."

"*August* 12.—A very good and attentive audience. Surely all will not be forgotten. How small their opportunity compared to ours who have been carefully instructed in the knowledge of divine truth from our earliest infancy! The Judge is just and merciful. He will deal fairly and kindly with all."

"*October* 15.—We had a good and very attentive audience yesterday, and I expatiated with great freedom on the love of Christ in dying, from his parting address in John xvi. It cannot be these precious truths will fall to the ground; but it is perplexing to observe no effects. They assent to the truth, but 'we don't know,' or 'you speak truly,' is all the response. In reading accounts of South Sea missions it is hard to believe the quickness of the vegetation of the good seed, but I know several of the men" [the South Sea missionaries], "and am sure they are of unimpeachable veracity. In trying to convey knowledge, and use the magic lantern, which is everywhere extremely popular, though they listen with apparent delight to what is said, questioning them on the following night reveals almost entire ignorance of the previous lesson. O that the Holy Ghost might enlighten them! To his soul-renewing influence my longing soul is directed. It is his word, and cannot die."

The long absence of Livingstone and the want of letters had caused great anxiety to his friends. The Moffats had been particularly concerned about him, and, in 1854, partly in the hope of hearing of him, Mr. Moffat undertook a visit to Mosilikatse, while a box of goods and comforts was sent to Linyanti to await his return, should that ever take place. A letter from Mrs. Moffat accompanied the box. It is amusing to read her motherly explanations about the white shirts, and the blue waistcoat, the woolen socks, lemon juice, quince jam, and tea and coffee, some of which had come all the way from Hamilton; but there are passages in that little note that make one's heart go with rapid beat:

"MY DEAR SON LIVINGSTON,—Your present position is almost too much for my weak nerves to suffer me to contemplate. Hitherto I have kept up my spirits, and been enabled to believe that our great Master may yet bring you out in safety, for though his ways are often inscrutable, I should have clung to the many precious promises made in his word as to temporal preservation, such as the 91st and 121st Psalms—but have been taught that we may not presume confidently to expect them to be fulfilled, and that every petition, however fervent, must be with devout submission to his will. My poor sister-in-law clung tenaciously to the 91st Psalm, and firmly believed that her dear husband would thus be preserved, and never indulged the idea that they should never meet on earth. But I apprehend submission was wanting. 'If it be Thy will,' I fancy she could not say—and, therefore, she was utterly confounded when the news came.[1] She had exercised strong faith, and was disappointed. Dear Livingstone, I have always endeavored to keep this in mind with regard to you. Since George [Fleming] came out it seemed almost hope against hope. Your having got so thoroughly feverised chills my expectations; still prayer, unceasing

---

[1] Rev. John Smith, missionary at Madras, had gone to Vizagapatam to the ordination of two native pastors, and when returning in a small vessel, a storm arose, when he and all on board perished.

prayer, is made for you. When I think of you my heart will go upward. 'Keep him as the apple of Thine eye,' 'Hold him in the hollow of Thy hand,' are the ejaculations of my heart."

In writing from Linyanti to his wife, Livingstone makes the best he can of his long detention. She seems to have put the matter playfully, wondering what the "source of attraction" had been. He says:

"Don't know what apology to make you for a delay I could not shorten. But as you are a mercifully kind-hearted dame, I expect you will write out an apology in proper form, and I shall read it before you with as long a face as I can exhibit. Disease was the chief obstacle. The repair of the wagon was the 'source of attraction' in Cape Town, and the settlement of a case of libel another 'source of attraction.' They tried to engulf me in a law-suit for simply asking the postmaster why some letters were charged double. They were so marked in my account. I had to pay £13 to quash it. They longed to hook me in, from mere hatred to London missionaries. I did not remain an hour after I could move. But I do not wonder at your anxiety for my speedy return. I am sorry you have been disappointed, but you know no mortal can control disease. The Makololo are wonderfully well pleased with the path we have already made, and if I am successful in going down to Quilimane, that will be still better. I have written you by every opportunity, and am very sorry your letters have been miscarried."

To his father-in-law he expresses his warm gratitude for the stores. It was feared by the natives that the goods were bewitched, so they were placed on an island, a hut was built over them, and there Livingstone found them on his arrival, a year after! A letter of twelve quarto pages to Mr. Moffat gives his impressions of his journey, while another of sixteen pages to Mrs. Moffat explains his "plans," about which she had asked more full information. He quiets her fears by his favorite texts for the present—"Commit thy way to the Lord," and "Lo, I am with you alway"; and his favorite vision of the future— the earth full of the knowledge of the Lord. He is somewhat cutting at the expense of so-called "missionaries to the heathen, who never march into real heathen territory,

17

and quiet their consciences by opposing their do-nothing-
ism to my blundering do-somethingism!" He is indignant
at the charge made by some of his enemies that no good
was done among the Bakwains. They were, in many
respects, a different people from before. Any one who
should be among the Makololo as he had been, would be
thankful for the state of the Bakwains. The seed would
always bear fruit, but the husbandman had need of great
patience, and the end was sure.

Sekelétu had not been behaving well in Livingstone's
absence. He had been conducting marauding parties
against his neighbors, which even Livingstone's men,
when they heard of it, pronounced to be "bad, bad."
Livingstone was obliged to reprove him. A new uniform
had been sent to the chief from Loanda, with which he
appeared at church, "attracting more attention than the
sermon." He continued, however, to show the same friendship for Livingstone, and did all he could for him when
he set out eastward. A new escort of men was provided,
above a hundred and twenty strong, with ten slaughter
cattle, and three of his best riding oxen; stores of food
were given, and a right to levy tribute over the tribes
that were subject to Sekelétu as he passed through their
borders. If Livingstone had performed these journeys
with some long-pursed society or individual at his back,
his feat even then would have been wonderful; but it
becomes quite amazing when we think that he went
without stores, and owed everything to the influence he
acquired with men like Sekelétu and the natives generally. His heart was much touched on one occasion by the
disinterested kindness of Sekelétu. Having lost their way
on a dark night in the forest, in a storm of rain and lightning, and the luggage having been carried on, they had to
pass the night under a tree. The chief's blanket had not
been carried on, and Sekelétu placed Livingstone under
it, and lay down himself on the wet ground. "If such

men must perish before the white by an immutable law of heaven," he wrote to the Geographical Society (25th January, 1856), " we must seem to be under the same sort of terrible necessity in our Caffre wars as the American Professor of Chemistry said he was under, when he dismembered the man whom he had murdered."

Again Livingstone sets out on his weary way, untrodden by white man's foot, to pass through unknown tribes, whose savage temper might give him his quietus at any turn of the road. There were various routes to the sea open to him. He chose the route along the Zambesi—though the the most difficult, and through hostile tribes—because it seemed the most likely to answer his desire to find a commercial highway to the coast. Not far to the east of Linyanti, he beheld for the first time those wonderful falls of which he had only heard before, giving an English name to them,—the first he had ever given in all his African journeys,—the Victoria Falls. We have seen how genuine his respect was for his Sovereign, and it was doubtless a real though quiet pleasure to connect her name with the grandest natural phenomenon in Africa. This is one of the discoveries[1] that have taken most hold on the popular imagination, for the Victoria Falls are like a second Niagara, but grander and more astonishing ; but except as illustrating his views of the structure of Africa, and the distribution of its waters, it had not much influence, and led to no very remarkable results. Right across the channel of the river was a deep fissure only eighty feet wide, into which the whole volume of the river, a thousand yards broad, tumbled to the depth of a hundred feet,[2] the fissure being continued in zigzag form for thirty miles, so that the stream had to change its course from right to left and left to right, and went through the hills boiling and roaring, sending up columns of steam, formed

---

[1] Virtually a discovery, though marked in an old map.

[2] Afterward ascertained by him to be 1800 yards and 320 feet respectively.

by the compression of the water falling into its narrow wedge-shaped receptacle.

A discovery as to the structure of the country, long believed in by him, but now fully verified, was of much more practical importance. It had been ascertained by him that skirting the central hollow there were two longitudinal ridges extremely favorable for settlements, both for missions and merchandise. We shall hear much of this soon.

Slowly but steadily the eastward tramp is continued, often over ground which was far from favorable for walking exercise. "Pedestrianism," said Livingstone, "may be all very well for those whose obesity requires much exercise; but for one who was becoming as thin as a lath through the constant perspiration caused by marching day after day in the hot sun, the only good I saw in it was that it gave an honest sort of a man a vivid idea of the tread-mill."

When Livingstone came to England, and was writing books, his tendency was rather to get stout than thin; and the disgust with which he spoke then of the "beastly fat" seemed to show that if for nothing else than to get rid of it he would have been glad to be on the tread-mill again. In one of his letters to Mr. Maclear he thus speaks of a part of this journey: "It was not likely that I should know our course well, for the country there is covered with shingle and gravel, bushes, trees, and grass, and we were without path. Skulking out of the way of villages where we were expected to pay after the purse was empty, it was excessively hot and steamy; the eyes had to be always fixed on the ground to avoid being tripped."

In the course of this journey he had even more exciting escapades among hostile tribes than those which he had encountered on the way to Loanda. His serious anxieties began when he passed beyond the tribes that owned the sovereignty of Sekelétu. At the union of the rivers

Loangwa and Zambesi, the suspicious feeling regarding him reached a climax, and he could only avoid the threatened doom of the Bazimka (*i.e.* Bastard Portuguese) who had formerly incurred the wrath of the chief, by showing his bosom, arms, and hair, and asking if the Bazimka were like that. Livingstone felt that there was danger in the air. In fact, he never seemed in more imminent peril:

"14*th January*, 1856.—At the confluence of the Loangwa and Zambesi. Thank God for his great mercies thus far. How soon I may be called to stand before Him, my righteous Judge, I know not. All hearts are in his hands, and merciful and gracious is the Lord our God. O Jesus, grant me resignation to Thy will, and entire reliance on Thy powerful hand. On Thy Word alone I lean. But wilt Thou permit me to plead for Africa? The cause is Thine. What an impulse will be given to the idea that Africa is not open if I perish now! See, O Lord, how the heathen rise up against me, as they did to Thy Son. I commit my way unto Thee. I trust also in Thee that Thou wilt direct my steps. Thou givest wisdom liberally to all who ask Thee—give it to me, my Father. My family is Thine. They are in the best hands. Oh! be gracious, and all our sins do Thou blot out.

'A guilty, weak, and helpless worm,
On Thy kind arms I fall.'

Leave me not, forsake me not. I cast myself and all my cares down at Thy feet. Thou knowest all I need, for time and for eternity.

"It seems a pity that the important facts about the two healthy longitudinal ridges should not become known in Christendom. Thy will be done! . . . They will not furnish us with more canoes than two. I leave my cause and all my concerns in the hands of God, my gracious Saviour, the Friend of sinners.

"*Evening.*—Felt much turmoil of spirit in view of having all my plans for the welfare of this great region and teeming population knocked on the head by savages to-morrow. But I read that Jesus came and said, 'All power is given unto me in heaven and in earth. Go ye therefore, and teach all nations—and lo, *I am with you alway, even unto the end of the world.*' It is the word of a gentleman of the most sacred and strictest honor, and there is an end on't. I will not cross furtively by night as I intended. It would appear as flight, and should such a man as I flee? Nay, verily, I shall take observations for latitude and longitude to-night, though they may be the last. I feel quite calm now, thank God.

"15*th January*, 1856.—Left bank of Loangwa. The natives of the surrounding country collected round us this morning all armed. Children and women were sent away, and Mburuma's wife who lives here was not allowed to approach, though she came some way from her village in order to pay me a visit. Only one canoe was lent, though we saw two tied to the bank. And the part of the river we crossed at, about a mile from the confluence, is a good mile broad. We passed all our goods first, to an island in the middle, then the cattle and men, I occupying the post of honor, being the last to enter the canoe. We had, by this means, an opportunity of helping each other in case of attack. They stood armed at my back for some time. I then showed them my watch, burning-glass, etc., etc., and kept them amused till all were over, except those who could go into the canoe with me. I thanked them all for their kindness and wished them peace."

Nine days later they were again threatened by Mpende:

"23*d January*, 1856.—At Mpende's this morning at sunrise, a party of his people came close to our encampment, using strange cries, and waving some red substance toward us. They then lighted a fire with charms in it, and departed uttering the same hideous screams as before. This is intended to render us powerless, and probably also to frighten us. No message has yet come from him, though several parties have arrived, and profess to have come simply to see the white man. Parties of his people have been collecting from all quarters long before daybreak. It would be considered a challenge—for us to move down the river, and an indication of fear and invitation to attack if we went back. So we must wait in patience, and trust in Him who has the hearts of all men in his hands. To Thee, O God, we look. And, oh! Thou who wast the man of sorrows for the sake of poor vile sinners, and didst not disdain the thief's petition, remember me and Thy cause in Africa. Soul and body, my family, and Thy cause, I commit all to Thee. Hear, Lord, for Jesus' sake."

In the entire records of Christian heroism, there are few more remarkable occasions of the triumph of the spirit of holy trust than those which are recorded here so quietly and modestly. We are carried back to the days of the Psalmist: "I will not be afraid of ten thousand of the people that have set themselves against me round about."

In the case of David Livingstone as of the other David, the triumph of confidence was not the less wonderful that it was preceded by no small inward tumult. Both were human creatures. But in both the flutter lasted only till the soul had time to rally its trust—to think of God as a living friend, sure to help in time of need. And how real is the sense of God's presence! The mention of the two longitudinal ridges, and of the refusal of the people to give more than two canoes, side by side with the most solemn appeals, would have been incongruous, or even irreverent, if Livingstone had not felt that he was dealing with the living God, by whom every step of his own career and every movement of his enemies were absolutely controlled.

A single text often gave him all the help he needed:

"It is singular," he says, "that the very same text which recurred to my mind at every turn of my course in life in this country and even in England, should be the same as Captain Maclure, the discoverer of the Northwest Passage, mentions in a letter to his sister as familiar in his experience: 'Trust in the Lord with all thine heart, and lean not to thine own understanding. In all thy ways acknowledge Him and He shall direct thy steps. Commit thy way unto thy Lord; trust also in Him and He shall bring it to pass.' Many more, I have no doubt, of our gallant seamen feel that it is graceful to acknowledge the gracious Lord in whom we live and move and have our being. It is an advance surely in humanity from that devilry which gloried in fearing neither God, nor man, nor Devil, and made our wooden walls floating hells."

His being enabled to reach the sanctuary of perfect peace in the presence of his enemies was all the more striking if we consider—what he felt keenly—that to live among the heathen is in itself very far from favorable to the vigor or the prosperity of the spiritual life. "Traveling from day to day among barbarians," he says in his Journal, "exerts a most benumbing effect on the religious feelings of the soul."

Among the subjects that occupied a large share of his thoughts in these long and laborious journeys, two appear to have been especially prominent: first, the configuration of the country; and second, the best way of conducting missions, and bringing the people of Africa to Christ. The configuration of intertropical South Africa had long been with him a subject of earnest study, and now he had come clearly to the conclusion that the middle part was a table-land, depressed, however, in the centre, and flanked by longitudinal ridges on the east and west; that originally the depressed centre had contained a vast accumulation of water, which had found ways of escape through fissures in the encircling fringe of mountains, the result of volcanic action or of earthquakes. The Victoria Falls presented the most remarkable of these fissures, and thus served to verify and complete his theory. The great lakes in the great heart of South Africa were the remains of the earlier accumulation before the fissures were formed. Lake 'Ngami, large though it was, was but a little fraction of the vast lake that had once spread itself over the south. This view of the structure of South Africa he now found, from a communication which reached him at Linyanti, had been anticipated by Sir Roderick Murchison, who in 1852 had propounded it to the Geographical Society. Livingstone was only amused at thus losing the credit of his discovery; he contented himself with a playful remark on his being "cut out" by Sir Roderick. But the coincidence of views was very remarkable, and it lay at the foundation of that brotherlike intimacy and friendship which ever marked his relation with Murchison. One important bearing of the geographical fact was this; it was evident that while the low districts were unhealthy, the longitudinal ridges by which they were fringed were salubrious. Another of its bearings was, that it would help them to find the course and perhaps the sources of the great rivers, and thus facilitate commercial and missionary operations. The discovery of the two healthy ridges, which made him so unwilling to die at the mouth of the Loangwa, gave him new hopes for missions and commerce.

These and other matters connected with the state of the country formed the subject of regular communications to the Geographical Society. Between Loanda and Quilimane, six despatches were written at different points.[1] Formerly, as we have seen, he had written through a Fellow of the Society, his friend and former fellow-traveler, Captain, now Colonel Steele; but as the Colonel had been called on duty to the Crimea, he now addressed his letters to his countryman, Sir Roderick Murchison. Sir Roderick was charmed with the compliment, and was not slow to turn it to account, as appears from the following letter, the first of very many communications which he addressed to Livingstone:

"16 BELGRAVE SQUARE, *October* 2, 1855.

"MY DEAR SIR,—Your most welcome letter reached me after I had made a tour in the Highlands, and just as the meeting of the British Association for the Advancement of Science commenced.

"I naturally communicated your despatch to the Geographical section of that body, and the reading of it called forth an unanimous expression of admiration of your labors and researches.

"In truth, you will long ago, I trust, have received the cordial thanks of all British geographers for your unparalleled exertions, and your successful accomplishment of the greatest triumph in geographical research which has been effected in our times.

"I rejoice that I was the individual in the Council of the British Geographical Society who proposed that you should receive our first gold medal of the past session, and I need not say that the award was made by an unanimous and cordial vote.

"Permit me to thank you sincerely for having selected me as your correspondent in the absence of Colonel Steele, and to assure you that I shall consider myself as much honored, as I shall certainly be gratified, by every fresh line which you may have leisure to write to me.

"Anxiously hoping that I may make your personal acquaintance,

---

[1] The dates were Pungo Andongo, 24th December, 1854; Cabango, 17th May, 1855; Linyanti, October 16, 1855; Chanyuni, 25th January, 1856; Tette, 4th March, 1856; Quilimane, 23d May, 1856.

and that you may return to us in health to receive the homage of all geographers,—I remain, my dear Sir, yours most faithfully,

"Rod^{ck} I. Murchison."

The other subject that chiefly occupied Livingstone's mind at this time was missionary labor. This, like all other labor, required to be organized, on the principle of making the very best use of all the force that was or could be contributed for missionary effort. With his fair, open mind, he weighed the old method of monastic establishments, and, *mutatis mutandis,* he thought something of the kind might be very useful. He thought it unfair to judge of what these monasteries were in their periods of youth and vigor, from the rottenness of their decay. Modern missionary stations, indeed, with their churches, schools, and hospitals, were like Protestant monasteries, conducted on the more wholesome principle of family life; but they wanted stability; they had not farms like monasteries, and hence they required to depend on the mother country. From infancy to decay they were pauper institutions. In Livingstone's judgment they needed to have more of the self-supporting element:

"It would be heresy to mention the idea of purchasing lands, like religious endowments, among the stiff Congregationalists; but an endowment conferred on a man who will risk his life in an unhealthy climate, in order, thereby, to spread Christ's gospel among the heathen, is rather different, I ween, from the same given to a man to act as pastor to a number of professed Christians. . . . Some may think it creditable to our principles that we have not a single acre of land, the gift of the Colonial Government, in our possession. But it does not argue much for our foresight that we have not farms of our own, equal to those of any colonial farmer."

Dr. Livingstone acknowledged the services of the Jesuit missionaries in the cause of education and literature, and even of commerce. But while conceding to them this meed of praise, he did not praise their worship. He was slow, indeed, to disparage any form of worship—any form in which men, however unenlightened, gave expression to their religious feelings; but he could not away with the sight of men of intelligence kissing the toe of an image of the Virgin, as he saw them doing in a Portuguese church, and taking part in services in which they did not, and could not, believe. If the missions of the Church of Rome had left good effects on some parts of Africa, how much greater blessing might not come from Protestant missions, with the Bible instead of the Syllabus as their basis, and animated with the spirit of freedom instead of despotism!

With regard to that part of Africa which he had been exploring, he gives his views at great length in a letter to the Directors, dated Linyanti, 12th October, 1855. After fully describing the physical features of the country, he fastens on the one element which, more than any other, was likely to hinder missions—fever. He does not deny that it is a serious obstacle. But he argues at great length that it is not insurmountable. Fever yields to proper treatment. His own experience was no rule to indicate what might be reckoned on by others. His journeys had been made under the worst possible conditions. Bad food, poor nursing, insufficient medicines, continual drenchings, exhausting heat and toil, and wearing anxiety had caused much of his illness. He gives a touching detail of the hardships incident to his peculiar case, from which other missionaries would be exempted, but with characteristic manliness he charges the Directors not to publish that part of his letter, lest he should appear to be making too much of his trials. "Sacrifices" he could never call them, because nothing could be worthy of that name in the service of Him who, though he was rich, for our sakes became poor. Two or three times every day he had been wet up to the waist in crossing streams and marshy ground. The rain was so drenching that he had often to put his watch under his arm-pit to keep it dry. His good ox Sindbad would never let him hold an umbrella. His bed was on grass, with only a horse-cloth between. His food often consisted of bird-seed, manioc-roots, and meal. No wonder if he suffered much. Others would not have all that to bear. Moreover, if the fever of the district was severe, it was almost the only disease. Consumption, scrofula, madness, cholera, cancer, delirium tremens, and certain contagious diseases of which much was heard in civilized countries, were hardly known. The beauty of some parts of the country could not be surpassed. Much of it was densely peopled, but in other parts the population was scattered. Many of the tribes were friendly, and, for reasons of their own, would welcome missionaries. The Makololo, for example, furnished an inviting field. The dangers he had encountered arose from the irritating treatment the tribes had received from half-cast traders and slave-dealers, in consequence of which they had imposed certain taxes on travelers, which, sometimes, he and his brother-chartists had refused to pay. They were mistaken for slave-dealers. But character was a powerful educator. A body of missionaries, maintaining everywhere the character of honest, truthful, kind-hearted Christian gentlemen, would scatter such prejudices to the winds.

In instituting a comparison between the direct and indirect results of missions, between conversion-work and the diffusion of better principles, he emphatically assigns the preference to the latter. Not that he undervalued the conversion of the most abject creature that breathed. To the man individually his conversion was of overwhelming consequence, but with relation to the final harvest, it was more important to sow the seed broadcast over a wide field than to reap a few heads of grain on a single spot. Concentration was not the true principle of missions. The Society itself had felt this, in sending Morrison and Milne to be lost among the three hundred millions of China; and the Church of England, in looking to the Antipodes, to Patagonia, to East Africa, with the full knowledge that charity began at home. Time was more essential than concentration. Ultimately there would be more conversions, if only the seed were now more widely spread.

He concludes by pointing out the difference between mere worldly enterprises and missionary undertakings for the good of the world. The world thought their mission schemes fanatical; the friends of missions, on the other hand, could welcome the commercial enterprises of the world as fitted to be useful. The Africans were all deeply imbued with the spirit of trade. Commerce was so far good that it taught the people their mutual dependence; but Christianity alone reached the centre of African wants. "Theoretically," he concludes, "I would pronounce the country about the junction of the Leeba and Leeambye or Kabompo, and river of the Bashukulompo, as a most desirable centre-point for the spread of civilization and Christianity; but unfortunately I must mar my report by saying I feel a difficulty as to taking my children there without their intelligent self-dedication. I can speak for my wife and myself only. WE WILL GO, WHOEVER REMAINS BEHIND."

Resuming the subject some months later, after he had got to the sea-shore, he dwells on the belt of elevated land eastward from the country of the Makololo, two degrees of longitude broad, and of unknown length, as remarkably suitable for the residence of European missionaries. It was formerly occupied by the Makololo, and they had a great desire to resume the occupation. One great advantage of such a locality was that it was on the border of the regions occupied by the true negroes, the real nucleus of the African population, to whom they owed a great debt, and who had shown themselves friendly and disposed to learn. It was his earnest hope that the Directors would plant a mission here, and his belief that they would thereby confer unlimited blessing on the regions beyond.

18

Some of the remarks in these passages, and also in the extracts which we have given from his Journals, are of profound interest, as indicating an important transition from the ideas of a mere missionary laborer to those of a missionary general or statesman. In the early part of his life he deemed it his joy and his honor to aim at the conversion of individual souls, and earnestly did he labor and pray for that, although his visible success was but small. But as he gets better acquainted with Africa, and reaches a more commanding point of view, he sees the necessity for other work. The continent must be surveyed, healthy localities for mission-stations must be found, the temptations to a cursed traffic in human flesh must be removed, the products of the country must be turned to account; its whole social economy must be changed. The accomplishment of such objects, even in a limited degree, would be an immense service to the missionary; it would be such a preparing of his way that a hundred years hence the spiritual results would be far greater than if all the effort now were concentrated on single souls. To many persons it appeared as if dealing with individual souls were the only proper work of a missionary, and as if one who had been doing such work would be lowering himself if he accepted any other. Livingstone never stopped to reason as to which was the higher or the more desirable work; he felt that Providence was calling him to be less of a missionary journeyman and more of a missionary statesman; but the great end was ever the same—

"THE END OF THE GEOGRAPHICAL FEAT IS ONLY
THE BEGINNING OF THE ENTERPRISE."

Livingstone reached the Portuguese settlement of Tette on the 3d March, 1856, and the "civilized breakfast" which the commandant, Major Sicard, sent forward to him, on his way, was a luxury like Mr. Gabriel's bed at Loanda, and made him walk the last eight miles without the least sensation of fatigue, although the road was so rough that, as a Portuguese soldier remarked, it was like "to tear a man's life out of him." At Loanda he had heard of the battle of the Alma; after being in Tette a short time he heard of the fall of Sebastopol and the end of the Crimean War. He remained in Tette till the 23d April, detained by an attack of fever, receiving extraordinary kindness from the Governor, and, among other tokens of affection, a gold chain for his daughter Agnes, the work of an inhabitant of the town. These gifts were duly acknowledged. It was at this place that Dr. Livingstone left his Makololo followers, with instructions to wait for him till he should return from England. Well entitled though he was to a long rest, he deliberately gave up the possibility of it, by engaging to return for his black companions.

In the case of Dr. Livingstone, rest meant merely change of employment, and while resting and recovering from fever, he wrote a large budget of long and interesting letters. One of these was addressed to the King of Portugal: it affords clear evidence that, however much Livingstone felt called to reprobate the deeds of some of his subordinates, he had a respectful feeling for the King himself, a grateful sense of the kindness received from his African subjects, and an honest desire to aid the wholesome development of the Portuguese colonies. It refutes, by anticipation, calumnies afterward circulated to the effect that Livingstone's real design was to wrest the Portuguese settlements in Africa from Portugal, and to annex them to the British Crown. He refers most gratefully to the great kindness and substantial aid he had received from His Majesty's subjects, and is emboldened thereby to address him on behalf of Africa. He suggests certain agricultural products—especially wheat and a species of wax—that might be cultivated with enormous profit. A great stimulus might be given to the cultivation of other products—coffee, cotton, sugar, and oil. Much had been done for Angola, but with little result, because the colonists leant on Government instead of trusting to themselves. Illegitimate traffic (the slave-trade) was not at present remunerative, and now was the time to make a great effort to revive wholesome enterprise. A good road into the interior would be a great boon. Efforts to provide roads and canals had failed for want of superintendents. Dr. Livingstone named a Portuguese engineer who would superintend admirably. The fruits of the Portuguese missions were still apparent, but there was a great want of literature, of books.

" It will not be denied," concludes the letter, "that those who, like your Majesty, have been placed over so many human souls, have a serious responsibility resting upon them in reference to their future welfare. The absence also of Portuguese women in the colony is a circumstance which seems to merit the attention of Government for obvious reasons. And if any of these suggestions should lead to the formation of a middle class of free laborers, I feel sure that Angola would have cause to bless your Majesty to the remotest time."

Dr. Livingstone has often been accused of claiming for himself the credit of discoveries made by others, of writing as if he had been the first to traverse routes in which he had really been preceded by the Portuguese. Even were it true that now and then an obscure Portuguese trader or traveler reached spots that lay in Dr. Livingstone's subsequent route, the fact would detract nothing from his merit, because he derived not a tittle of benefit from their experience, and what he was concerned about was, not the mere honor of being first at a place, as if he had been running a race, but to make it known to the world, to bring it into the circuit of commerce and Christianity, and thus place it under the influence of the greatest blessings. But even as to being first, Livingstone was careful not to claim anything that was really due to others. Writing from Tette to Sir Roderick in March, 1856, he says: " It seems proper to mention what has been done in former times in the way of traversing the continent, and the result of my inquiries leads to the belief that the honor belongs to our country." He refers to the brave attempt of Captain José da Roga, in 1678, to penetrate from Benguela to the Rio da Senna, in which attempt, however, so much opposition was encountered that he was compelled to return. In 1800, Lacerda revived the project by proposing a chain of forts along the banks of the Coanza. In 1815, two black traders showed the possibility of communication from east to west, by bringing to Loanda communications from the Governor of Mozambique. Some Arabs and Moors went from the East Coast to Benguela, and with a view to improve the event, "a million of Reis (£142) and an honorary captaincy in the Portuguese army was offered to any one who would accompany them back—but none went." The journey had several times been performed by Arabs.

"I do not feel so much elated," continued Dr. Livingstone, "by the prospect of accomplishing this feat. I feel most thankful to God for preserving my life, where so many, who by superior intelligence would have done more good, have been cut off. But it does not look as if I had reached the goal. Viewed in relation to my calling, the end of the geographical feat is only the beginning of the enterprise. Apart from family longings, I have a most intense longing to hear how it has fared with our brave men at Sebastopol. My last scrap of intelligence was the *Times*, 17th November, 1855, after the terrible affair of the Light Cavalry. The news was not certain about a most determined attack to force the way to Balaclava, and Sebastopol expected every day to fall, and I have had to repress all my longings since, except in a poor prayer to prosper the cause of justice and right, and cover the heads of our soldiers in the day of battle." [A few days later he heard the news.] "We are all engaged in very much the same cause. Geographers, astronomers, and mechanicians, laboring to make men better acquainted with each other; sanitary reformers, prison reformers, promoters of ragged schools and Niger Expeditions; soldiers fighting for right against oppression, and sailors rescuing captives in deadly climes, as well as missionaries, are all aiding in hastening on a glorious consummation to all God's dealings with our race. In the hope that I may yet be honored to do some good to this poor long downtrodden Africa, the gentlemen over whom you have the honor to preside will, I believe, cordially join."

From Tette he went on to Senna. Again he is treated with extraordinary kindness by Lieutenant Miranda, and others, and again he is prostrated by an attack of fever. Provided with a comfortable boat, he at last reaches Quilimane on the 20th May, and is most kindly received by Colonel Nunes, " one of the best men in the country." Dr. Livingstone has told us in his book how his joy in reaching Quilimane was embittered on his learning that Captain Maclure, Lieutenant Woodruffe, and five men of H. M. S. " Dart," had been drowned off the bar in coming to Quilimane to pick him up, and how he felt as if he would rather have died for them.[1]

News from across the Atlantic likewise informed him that his nephew and namesake, David Livingston, a fine lad eleven years of age, had been drowned in Canada. All the deeper was his gratitude for the goodness and mercy that had followed him and preserved him, as he says in his private Journal, from " many dangers not recorded in this book."

The retrospect in his *Missionary Travels* of the manner in which his life had been ordered up to this point, is so striking that our narrative would be deficient if it did not contain it:

" If the reader remembers the way in which I was led, while teaching the Bakwains, to commence exploration, he will, I think, recognize the hand of Providence. Anterior to that, when Mr. Moffat began to give the Bible—the Magna Charta of all the rights and privileges of modern civilization—to the Bechuanas, Sebituane went north, and spread the language into which he was translating the sacred oracles,

---

[1] Among Livingstone's papers we have found draft letter to the Admiralty, earnestly commending to their Lordship's favorable consideration a petition from the widow of one of the men. He had never seen her, he said, but he had been the unconscious cause of her husband's death, and all the joy he felt in crossing the continent was embittered when the news of the sad catastrophe reached him.

in a new region larger than France. Sebituane, at the same time, rooted out hordes of bloody savages, among whom no white man could have gone without leaving his skull to ornament some village. He opened up the way for me—let us hope also for the Bible. Then, again, while I was laboring at Kolobeng, seeing only a small arc of the cycle of Providence, I could not understand it, and felt inclined to ascribe our successive and prolonged droughts to the wicked one. But when forced by these, and the Boers, to become explorer, and open a new country in the north rather than set my face southward, where missionaries are not needed, the gracious Spirit of God influenced the minds of the heathen to regard me with favor, the Divine hand is again perceived. Then I turned away westward, rather than in the opposite direction, chiefly from observing that some native Portuguese, though influenced by the hope of a reward from their Government to cross the continent, had been obliged to return from the east without accomplishing their object. Had I gone at first in the eastern direction, which the course of the great Leeambye seemed to invite, I should have come among the belligerents near Tette when the war was raging at its height, instead of, as it happened, when all was over. And again, when enabled to reach Loanda, the resolution to do my duty by going back to Linyanti probably saved me from the fate of my papers in the 'Forerunner.' And then, last of all, this new country is partially opened to the sympathies of Christendom, and I find that Sechéle himself has, though unbidden by man, been teaching his own people. In fact, he has been doing all that I was prevented from doing, and I have been employed in exploring—a work I had no previous intention of performing. I think that I see the operation of the Unseen Hand in all this, and I humbly hope that it will still guide me to do good in my day and generation in Africa."

In looking forward to the work to which Providence seemed to be calling him, a communication received at Quilimane disturbed him not a little. It was from the London Missionary Society. It informed him that the Directors were restricted in their power of aiding plans connected only remotely with the spread of the gospel, and that even though certain obstacles (from tsetse, etc.) should prove surmountable, "the financial circumstances of the Society are not such as to afford any ground of hope that it would be in a position within any definite period to undertake untried any remote and difficult fields of labor." Dr. Livingstone very naturally understood this as a declinature of his proposals. Writing on the subject to Rev. William Thompson, the Society's agent at Cape Town, he said:

"I had imagined in my simplicity that both my preaching, conversation, and travel were as nearly connected with the spread of the gospel as the Boers would allow them to be. A plan of opening up a path from either the East or West Coast for the teeming population of the interior was submitted to the judgment of the Directors, and received their formal approbation.

"I have been seven times in peril of my life from savage men while laboriously and without swerving pursuing that plan, and never doubting that I was in the path of duty.

"Indeed, so clearly did I perceive that I was performing good service to the cause of Christ, that I wrote to my brother that I would perish rather than fail in my enterprise. I shall not boast of what I have done, but the wonderful mercy I have received will constrain me to follow out the work in spite of the veto of the Board.

"If it is according to the will of God, means will be provided from other quarters."

A long letter to the Secretary gives a fuller statement of his views. It is so important as throwing light on his missionary consistency, that we give it in full in the Appendix.[1]

The Directors showed a much more sympathetic spirit when Livingstone came among them, but meanwhile, as he tells us in his book, his old feeling of independence had returned, and it did not seem probable that he would remain in the same relation to the Society.

After Livingstone had been six weeks at Quilimane, H.M. brig "Frolic" arrived, with ample supplies for all his need, and took him to the Mauritius, where he arrived on 12th August, 1856. It was during this voyage that the lamentable insanity and suicide of his native attendant Sekwebu occurred, of which we have an account in the *Missionary Travels.* At the Mauritius he was the guest of

---

Appendix No. III.

General Hay, from whom he received the greatest kindness, and so rapid was his recovery from an affection of
the spleen which his numerous fevers had bequeathed,
that before he left the island he wrote to Commodore
Trotter and other friends that he was perfectly well, and
" quite ready to go back to Africa again." This, however,
was not to be just yet. In November he sailed through
the Red Sea, on the homeward route. He had expected
to land at Southampton, and there Mrs. Livingstone and
other friends had gone to welcome him. But the perils
of travel were not yet over. A serious accident befell the
ship, which might have been followed by fatal results but
for that good Providence that held the life of Livingstone
so carefully. Writing to Mrs. Livingstone from the Bay
of Tunis (27th November, 1856), he says:

" We had very rough weather after leaving Malta, and yesterday ฀.฀
midday the shaft of the engine—an enormous mass of malleable iron—
broke with a sort of oblique fracture, evidently from the terrific strains
which the tremendous seas inflicted as they thumped and tossed this
gigantic vessel like a plaything. We were near the island called
Zembra, which is in sight of the Bay of Tunis. The wind, which had
been a full gale ahead when we did not require it, now fell to a dead
calm, and a current was drifting our gallant ship, with her sails flap-
ping all helplessly, against the rocks; the boats were provisioned,
watered, and armed, the number each was to carry arranged (the
women and children to go in first, of course), when most providentially
a wind sprung up and carried us out of danger into the Bay of Tunis,
where I now write. The whole affair was managed by Captain Powell
most admirably. He was assisted by two gentlemen whom we all
admire—Captain Tregear of the same Company, and Lieutenant Chim-
nis of the Royal Navy, and though they and the sailors knew that the
vessel was so near destruction as to render it certain that we should
scarcely clear her in the boats before the swell would have overwhelmed
her, all was managed so quietly that none of us passengers knew much
about it. Though we saw the preparation, no alarm spread among us.
The Company will do everything in their power to forward us quickly
and safely. I'm only sorry for your sake, but patience is a great virtue,
you know. Captain Tregear has been six years away from his family,
I only four and a half."

The passengers were sent on *viâ* Marseilles, and Livingstone proceeded homeward by Paris and Dover.

At last he reached "dear old England" on the 9th of December, 1856. Tidings of a great sorrow had reached him on the way. At Cairo he heard of the death of his father. He had been ill a fortnight, and died full of faith and peace. "You wished so much to see David," said his daughter to him as his life was ebbing away. "Ay, very much, very much; but the will of the Lord be done." Then after a pause he said, "But I think I'll know whatever is worth knowing about him. When you see him, tell him I think so." David had not less eagerly desired to sit once more at the fireside and tell his father of all that had befallen him on the way. On both sides the desire had to be classed among hopes unfulfilled. But on both sides there was a vivid impression that the joy so narrowly missed on earth would be found in a purer form in the next stage of being.

CHAPTER X.

FIRST VISIT HOME.

A.D. 1856–1857.

Mrs. Livingstone—Her intense anxieties—Her poetical welcome—Congratulatory letters from Mrs. and Dr. Moffat—Meeting of welcome of Royal Geographical Society—of London Missionary Society—Meeting in Mansion House—Enthusiastic public meeting at Cape Town—Livingstone visits Hamilton—Returns to London to write his book—Letter to Mr. Maclear.—Dr. Risdon Bennett's reminiscences of this period—Mr. Frederick Fitch's—Interview with Prince Consort—Honors—Publication and great success of *Missionary Travels*—Character and design of the book—Why it was not more of a missionary record—Handsome conduct of publisher—Generous use of the profits—Letter to a lady in Carlisle vindicating the character of his speeches.

THE years that had elapsed since Dr. Livingstone bade his wife farewell at Cape Town had been to her years of deep and often terrible anxiety. Letters, as we have seen, were often lost, and none seem more frequently to have gone missing than those between him and her. A stranger in England, without a home, broken in health, with a family of four to care for, often without tidings of her husband for great stretches of time, and harassed with anxieties and apprehensions that sometimes proved too much for her faith, the strain on her was very great. Those who knew her in Africa, when, "queen of the wagon," and full of life, she directed the arrangements and sustained the spirits of a whole party, would hardly have thought her the same person in England. When Livingstone had been longest unheard of, her heart sank altogether; but through prayer, tranquillity of mind returned, even before the arrival of any letter announcing his safety. She had been waiting for him at Southampton, and, owing to the casualty in the Bay of Tunis, he arrived at Dover, but as soon as possible he was with her, reading the poetical welcome which she had prepared in the hope that they would never part again:

> " A hundred thousand welcomes, and it's time for you to come
> From the far land of the foreigner, to your country and your home.
> O long as we were parted, ever since you went away,
> I never passed a dreamless night, or knew an easy day.
>
> Do you think I would reproach you with the sorrows that I bore?
> Since the sorrow is all over, now I have you here once more,
> And there's nothing but the gladness, and the love within my heart,
> And the hope so sweet and certain that again we'll never part.
> *   *   *   *   *   *   *   *   *
> A hundred thousand welcomes! how my heart is gushing o'er
> With the love and joy and wonder thus to see your face once more.
> How did I live without you these long long years of woe?
> It seems as if 'twould kill me to be parted from you now.
>
> You'll never part me, darling, there's a promise in your eye;
> I may tend you while I'm living, you may watch me when I die;
> And if death but kindly lead me to the blessed home on high,
> What a hundred thousand welcomes will await you in the sky!
> " MARY."

Having for once lifted the domestic veil, we cannot resist the temptation to look into another corner of the home circle. Among the letters of congratulation that poured in at this time, none was more sincere or touching than that which Mrs. Livingstone received from her mother, Mrs. Moffat.[1] In the fullnes of her congratulations she does not forget the dark shadow that falls on the missionary's wife when the time comes for her to go back

---

[1] We have been greatly impressed by Mrs. Moffat's letters. She was evidently a woman of remarkable power. If her life had been published, we are convinced that it would have been a notable one in missionary biography. Heart and head were evidently of no common calibre. Perhaps it is not yet too late for some friend to think of this.

with her husband to their foreign home, and requires her to part with her children; tears and smiles mingle in Mrs. Moffat's letter as she reminds her daughter that they that rejoice need to be as though they rejoiced not:

"*Kuruman, December* 4, 1856.—MY DEAREST MARY,—In proportion to the anxiety I have experienced about you and your dear husband for some years past, so now is my joy and satisfaction; even though we have not yet heard the glad tidings of your having really met, but this for the present we take for granted. Having from the first been in a subdued and chastened state of mind on the subject, I endeavor still to be moderate in my joy. With regard to you both ofttimes has the sentence of death been passed in my mind, and at such seasons I dared not, desired not, to rebel, submissively leaving all to the Divine disposal; but I now feel that this has been a suitable preparation for what is before me, having to contemplate a complete separation from you till that day when we meet with the spirits of just men made perfect in the kingdom of our Father. Yes, I do feel solemn at death, but there is no melancholy about it, for what is our life, so short and so transient? And seeing it is so, we should be happy to do or to suffer as much as we can for him who bought us with his blood. Should you go to those wilds which God has enabled your husband, through numerous dangers and deaths, to penetrate, there to spend the remainder of your life, and as a consequence there to suffer manifold privations, in addition to those trials through which you have already passed—and they have not been few (for you had a hard life in this interior)—you will not think all *too much*, when you stand with that multitude who have washed their robes in the blood of the Lamb!

" Yet, my dear Mary, while we are yet in the flesh my heart will yearn over you. You are my own dear child, my first-born, and recent circumstances have had a tendency to make me feel still more tenderly toward you, and deeply as I have sympathized with you for the last few years, I shall not cease to do so for the future. Already is my imagination busy picturing the various scenes through which you must pass, from the first transport of joy on meeting till that painful anxious hour when you must bid adieu to your darlings, with faint hopes of ever seeing them again in this life; and then, what you may both have to pass through in those inhospitable regions. . . .

" From what I saw in Mr. Livingston's letter to Robert, I was shocked to think that that poor head, in the prime of manhood, was so like my own, who am literally worn out. The symptoms he describes are so like my own. Now, with a little rest and relaxation, having youth on his side, he might regain all, but I cannot help fearing for

19

him if he dashes at once into hardships again. He is certainly the wonder of his age, and with a little prudence as regards his health, the stores of information he now possesses might be turned to a mighty account for poor wretched Africa. . . . We do not yet see how Mr. L. will get on—the case seems so complex. I feel, as I have often done, that as regards ourselves it is a subject more for prayer than for deliberation, separated as we are by such distances, and such a tardy and eccentric post. I used to imagine that when he was once got out safely from this dark continent we should only have to praise God for all his mercies to him and to us all, and for what He had effected by him; but now I see we must go on seeking the guidance and direction of his providential hand, and sustaining and preventing mercy. We cannot cease to remember you daily, and thus our sympathy will be kept alive with you. . . ."

Dr. Moffatt's congratulation to his son-in-law was calm and hearty:

"Your explorations have created immense interest, and especially in England, and that man must be made of bend-leather who can remain unmoved at the rehearsal even of a tithe of your daring enterprises. The honors awaiting you at home would be enough to make a score of light heads dizzy, but I have no fear of their affecting your upper story, beyond showing you that your labors to lay open the recesses of the vast interior have been appreciated. It will be almost too much for dear Mary to hear that you are verily unscathed. She has had many to sympathize with her, and I daresay many have called you a very naughty man for thus having exposed your life a thousand times. Be that as it may, you have succeeded beyond the most sanguine expectations in laying open a world of immortal beings, all needing the gospel, and at a time, now that war is over, when people may exert their energies on an object compared with which that which has occupied the master minds of Europe, and expended so much money, and shed so much blood, is but a phantom."

On the 9th of December, as we have seen, Livingstone arrived at London. He went first to Southampton, where his wife was waiting for him, and on his return to London was quickly in communication with Sir Roderick Murchison. On the 15th December the Royal Geographic Society held a special meeting to welcome him. Sir Roderick was in the chair; the attendance was numerous and distinguished, and included some of Livingstone's previous fellow-travelers, Colonel Steele, Captain Vardon, and Mr. Oswell. The President referred to the meeting of May, 1855, when the Victoria or Patron's medal had been awarded to Livingstone for his journey from the Cape to Linyanti and Loanda. Now Livingstone had added to that feat the journey from the Atlantic Ocean at Loanda to the Indian Ocean at Quilimane, and during his several journeys had traveled over not less than eleven thousand miles of African ground. Surpassing the French missionary travelers, Huc and Gabet, he had determined, by astronomical observations, the site of numerous places, hills, rivers, and lakes, previously unknown. He had seized every opportunity of describing the physical structure, geology, and climatology of the countries traversed, and making known their natural products and capabilities. He had ascertained by experience, what had been only conjectured previously, that the interior of Africa was a plateau intersected by various lakes and rivers, the waters of which escaped to the Eastern and Western oceans by deep rents in the flanking hills. Great though these achievements were, the most honorable of all Livingstone's acts had yet to be mentioned—the fidelity that kept his promise to the natives, who, having accompanied him to St. Paul de Loanda, were reconducted by him from that city to their homes.

" Rare fortitude and virtue must our medalist have possessed, when, having struggled at the imminent risk of his life through such obstacles, and when, escaping from the interior, he had been received with true kindness by our old allies, the Portuguese at Angola, he nobly resolved to redeem his promise and retrace his steps to the interior of the vast continent! How much indeed must the influence of the British name be enhanced throughout Africa, when it has been promulgated that our missionary has thus kept his plighted word to the poor natives who faithfully stood by him !"

On receiving the medal, Livingstone apologized for his

rustiness in the use of his native tongue; said that he had only done his duty as a Christian missionary in opening up a part of Africa to the sympathy of Christendom : that Steele, Vardon, or Oswell might have done all that he had done; that as yet he was only buckling on his armor, and therefore in no condition to speak boastfully; and that the enterprise would never be complete till the slave-trade was abolished, and the whole country opened up to commerce and Christianity.

Among the distinguished men who took part in the conversation that followed was Professor Owen. He bore testimony to the value of Livingstone's contributions to zoology and palæontology, not less cordial than Sir Roderick Murchison had borne to his service to geography. He had listened with very intense interest to the sketches of these magnificent scenes of animal life that his old and most esteemed friend had given them. He cordially hoped that many more such contributions would follow, and expressed his admiration of the moral qualities of the man who had taken such pains to keep his word.

In the recognition by other gentlemen of Dr. Livingstone's labors, much stress was laid on the scientific accuracy with which he had laid down every point over which he had traveled. Thanks were given to the Portuguese authorities in Africa for the remarkable kindness which they had invariably shown him. Mr. Consul Brand reported tidings from Mr. Gabriel at Loanda, to the effect that a company of Sekelétu's people had arrived at Loanda, with a cargo of ivory, and though they had not been very successful in business, they had shown the practicability of the route. He added, that Dr. Livingstone, at Loanda, had written some letters to a newspaper, which had given such an impetus to literary taste there, that a new journal had been started—the *Loanda Aurora.*

On one other point there was a most cordial expression of feeling, especially by those who had themselves been in

South Africa,—gratitude for the unbounded kindness and
hospitality that Dr. and Mrs. Livingstone had shown to
South African travelers in the neighborhood of their
home. Happily Mrs. Livingstone was present, and heard
this acknowledgment of her kinduess.

Next day, 16th December, Dr. Livingstone had his
reception from the London Missionary Society in Freemason's Hall. Lord Shaftesbury was in the chair:

"What better thing can we do," asked the noble Earl, "than to
welcome such a man to the shores of our country? What better than
to receive him with thanksgiving and rejoicings that he is spared to
refresh us with his presence, and give his strength to future exertions?
What season more appropriate than this, when at every hearth, and in
every congregation of worshipers, the name of Christ will be honored
with more than ordinary devotion, to receive a man whose life and
labors have been in humble, hearty, and willing obedience to the
angels' song, 'Glory to God in the highest, on earth peace, good-will
toward men.'"

In reply, Livingstone acknowledged the kindness of the
Directors, with whom, for sixteen years, he had never had
a word of difference. He referred to the slowness of the
African tribes, in explanation of the comparatively small
progress of the gospel among them. He cordially acknowledged the great services of the British squadron on the
West Coast in the repressing of the slave-trade. He had
been told that to make such explorations as he was
engaged in was only a tempting of Providence, but such
ridiculous assertions were only the utterances of the weaker
brethren.

Lord Shaftesbury's words at the close of this meeting, in
honor of Mrs. Livingstone, deserve to be perpetuated:

"That lady," he said, "was born with one distinguished name,
which she had changed for another. She was born a Moffat, and she
became a Livingstone. She cheered the early part of our friend's
career by her spirit, her counsel, and her society. Afterward, when
she reached this country, she passed many years with her children in

solitude and anxiety, suffering the greatest fears for the welfare of her husband, and yet enduring all with patience and resignation, and even joy, because she had surrendered her best feelings, and sacrificed her own private interests, to the advancement of civilization and the great interests of Christianity."

A more general meeting was held in the Mansion House on the 5th of January, to consider the propriety of presenting a testimonial to Dr. Livingstone. It was addressed ˉy the Bishop of London, Mr. Raikes Currie, and others.

Meanwhile, a sensible impulse was given to the *scientific* enthusiasm for Livingstone by the arrival of the report of a great meeting held in Africa itself, in honor of the missionary explorer. At Cape Town, on 12th November, 1856, His Excellency the Governor, Sir George Grey, the Colonial Secretary, the Astronomer-Royal, the Attorney-General, Mr. Rutherfoord, the Bishop, the Rev. Mr. Thompson, and others, vied with each other in expressing their sense of Livingstone's character and work. The testimony of the Astronomer-Royal to Livingstone's eminence as an astronomical observer was even more emphatic than Murchison's and Owen's to his attainments in geography and natural history. Going over his whole career, Mr. Maclear showed his unexampled achievements in accurate lunar observation. "I never knew a man," he said, "who, knowing scarcely anything of the method of making geographical observations, or laying down positions, became so soon an adept, that he could take the complete lunar observation, and altitudes for time, within fifteen minutes." His observations of the course of the Zambesi, from Sesheke to its confluence with the Lonta, were considered by the Astronomer-Royal to be "the finest specimens of sound geographical observation he ever met with."

"To give an idea of the laboriousness of this branch of his work," he adds, "on an average each lunar distance consists of five partial observations, and there are 148 sets of distances, being 740 contacts,—

and there are two altitudes of each object before, and two after, which, together with altitudes for time, amount to 2812 partial observations. But that is not the whole of his observations. Some of them intrusted to an Arab have not been received, and in reference to those transmitted he says, 'I have taken others which I do not think it necessary to send.' How completely all this stamps the impress of Livingstone on the interior of South Africa! . . . I say, what that man has done is unprecedented. . . . You could go to any point across the entire continent, along Livingstone's track, and feel certain of your position." [1]

Following this unrivaled eulogium on the scientific powers of Livingstone came the testimony of Mr. Thompson to his missionary ardor:

'I am in a position to express my earnest conviction, formed in long, intimate, unreserved communications with him, personally and by letter, that in the privations, sufferings, and dangers he has passed through, during the last eight years, he has not been actuated by mere curiosity, or the love of adventure, or the thirst for applause, or by any other object, however laudable in itself, less than his avowed one as a messenger of Christian love from the Churches. If ever there was a man who, by realizing the obligations of his sacred calling as a Christian missionary, and intelligently comprehending its object, sought to pursue it to a successful issue, such a man is Dr. Livingstone. The spirit in which he engages in his work may be seen in the following extract from one of his letters: 'You kindly say you fear for the result of my going in alone. I hope I am in the way of duty; my own conviction that such is the case has never wavered. I am doing something for God. I have preached the gospel in many a spot where the name of Christ has never been heard, and I would wish to do still more in the way of reducing the Barotse language, if I had not suffered so

---

It seems unaccountable that in the face of such unrivaled testimonies, reflections should continue to be cast on Livingstone's scientific accuracy, even so late as the meeting of the British Association at Sheffield in 1879. The family of the late Sir Thomas Maclear have sent home his collection of Livingstone's papers. They fill a box which one man could with difficulty carry. And their mass is far from their most striking quality. The evidence of laborious, painstaking care to be accurate is almost unprecedented. Folio volumes of pages covered with figures show how much time and labor must have been spent in these computations. Explanatory remarks often indicate the particulars of the observation.

severely from fever. Exhaustion produced vertigo, causing me, if 1 looked suddenly up, almost to lose consciousness; this made me give up sedentary work; but I hope God will accept of what I can do.' "

A third gentleman at this meeting, Mr. Rutherfoord, who had known Livingstone for many years, besides describing him as "one of the most honorable, benevolent, conscientious men I ever met with," bore testimony to his capacity in mercantile affairs; not exercised in his own interest, but in that of others. It was Mr. Rutherfoord who, when Livingstone was at the Cape in 1852, entered into his plans for supplanting the slave-trade by lawful traffic, and at his suggestion engaged George Fleming to go north with him as a trader, and try the experiment. The project was not very successful, owing to innumerable unforeseen worries, and especially the rascality of Fleming's men. Livingstone found it impossible to take Fleming to the coast, and had therefore to send him back, but he did his utmost to prevent loss to his friend; and thus, as Mr. Rutherfoord said, "at the very time that he was engaged in such important duties, and exposed to such difficulties, he found time to fulfill his promise to do what he could to save me from loss, to attend to a matter quite foreign to his usual avocations, and in which he had no personal interest; and by his energy and good sense, and self-denying exertions, to render the plan, if not perfectly successful, yet by no means a failure."

Traveler, geographer, zoologist, astronomer, missionary, physician, and mercantile director, did ever man sustain so many characters at once? Or did ever man perform the duties of each with such painstaking accuracy and so great success?

As soon as he could tear himself from his first engagements, he ran down to Hamilton to see his mother, children, and other relatives. His father's empty chair deeply affected him. "The first evening," writes one of his sisters, "he asked all about his illness and death. One of us remarking that after he knew he was dying his spirits seemed to rise, David burst into tears. At family worship that evening he said with deep feeling—'We bless thee, O Lord, for our parents; we give thee thanks for the dead who has died in the Lord.'"

At first Livingstone thought that his stay in this country could be only for three or four months, as he was eager to be at Quilimane before the unhealthy season set in, and thus fulfill his promise to return to his Makololo at Tette. But on receiving an assurance from the Portuguese Government (which, however, was never fulfilled *by them*) that his men would be looked after, he made up his mind for a somewhat longer stay. But it could not be called rest. As soon as he could settle down he had to set to work with a book. So long before as May, 1856, Sir Roderick Murchison had written to him that "Mr. John Murray, the great publisher, is most anxious to induce you to put together all your data, and to make a good book," adding his own strong advice to comply with the request. If he ever doubted the propriety of writing the book, the doubt must have vanished, not only in view of the unequaled interest excited by the subject, but also of the readiness of unprincipled adventurers, and even some respectable publishers, to circulate narratives often mythical and quite unauthorized.

The early part of the year 1857 was mainly occupied with the labor of writing. For this he had materials in the Journals which he had kept so carefully; but the business of selection and supplementing was laborious, and the task of arrangement and transcription very irksome. In fact, this task tried the patience of Livingstone more than any which he had yet undertaken, and he used to say that he would rather cross Africa than write another book. His experience of book-making increased his respect for authors and authoresses a hundred-fold!

We are not, however, inclined to think that this trial was due to the cause which Livingstone assigned,—his want of experience, and want of command over the English tongue. He was by no means an inexperienced writer. He had written large volumes of Journals, memoirs for the Geographical Society, articles on African Missions, letters for the Missionary Society, and private letters without end, each usually as long as a pamphlet. He was master of a clear, simple, idiomatic style, well fitted to record the incidents of a journey—sometimes poetical in its vivid pictures, often brightening into humor, and sometimes deepening into pathos. Viewing it page by page, the style of the *Missionary Travels* is admirable, the chief defect being want of perspective; the book is more a collection of pieces than an organized whole: a fault inevitable, perhaps, in some measure, from its nature, but aggravated, as we believe, by the haste and pressure under which it had to be written. In his earlier private letters, Livingstone, in his single-hearted desire to rouse the world on the subject of Africa, used to regret that he could not write in such a way as to command general attention: had he been master of the flowing periods of the *Edinburgh Review,* he thought he could have done much more good. In point of fact, if he had had the pen of Samuel Johnson, or the tongue of Edmund Burke, he would not have made the impression he did. His simple style and plain speech were eminently in harmony with his truthful, unexaggerating nature, and showed that he neither wrote nor spoke for effect, but simply to utter truth. What made his work of composition irksome was, on the one hand, the fear that he was not doing it well, and on the other, the necessity of doing it quickly. He had always a dread that his English was not up to the critical mark, and yet he was obliged to hurry on, and leave the English as it dropped from his pen. He had no time to plan, to shape, to organize; the architectural talent could not be brought into play. Add to this that he had been so accustomed to open-air life and physical exercise, that the close air and sedentary attitude of the study must have been exceedingly irksome; so that it is hardly less wonderful that his health stood the confinement of book-making in England, than that it survived the tear and wear, labor and sorrow, of all his journeys in Africa.

An extract from a letter to Mr. Maclear, on the eve of his beginning his book (21st January, 1857), will show how his thoughts were running:

"I begin to-morrow to write my book, and as I have a large party of men (110) waiting for me at Tette, and I promised to join them in April next, you will see I shall have enough to do to get over my work here before the end of the month. . . . Many thanks for all the kind things you said at the Cape Town meeting. Here they laud me till I shut my eyes, for only trying to do my duty. They ought to vote thanks to the Boers who set me free to discover the fine new country. They were determined to shut the country, and I was determined to open it. They boasted to the Portuguese that they had expelled two missionaries, and outwitted themselves rather. I got the gold medal, as you predicted, and the freedom of the town of Hamilton, which insures me protection from the payment of jail fees if put in prison!"

In writing his book, he sometimes worked in the house of a friend, but generally in a London or suburban lodging, often with his children about him, and all their noise; for, as in the Blantyre mill, he could abstract his attention from sounds of whatever kind, and go on calmly with his work. Busy though he was, this must have been one of the happiest times in his life. Some of his children still remember his walks and romps with them in the Barnet woods, near which they lived part of the time—how he would suddenly plunge into the ferny thicket, and set them looking for him, as people looked for him afterward when he disappeared in Africa, coming out all at once at some unexpected corner of the thicket. One of his greatest troubles was the penny post. People used to ask him the most frivolous questions. At first he struggled to answer them, but in a few weeks he had to give this up in despair. The simplicity of his heart is seen in the child-like joy with which he welcomes the early products of the spring. He writes to Mr. Maclear that, one day at Professor Owen's, they had "seen daisies, primroses, hawthorns, and robin-redbreasts. Does not Mrs. Maclear envy us? It was so pleasant."

But a better idea of his mode of life at home will be conveyed by the notes of some of the friends with whom he stayed. For that purpose, we resume the recollections of Dr. Risdon Bennett:

"On returning to England, after his first great journey of discovery, he and Mrs. Livingstone stayed in my house for some time, and I had frequent conversations with him on subjects connected with his African life, especially on such as related to natural history and medicine, on which he had gathered a fund of information. His observation of malarious diseases, and the methods of treatment adopted by both the natives and Europeans, had led him to form very definite and decided views, especially in reference to the use of purgatives, preliminary to, and in conjunction with, quinine and other acknowledged febrifuge medicines. He had, while staying with me, one of those febrile attacks to which persons who have once suffered from malarious disease are so liable, and I could not fail to remark his sensible observations thereon, and his judicious management of his sickness. He had a great natural predilection for medical science, and always took great interest in all that related to the profession. I endeavored to persuade him to commit to writing the results of his medical observations and experience among the natives of Africa, but he was too much occupied with the preparation of his Journal for the press to enable him to do this. Moreover, as he often said, writing was a great drudgery to him. He, however, attended with me the meetings of some of the medical societies, and gave some verbal accounts of his medical experience which greatly interested his audience. His remarks on climates, food, and customs of the natives, in reference to the origin and spread of disease, evinced the same acuteness of observation which characterized all the records of his life. He specially commented on the absence of consumption and all forms of tubercular disease among the natives, and connected this with their constant exposure and out-of-door life.

"After leaving my house he had lodgings in Chelsea, and used frequently to come and spend the Sunday afternoon with my family, often bringing his sister, who was staying with him, and his two elder children. It was beautiful to observe how thoroughly he enjoyed domestic life and the society of children, how strong was his attachment to his own family after his long and frequent separations from them, and how entirely he had retained his simplicity of character.

"Like so many of his countrymen, he had a keen sense of humor, which frequently came into play when relating his many adventures and hardships. On the latter he never dilated in the way of complaint, and he had little sympathy with, or respect for, those travelers who did so. Nor was he apt to say much on direct religious topics, or on the results of his missionary efforts as a Christian teacher. He had unbounded confidence in the influence of Christian character and principles, and gave many illustrations of the effect produced on the minds and conduct of the benighted and savage tribes with whom he was brought into contact by his own unvarying uprightness of conduct and self-denying labor. The fatherly character of God, his never-failing goodness and mercy, and the infinite love of the Lord Jesus Christ, and efficacy of his atoning sacrifice, appeared to be the topics on which he loved chiefly to dwell. The all-pervading deadly evils of slavery, and the atrocities of the slave-trade, never failed to excite his righteous indignation. If ever he was betrayed into unmeasured language, it was when referring to these topics, or when speaking of the injurious influence exerted on the native mind by the cruel and unprincipled conduct of wicked and selfish traders. His love for Africa, and confidence in the steady dawn of brighter days for its oppressed races, were unbounded."

From a member of another family, that of Mr. Frederick Fitch, of Highbury New Park, with whom also the Livingstones spent part of their time, we have some homely but graphic reminiscences:

"Dr. Livingstone was very simple and unpretending, and used to be annoyed when he was made a lion of. Once a well-known gentleman, who was advertised to deliver a lecture next day, called on him to pump him for material. The Doctor sat rather quiet, and, without being rude, treated the gentleman to monosyllabic answers. He could do that—could keep people at a distance when they wanted to make capital out of him. When the stranger had left, turning to my mother, he would say, 'I'll tell *you* anything you like to ask.'

"He never liked to walk in the streets for fear of being mobbed. Once he was mobbed in Regent street, and did not know how he was to escape, till he saw a cab, and took refuge in it. For the same reason it was painful for him to go to church. Once, being anxious to go with

20

us, my father persuaded him that, as the seat at the top of our pew was under the gallery, he would not be seen. As soon as he entered, he held down his head, and kept it covered with his hands all the time, but the preacher somehow caught sight of him, and rather unwisely, in his last prayer, adverted to him. This gave the people the knowledge that he was in the chapel, and after the service they came trooping toward him, even over the pews, in their anxiety to see him and shake hands.[1]

"Dr. Livingstone usually conducted our family worship. On Sunday morning he always gave us a text for the day. His prayers were very direct and simple, just like a child asking his Father for what he needed.

"He was always careful as to dress and appearance. This was his habit in Africa, too, and with Mrs. Livingstone it was the same. They thought that this was fitted to secure respect for themselves, and that it was for the good of the natives too, as it was so difficult to impress them with proper ideas on the subject of dress.

"Dr. and Mrs. Livingstone were much attached, and thoroughly understood each other. The doctor was sportive and fond of a joke, and Mrs. Livingstone entered into his humor. Mrs. Livingstone was terribly anxious about her husband when he was in Africa, but before others she concealed her emotion. In society both were reserved and quiet. Neither of them cared for grandeur; it was a great trial to Dr. Livingstone to go to a grand dinner. Yet in his quiet way he would exercise an influence at the dinner-table. He told us that once at a dinner at Lord ——'s, every one was running down London tradesmen. Dr. Livingstone quietly remarked that though he was a stranger in London, he knew one tradesman of whose honesty he was thoroughly assured; and if there was one such in his little circle, surely there must be many more.

"He used to rise early: about seven he had a cup of tea or coffee, and then he set to work with his writing. He had not the appearance of a very strong man."

In spite of his literary work, the stream of public honors and public engagements began to flow very strongly. The Prince Consort granted him an interview, soon after his arrival, in presence of some of the younger members of the Royal Family. In March it was agreed to present him with the freedom of the City of London, in a box of the value of fifty guineas, and in May the presen-

---

[1] A similar occurrence took place in a church at Bath during the meetings of the British Association in 1864.

tation took place. Most of his public honors, however, were reserved till the autumn.

The *Missionary Travels* was published in November, 1857, and the success of the book was quite remarkable. Writing to Mr. Maclear, 10th November, 1857, he says, after an apology for delay:

"You must ascribe my culpable silence to 'aberration.' I am out of my orbit, rather, and you must have patience till I come in again. The book is out to-day, and I am going to Captain Washington to see about copies to yourself, the Governor, the Bishop, Fairbairn, Thompson, Rutherfoord, and Saul Solomon.[1] Ten thousand were taken by the London trade alone. Thirteen thousand eight hundred have been ordered from an edition of twelve thousand, so the printers are again at work to supply the demand. Sir Roderick gave it a glowing character last night at the Royal Geographical Society, and the *Athenæum* has come out strongly on the same side. This is considered a successful launch for a guinea book."

It has sometimes been a complaint that so much of the book is occupied with matters of science, geographical inquiries, descriptions of plants and animals, accounts of rivers and mountains, and so little with what directly concerns the work of the missionary. In reply to this, it may be stated, in the first place, that if the information given and the views expressed on missionary topics were all put together, they would constitute no insignificant contribution to missionary literature. But there was another consideration. Livingstone regarded himself as but a pioneer in missionary enterprise. During sixteen years he had done much to bring the knowledge of Christ to tribes that had never heard of Him—probably no missionary in Africa had ever preached to so many blacks. In some instances he had been successful in the highest sense—he had been the instrument of turning men from darkness to light; but he did not think it right to dwell on these cases, because

---

[1] Livingstone was quite lavish with presentation copies; every friend on the list seemed to be included in his list. He tried to remember every one who had shown kindness to himself, and particularly to his wife and children.

the converts were often inconsistent, and did not exemplify a high moral tone. In most cases, however, he had been a sower of seed, and not a reaper of harvests. He had no triumphs to record, like those which had gladdened the hearts of some of his missionary brethren in the South Sea Islands. He wished his book to be a record of facts, not a mere register of hopes. The missionary work was yet to be done. It belonged to the future, not to the past. By showing what vast fields there were in Africa ripe for the harvest, he sought to stimulate the Christian enterprise of the Churches, and lead them to take possession of Africa for Christ. He would diligently record facts which he had ascertained about Africa, facts that he saw had some bearing on its future welfare, but whose full significance in that connection no one might yet be able to perceive. In a sense, the book was a work of faith. He wished to interest men of science, men of commerce, men of philanthropy, ministers of the Crown, men of all sorts, in the welfare of Africa. Where he had so varied a constituency to deal with, and where the precise method by which Africa would be civilized was yet so indefinite, he would faithfully record what he had come to know, and let others build as they might with his materials. Certainly, in all that Livingstone has written, he has left us in no doubt as to the consummation to which he ever looked. His whole writings and his whole life are a commentary on his own words—"The end of the geographical feat is only the beginning of the enterprise."

Through the great success of the volume and the handsome conduct of the publishers, the book yielded him a little fortune. We shall see what generous use he made of it—how large a portion of the profits went to forward directly the great object to which his heart and his life were so cordially given. More than half went to a single object connected with the Zambesi Expedition, and of the remainder he was ready to devote a half to another favorite project. All that he thought it his duty to reserve for his children was enough to educate them, and prepare them for their part in life. Nothing would have seemed less desirable or less for their good than to found a rich family to live in idleness. It was and is a common impression that Livingstone received large sums from friends to aid him in his work. For the most part these impressions were unfounded; but his own hard-earned money was bestowed freely and cheerfully wherever it seemed likely to do good. The complaint that he was not sufficiently a missionary was sometimes made of his speeches as well as his book. At Carlisle, a lady wrote to him in this strain. A copy of his reply is before us. After explaining that reporters were more ready to report his geography than his missionary views, he says:

"Nowhere have I ever appeared as anything else but a servant of God, who has simply followed the leadings of his hand. My views of what is *missionary* duty are not so contracted as those whose ideal is a dumpy sort of man with a Bible under his arm. I have labored in bricks and mortar, at the forge and carpenter's bench, as well as in preaching and medical practice. I feel that I am 'not my own.' I am serving Christ when shooting a buffalo for my men, or taking an astronomical observation, or writing to one of his children who forget, during the little moment of penning a note, that charity which is eulogized as 'thinking no evil'; and after having by his help got information, which I hope will lead to more abundant blessing being bestowed on Africa than heretofore, am I to hide the light under a bushel, merely because some will consider it not sufficiently, or even at all, *missionary?* Knowing that some persons do believe that opening up a new country to the sympathies of Christendom was not a proper work for an agent of a missionary society to engage in, I now refrain from taking any salary from the Society with which I was connected; so no pecuniary loss is sustained by any one."

Subsequently, when detained in Manyuema, and when his immediate object was to determine the water-shed, Dr. Livingstone wrote: "I never felt a single pang at having left the Missionary Society. I acted for my Master, and believe that all ought to devote their special faculties to Him. I regretted that unconscientious men took occasion to prevent many from sympathizing with me."

CHAPTER XI.

FIRST VISIT HOME—*continued.*

A.D. 1857–1858.

Livingstone at Dublin, at British Association—Letter to his wife—He meets the Chamber of Commerce at Manchester—At Glasgow, receives honors from Corporation, University, Faculty of Physicians and Surgeons, United Presbyterians, Cotton-spinners—His speeches in reply—His brother Charles joins him—Interesting meeting and speech at Hamilton—Reception from "Literary and Scientific Institute of Blantyre"—Sympathy with operatives—Quick apprehension of all public questions—His social views in advance of the age—He plans a People's Café—Visit to Edinburgh—More honors—Letter to Mr. Maclear—Interesting visit to Cambridge—Lectures there—Professor Sedgwick's remarks on his visit—Livingstone's great satisfaction—Relations to London Missionary Society—He severs his connection—Proposal of Government expedition—He accepts consulship and command of expedition—Kindness of Lords Palmerston and Clarendon—The Portuguese Ambassador—Livingstone proposes to go to Portugal—Is dissuaded—Lord Clarendon's letter to Sekelétu—Results of Livingstone's visit to England—Farewell banquet, Feb., 1858—Interview with the Queen—Valedictory letters—Professor Sedgwick and Sir Roderick Murchison—Arrangements for expedition—Dr., Mrs., and Oswell Livingstone set sail from Liverpool—Letters to children.

FINDING himself, in the autumn, free of the toil of book-making, Dr. Livingstone moved more freely through the country, attended meetings, and gave addresses. In August he went to Dublin, to the meeting of the British Association for the Advancement of Science, and gave an interesting lecture. Mrs. Livingstone did not accompany him. In a letter to her we have some pleasant notes of his Dublin visit:

"*Dublin, 29th August,* 1857.—I am very sorry now that I did not bring you with me, for all inquired after you, and father's book is better known here than anywhere else I have been. But it could scarcely have been otherwise. I think the visit to Dublin will be beneficial to our cause, which, I think, is the cause of Christ in Africa. Lord Radstock is much interested in it, and seems willing and anxious to promote it. He was converted out at the Crimea, whither he had gone as an amateur. His lady is a beautiful woman, and I think, what is far better, a good, pious one. The Archbishop's daughters asked me if they could be of any use in sending out needles, thread, etc., to your school. I, of course, said Yes. His daughters are devotedly missionary, and work hard in ragged schools, etc. One of them nearly remained in Jerusalem as a missionary, and is the same in spirit here. It is well to be servants of Christ everywhere, at home or abroad, wherever He may send us or take us. . . . I hope I may be enabled to say a word for Him on Monday. There is to be a grand dinner and soiree at the Lord-Lieutenant's on Monday, and I have got an invitation in my pocket, but will have to meet Admiral Trotter on Tuesday. I go off as soon as my lecture is over. . . . Sir Duncan Macgregor is the author of *The Burning of the Kent East Indiaman.* His son, the only infant saved, is now a devoted Christian, a barrister." [1]

In September we find him in Manchester, where the Chamber of Commerce gave him a hearty welcome, and entered cordially into his schemes for the commercial development of Africa. He was subjected to a close cross-examination regarding the products of the country, and the materials it contained for commerce; but here, too, the missionary was equal to the occasion. He had brought home five or six and twenty different kinds of fruit; he told them of oils they had never heard of—dyes that were kept secret by the natives—fibres that might be used for the manufacture of paper—sheep that had hair instead of wool—honey, sugar-cane, wheat, millet, cotton, and

---

[1] Dr. Livingstone always liked that style of earnest Christianity which he notices in this letter. In November of the same year, after he had resigned his connection with the London Missionary Society, and was preparing to return to Africa as H.M. Consul and head of the Zambesi Expedition, he writes thus to his friend Mr. James Young: "I read the life of Hedley Vicars for the first time through, when down at Rugby. It is really excellent, and makes me ashamed of the coldness of my services in comparison. That was his sister you saw me walking with in Dublin at the Gardens (Lady Rayleigh). If you have not read it, the sooner you dip into it the better. You will thank me for it."

iron, all abounding in the country. That all these should abound in what used to be deemed a sandy desert appeared very strange. A very cordial resolution was unanimously agreed to, and a strong desire expressed that Her Majesty's Government would unite with that of Portugal in giving Dr. Livingstone facilities for further exploration in the interior of Africa, and especially in the district around the river Zambesi and its tributaries, which promised to be the most suitable as a basis both for commercial and missionary settlements.

In the course of the same month his foot was again on his native soil, and there his reception was remarkably cordial. In Glasgow, the University, the Corporation, the Faculty of Physicians and Surgeons, the United Presbyterians, and the Associated Operative Cotton-spinners of Scotland came forward to pay him honor. A testimonial of £2000 had been raised by public subscription. The Corporation presented him with the freedom of the city in a gold box, in acknowledging which he naturally dwelt on some of the topics that were interesting to a commercial community. He gave a somewhat new view of "Protection" when he called it a remnant of heathenism. The heathen would be dependent on no one; they would depress all other communities. Christianity taught us to be friends and brothers, and he was glad that all restrictions on the freedom of trade were now done away with. He dwelt largely on the capacity of Africa to furnish us with useful articles of trade, and especially cotton.

His reception by the Faculty of Physicians and Surgeons had a special interest in relation to his medical labors. For nearly twenty years he had been a licentiate of this Faculty, one of the oldest medical institutions of the country, which for two centuries and a half had exerted a great influence in the west of Scotland. He was now admitted an honorary Fellow—an honor rarely conferred, and only on pre-eminently distinguished men

The President referred to the benefit which he had found from his scientific as well as his more strictly medical studies, pursued under their auspices, and Livingstone cordially echoed the remark, saying he often hoped that his sons might follow the same course of study and devote themselves to the same noble profession:

"In the country to which I went," he continued, "I endeavored to follow the footsteps of my Lord and Master. Our Saviour was a physician; but it is not to be expected that his followers should perform miracles. The nearest approach which they could expect to make was to become acquainted with medical science, and endeavor to heal the diseases of man. . . . One patient expressed his opinion of my religion to the following effect: "We like you very much; you are the only white man we have got acquainted with. We like you because you aid us whilst we are sick, but we don't like your everlasting preaching and praying. We can't get accustomed to that!""

To the United Presbyterians of Glasgow he spoke of mission work in Africa. At one time he had been somewhat disappointed with the Bechuana Christians, and thought the results of the mission had been exaggerated, but when he went into the interior and saw heathenism in all its unmitigated ferocity, he changed his opinion, and had a higher opinion than ever of what the mission had done. Such gatherings as the present were very encouraging; but in Africa mission work was hard work without excitement; and they had just to resolve to do their duty without expecting to receive gratitude from those whom they labored to serve. When gratitude came, they were thankful to have it; but when it did not come they must go on doing their duty, as unto the Lord.

His reply to the cotton-spinners is interesting as showing how fresh his sympathy still was with the sons of toil, and what respect he had for their position. He congratulated himself on the Spartan training he had got at the Blantyre mill, which had really been the foundation of all the work he had done. Poverty and hard work were often looked down on,—he did not know why,—for wickedness was the only thing that ought to be a reproach to any man. Those that looked down on cotton-spinners with contempt were men who, had they been cotton-spinners at the beginning, would have been cotton-spinners to the end. The life of toil was what belonged to the great majority of the race, and to be poor was no reproach. The Saviour occupied the humble position that they had been born in, and he looked back on his own past life as having been spent in the same position in which the Saviour lived.

"My great object," he said, " was to be like Him—to imitate Him as far as He could be imitated. We have not the power of working miracles, but we can do a little in the way of healing the sick, and I sought a medical education in order that I might be like Him. In Africa I have had hard work. I don't know that any one in Africa despises a man who works hard. I find that all eminent men work hard. Eminent geologists, mineralogists, men of science in every department, if they attain eminence, work hard, and that both early and late. That is just what we did. Some of us have left the cotton-spinning, but I think that all of us who have been engaged in that occupation look back on it with feelings of complacency, and feel an interest in the course of our companions. There is one thing in cotton-spinning that I always felt to be a privilege. We were confined through the whole day, but when we got out to the green fields, and could wander through the shady woods, and rove about the whole country, we enjoyed it immensely. We were delighted to see the flowers and the beautiful scenery. We were prepared to admire. We were taught by our confinement to rejoice in the beauties of nature, and when we got out we enjoyed ourselves to the fullest extent."

At Hamilton an interesting meeting took place in the Congregational Chapel where he had been a worshiper in his youth. Here he was emphatically at home; and he took the opportunity (as he often did) to say how little he liked the lionizing he was undergoing, and how unexpected all the honors were that had been showered upon him. He had hoped to spend a short and quiet visit, and then return to his African work. It was his sense of the kindness shown him, and the desire not to be disobliging, that made him accept the public invitations he was receiving. But he did not wish to take the honor to himself, as if he had achieved anything by his own might or wisdom. He thanked God sincerely for employing him as an instrument in his work. One of the greatest honors was to be employed in winning souls to Christ, and proclaiming to the captives of Satan the liberty with which he had come to make them free. He was thankful that to him, "the least of all saints," this honor had been given. He then proceeded to notice the presence of members of various Churches, and to advert to the broadening process that had been going on in his own mind while in Africa, which made him feel himself more than ever the brother of all:

"In going about we learn something, and it would be a shame to us if we did not; and we look back to our own country and view it as a whole, and many of the little feelings we had when immersed in our own denominations we lose, and we look to the whole body of Christians with affection. We rejoice to see them advancing. I believe that every Scotch Christian abroad rejoiced in his heart when he saw the Free Church come boldly out on principle, and I may say we shall rejoice very much when we see the Free Church and the United Presbyterian Church one, as they ought to be. . . . I am sure I look on all the different denominations in Hamilton and in Britain with feelings of affection. I cannot say which I love most. I am quite certain I ought not to dislike any of them. Really, perhaps I may be considered a little heterodox, if I were living in this part of the country, I could not pass one Evangelical Church in order to go to my own denomination beyond it.[1] I still think that the different denominational peculiarities have, to a certain degree, a good effect in this country, but I think we ought to be much more careful lest we should appear to our fellow-Christians unchristian, than to appear inconsistent with the denominational principles we profess. . . . Let this meeting be the ratification of the

---

[1] Dr. Livingstone gave practical evidence of his sincerity in these remarks in the case of his elder daughter, saying, in reply to one of her guardians with whom she was residing, that he had no objections to her joining the Church of Scotland. This, however, she did not do; but afterward, when at Newstead Abbey, she was confirmed by the Bishop of Lincoln, and received the Communion along with her father, who helped to prepare her.

bond of union between my brother[2] and me, and all the denominations of Hamilton. Remember us in your prayers. Bear us on your spirits when we are far away, for when abroad we often feel as if we were forgot by every one. My entreaty to all the Christians of Hamilton is to pray that grace may be given to us to be faithful to our Saviour even unto death."

At Blantyre, his native village, the Literary and Scientific Institute gave him a reception, Mr. Hannan, one of the proprietors of the works, a magistrate of Glasgow, and an old acquaintance of Livingstone's, being in the chair. The Doctor was laboring under a cold, the first he had had for sixteen years. He talked to them of his travels, and by particular request gave an account of his encounter with the Mabotsa lion. He ridiculed Mrs. Beecher Stowe's notion that factory-workers were slaves. He counseled them strongly to put more confidence than workmen generally did in the honest good intentions of their employers, reminding them that some time ago, when the Blantyre proprietors had wished to let every workman have a garden, it was said by some that they only wished to bring the ground into good order, and then they would take the garden away. That was nasty and suspicious. If masters were more trusted, they would do more good. Finally, he exhorted them cordially to accept God's offers of mercy to them in Christ, and give themselves wholly to Him. To bow down before God was not mean; it was manly. His one wish for them all was that they might have peace with God, and rejoice in the hope of the eternal inheritance.

His remarks to the operatives show how sound and sagacious his views were on social problems; in this sphere, indeed, he was in advance of the age. The quickness and correctness with which he took up matters of public interest in Britain, mastered facts, and came to clear, intelligent conclusions on them, was often the astonishment of his

---

[2] Dr. Livingstone had been joined by his brother Charles, who was present on this occasion.

friends. It was as if, instead of being buried in Africa, he had been attending the club and reading the daily newspapers for years,—this, too, while he was at work writing his book, and delivering speeches almost without end. We find him at this time anticipating the temperance coffee-house movement, now so popular and successful. On 11th July, 1857, he wrote on this subject to a friend, in reference to a proposal to deliver a lecture in Glasgow. It should be noticed that he never lectured for money, though he might have done so with great pecuniary benefit:

"I am thinking of giving, or trying to give, a lecture by invitation at the Athenæum. I am offered thirty guineas, and as my old friends the cotton-spinners have invited me to meet them, I think of handing the sum, whatever it may be, to them, or rather letting them take it and fit up a room as a coffee-room on the plan of the French cafés, where men, women, and children may go, instead of to whisky-shops. There are coffee-houses already, but I don't think there are any where they can laugh and talk and read papers just as they please. The sort I contemplate would suit poor young fellows who cannot have a comfortable fire at home. I have seen men dragged into drinking ways from having no comfort at home, and women also drawn to the dram-shop from the same cause. Don't you think something could be done by setting the persons I mention to do something for themselves?"

Edinburgh conferred on Livingstone the freedom of the city, besides entertaining him at a public breakfast and hearing him at another meeting. We are not surprised to find him writing to Sir Roderick Murchison from Rossie Priory, on the 27th September, that he was about to proceed to Leeds, Liverpool, and Birmingham, "and then farewell to public spouting for ever. I am dead tired of it. The third meeting at Edinburgh quite knocked me up." It was generally believed that his appearances at Edinburgh were not equal to some others; and probably there was truth in the impression, for he must have come to it exhausted; and besides, at a public breakfast, he was put out by a proposal of the chairman, that they should try to get him a pension. Yet some who heard him in Edinburgh received

21

impressions that were never effaced, and it is probable that seed was silently sown which led afterward to the Scotch Livingstonia Mission—one of the most hopeful schemes for carrying out Livingstone's plans that have yet been organized.

Among the other honors conferred on him during this visit to Britain was the degree of D.C.L. from the University of Oxford. Some time before, Glasgow had given him the honorary degree of LL.D. In the beginning of 1858, when he was proposed as a Fellow of the Royal Society, the certificate on his behalf was signed, among others, by the Earl of Carlisle, then Lord-Lieutenant of Ireland, who after his signature added P.R. (*pro Regina*), a thing that had never been done before.[1]

The life he was now leading was rather trying. He writes to his friend Mr. Maclear on the 10th November:

"I finish my public spouting next week at Oxford. It is really very time-killing, this lionizing, and I am sure you pity me in it. I hope to leave in January. Wonder if the Portuguese have fulfilled the intention of their Government in supporting my men. . . . I shall rejoice when I see you again in the quiet of the Observatory. It is more satisfactory to serve God in peace. May He give his grace and blessing to us all! I am rather anxious to say something that will benefit the young men at Oxford. They made me a D.C.L. There!! Wonder if they would do so to the Editor of the *Grahamstown Journal?*"

Livingstone was not yet done with "public spouting," even after his trip to Oxford. Among the visits paid by him toward the end of 1857, none was more interesting or led to more important results than that to Cambridge. It was on 3d December he arrived there, becoming the guest of the Rev. Wm. Monk, of St. John's. Next morning, in the senate-house, he addressed a very large audience, consisting of graduates and undergraduates and many visitors from the town and neighborhood. The Vice-Chancellor presided and introduced the stranger. Dr. Livingstone's

---

[1] For list of Dr. Livingstone's honors, see Appendix No. **V.**

lecture consisted of facts relating to the country and its people, their habits and religious belief, with some notices of his travels, and an emphatic statement of his great object—to promote commerce and Christianity in the country which he had opened. The last part of his lecture was an earnest appeal for missionaries.

"It is deplorable to think that one of the noblest of our missionary societies, the Church Missionary Society, is compelled to send to Germany for missionaries, whilst other Societies are amply supplied. Let this stain be wiped off. The sort of men who are wanted for missionaries are such as I see before me; men of education, standing, enterprise, zeal, and piety. . . . I hope that many whom I now address will embrace that honorable career. Education has been given us from above for the purpose of bringing to the benighted the knowledge of a Saviour. If you knew the satisfaction of performing such a duty, as well as the gratitude to God which the missionary must always feel, in being chosen for so noble, so sacred a calling, you would have no hesitation in embracing it.

"For my own part, I have never ceased to rejoice that God has appointed me to such an office. People talk of the sacrifice I have made in spending so much of my life in Africa. Can that be called a sacrifice which is simply paid back as a small part of a great debt owing to our God, which we can never repay? Is that a sacrifice which brings its own blest reward in healthful activity, the consciousness of doing good, peace of mind, and a bright hope of a glorious destiny hereafter? Away with the word in such a view, and with such a thought! It is emphatically no sacrifice. Say rather it is a privilege. Anxiety, sickness, suffering, or danger, now and then, with a foregoing of the common conveniences and charities of this life, may make us pause, and cause the spirit to waver, and the soul to sink; but let this only be for a moment. All these are nothing when compared with the glory which shall hereafter be revealed in and for us. I never made a sacrifice. Of this we ought not to talk when we remember the great sacrifice which He made who left his father's throne on high to give himself for us; 'who being the brightness of that Father's glory, and the express image of his person, and upholding all things by the word of his power, when he had by himself purged our sins, sat down on the right hand of the Majesty on high.' . . .

"I beg to direct your attention to Africa: I know that in a few years I shall be cut off in that country, which is now open; do not let it be shut again! I go back to Africa to t~~ to make an open path for com-

merce and Christianity; do you carry out the work which I have begun.
I LEAVE IT WITH YOU!"

In a prefatory letter prefixed to the volume entitled
Dr. Livingstone's Cambridge Lectures, the late Professor
Sedgwick remarked, in connection with this event, that
in the course of a long academic life he had often been
present in the senate-house on exciting occasions; in the
days of Napoleon he had heard the greetings given to our
great military heroes; he had been present at four
installation services, the last of which was graced by the
presence of the Queen, when her youthful husband was
installed as Chancellor, amid the most fervent gratulations
that subjects are permitted to exhibit in the presence of
their Sovereign. But on none of these occasions "were the
gratulations of the University more honest and true-
hearted than those which were offered to Dr. Livingstone.
He came among us without any long notes of preparation,
without any pageant or eloquence to charm and captivate
our senses. He stood before us, a plain, single-minded man,
somewhat attenuated by years of toil, and with a face tinged
by the sun of Africa. . . . While we listened to the
tale he had to tell, there arose in the hearts of all the
listeners a fervent hope that the hand of God which had so
long upheld him would uphold him still, and help him to
carry out the great work of Christian love that was still
before him."

Next day, December 5th, Dr. Livingstone addressed a
very crowded audience in the Town Hall, the Mayor pre-
siding. Referring to his own plans, he said:

" I contend that we ought not to be ashamed of our religion, and had
we not kept this so much out of sight in India, we should not now be in
such straits in that country " [referring to the Indian Mutiny]. " Let
us appear just what we are. For my own part, I intend to go out as a
missionary, and hope boldly, but with civility, to state the truth of
Christianity, and my belief that those who do not possess it are in error.
My object in Africa is not only the elevation of man, but that the

country might be so opened that man might see the need of his soul's
salvation. I propose in my next expedition to visit the Zambesi, and
propitiate the different chiefs along its banks, endeavoring to induce
them to cultivate cotton, and to abolish the slave-trade : already they
trade in ivory and gold-dust, and are anxious to extend their commercial
operations. There is thus a probability of their interests being linked
with ours, and thus the elevation of the African would be the result.
"I believe England is alive to her duty of civilizing and Christianizing
the heathen. We cannot all go out as missionaries, it is true ; but we
may all do something toward providing a substitute. Moreover, all may
especially do that which every missionary highly prizes, viz.—COMMEND
THE WORK IN THEIR PRAYERS. I HOPE THAT THOSE WHOM I NOW ADDRESS
WILL BOTH PRAY FOR AND HELP THOSE WHO ARE THEIR SUBSTITUTES."

Dr. Livingstone was thoroughly delighted with his reception at Cambridge. Writing to a friend, on 6th December
1857, he says: "Cambridge, as Playfair would say, was
grand. It beat Oxford hollow. To make up my library
again they subscribed at least forty volumes at once. I
shall have reason soon to bless the Boers."

Referring to his Cambridge visit a few weeks afterward,
in a letter to Rev. W. Monk, Dr. Livingstone said: "I look
back to my visit to Cambridge as one of the most pleasant
episodes of my life. I shall always revert with feelings of
delight to the short intercourse I enjoyed with such noble
Christian men as Sedgwick, Whewell, Selwyn, etc. etc., as
not the least important privilege conferred on me by my
visit to England. It is something inspiriting to remember
that the eyes of such men are upon one's course. May
blessings rest upon them all, and on the seat of learning
which they adorn!"

Among the subjects that had occupied Dr. Livingstone's
attention most intensely during the early part of the year
1857 was that of his relation to the London Missionary
Society. The impression caused by Dr. Tidman's letter
received at Quilimane had been quite removed by personal
intercourse with the Directors, who would have been
delighted to let Livingstone work in their service in his

own way. But with the very peculiar work of exploration and inquiry which he felt that his Master had now placed in his hands, Dr. Livingstone was afraid that his freedom would be restricted by his continuing in the service of the Society, while the Society itself would be liable to suffer from the handle that might be given to contributors to say that it was departing from the proper objects of a missionary body. That in resigning his official connection he acted with a full knowledge of the effect which this might have upon his own character, and his reputation before the Church and the world, is evident from his correspondence with one of his most intimate friends and trusted counselors, Mr. J. B. Braithwaite, of Lincoln's Inn. Though himself a member of the Society of Friends, Mr. Braithwaite was desirous that Dr. Livingstone should continue to appear before the public as a Christian minister:

"To dissolve thy connection with the Missionary Society would at once place thee before the public in an aspect wholly distinct from that in which thou art at present, and, what is yet more important. would in a greater or less degree, and, perhaps, very gradually and almost insensibly to thyself, turn the current of thy own thoughts and feelings away from those channels of usefulness and service, as a minister of the gospel, with which I cannot doubt thy deepest interest and highest aspirations are inseparably associated."

On Dr. Livingstone explaining that, while he fully appreciated these views, it did not appear to him consistent with duty to be receiving the pay of a working missionary while engaged to a considerable extent in scientific exploration, Mr. Braithwaite expressed anew his sympathy for his feelings, and respect for his decision, but not as one quite convinced:

"Thy heart is bound, as I truly believe, in its inmost depths to the service of Christ. This is the 'one thing' which, through all, it is thy desire to keep in view. And my fear has been lest the severing of thy connection with a recognized religious body should lead any to suppose that thy Christian interests were in the least weakened; or that thou wast now going forth with any lower aim than the advancement of the Redeemer's kingdom. Such a circumstance would be deeply to be regretted, for thy character is now, if I may so speak, not thy own, but the common property, in a certain sense, of British Christianity, and anything which tended to lower thy high standing would cast a reflection on the general cause."

The result showed that Mr. Braithwaite was right as to the impression likely to be made on the public; but the contents of this volume amply prove that the impression was wrong.

Dr. Livingstone had said at Quilimane that if it were the will of God that he should do the work of exploration and settlement of stations which was indispensable to the opening up of Africa, but which the Directors did not then seem to wish him to undertake, the means would be provided from some other quarter. At the meeting of the British Association in Dublin, a movement was begun for getting the Government to aid him. The proposal was entertained favorably by the Government, and practically settled before the end of the year. In February, 1858, Dr. Livingstone received a formal commission, signed by Lord Clarendon, Foreign Secretary, appointing him Her Majesty's Consul at Quilimane for the Eastern Coast and the independent districts in the interior, and commander of an expedition for exploring Eastern and Central Africa. Dr. Livingstone accepted the appointment, and during the last part of his stay in England was much engaged in arranging for the expedition. A paddle steamer of light draught was procured for the navigation of the Zambesi, and the various members of the expedition received their appointments. These were—Commander Bedingfield, R.N., Naval Officer; John Kirk, M.D., Botanist and Physician; Mr. Charles Livingstone, brother of Dr. Livingstone, General Assistant and Secretary; Mr. Richard Thornton, Practical Mining Geologist; Mr. Thomas Baines, Artist and Storekeeper; and Mr. George Rae, Ship Engineer; and whoever afterward might join the expedition were required to obey Dr. Livingstone's directions as leader.

"We managed your affair very nicely," Lord Palmerston said to Livingstone at a reception at Lady Palmerston's on the 12th December. "Had we waited till the usual time when Parliament should be asked, it would have been too late." Lord Shaftesbury, at the reception, assured him that the country would do everything for him, and congratulated him on going out in the way now settled. So did the Lord Chancellor (Cranworth), Sir Culling Eardley, and Mr. Calcraft, M.P.

Dr. Livingstone was on the most friendly terms with the Portuguese Ambassador, the Count de Lavradio, who ever avowed the highest respect for himself, and a strong desire to help him in his work. To get this assurance turned into substantial assistance appeared to Livingstone to be of the very highest importance. Unless strong influence were brought to bear on the local Portuguese Governors in Africa, his scheme would be wrecked. The Portuguese Ambassador was then at Lisbon, and Livingstone had resolved to go there, to secure the influence from headquarters which was so necessary. The Prince Consort had promised to introduce him to his cousin, the King of Portugal. There were, however, some obstacles to his going. Yellow fever was raging at Lisbon, and moreover, time was precious, and a little delay might lead to the loss of a season on the Zambesi. At Lady Palmerston's reception, Lord Palmerston had said to him that Lord Clarendon might manage the Portuguese affair without his going to Lisbon. A day or two after, Livingstone saw Lord Clarendon, who confirmed Lord Palmerston's opinion, and assured him that when Lavradio returned, the affair would be settled. The Lisbon journey was accordingly given up. The Count returned to London before Livingstone left, and expressed a wish to send a number of Portuguese agents along with him. But to this both Lord Clarendon and he had the strongest objections, as complicating the expedition. Livingstone was furnished with letters from the Portuguese Government to the local Governors, instructing them to give him all needful help. But when he returned to the Zambesi he found that these public instructions were strangely neutralized and reversed by some unseen process. He himself believed to the last in the honest purpose of the King of Portugal, but he had not the same confidence in the Government. From some of the notes written to him at this time by friends who understood more of diplomacy than he did, we can see that little actual help was expected from the local Governors in the Portuguese settlements, one of these friends expressing the conviction that "the sooner those Portuguese dogs-in-the-manger are eaten up, body and bones, by the Zulu Caffres, the better."

The co-operation of Lord Clarendon was very cordial. " He told me to go to Washington (of the Admiralty) as if all had been arranged, and do everything necessary, and come to him for everything I needed. He repeated, ' Just come here and tell me what you want, and I will give it you.' He was wonderfully kind. I thank God who gives the influence." Among other things, Lord Clarendon wrote an official letter to the chief Sekelétu, thanking him, in the name of the Queen, for his kindness and help to her servant, Dr. Livingstone, explaining the desire of the British nation, as a commercial and Christian people, to live at peace with all and to benefit all; telling him, too, what they thought of the slave-trade; hoping that Sekelétu would help to keep "God's highway," the river Zambesi, as a free pathway for all nations; assuring him of friendship and good-will; and respectfully hinting that, " as we have derived all our greatness from the divine religion we received from heaven, it will be well if you consider it carefully when any of our people talk to you about it."[1]

Most men, after receiving such *carte blanche* as Lord

---

[1] See Appendix No. IV.

Clarendon had given to Livingstone, would have been drawing out plans on a large scale, regardless of expense. Livingstone's ideas were quite in the opposite direction. Instead of having to press Captain Washington, he had to restrain him. The expedition as planned by Washington, with commander and assistant, and a large staff of officers, was too expensive. All that Livingstone wished was a steam launch, with an economic botanist, a practical mining geologist, and an assistant. All was to be plain and practical; nothing was wished for ornament or show.

Before we come to the last adieus, it is well to glance at the remarkable effect of Dr. Livingstone's short visit, in connection with his previous labors, on the public opinion of the country in regard to Africa. In the first place, as we have already remarked, there was quite a revolution of ideas as to the interior of the country. It astonished men to find that, instead of a vast sandy desert, it was so rich and productive a land, and merchants came to see that if only a safe and wholesome traffic could be introduced, the result would be hardly less beneficial to them than to the people of Africa. In the second place, a new idea was given of the African people. Caffre wars and other mismanaged enterprises had brought out the wildest aspects of the native character, and had led to the impression that the blacks were just as brutish and ferocious as the tigers and crocodiles among which they lived. But Livingstone showed, as Moffat had showed before him, that, rightly dealt with, they were teachable and companionable, full of respect for the white man, affectionate toward him when he treated them well, and eager to have him dwelling among them. On the slave-trade of the interior he had thrown a ghastly light, although it was reserved to him in his future journeys to make a full exposure of the devil's work in that infamous traffic. He had thrown light, too, on the structure of Africa, shown where healthy localities were to be found, copiously illustrated its fauna and flora, discovered great rivers and lakes, and laid them down on its map with the greatest accuracy; and he had shown how its most virulent disease might be reduced to the category of an ordinary cold. In conjunction with other great African travelers, he had contributed not a little to the great increase of popularity which had been acquired by the Geographical Society. He had shown abundance of openings for Christian missions from Kuruman to the Zambesi, and from Loanda to Quilimane. He had excited no little compassion for the negro, by vivid pictures of his dark and repulsive life, with so much misery in it and so little joy. In the cause of missions he did not appeal in vain. At the English Universities, young men of ability and promise got new light on the purposes of life, and wondered that they had not thought sooner of offering themselves for such noble work. In Scotland, men like James Stewart, now of Lovedale, were set thinking whether they should not give themselves to Africa, and older men, like Mr. R. A. Macfie and the late Mr. James Cunningham, of Edinburgh, were pondering in what manner the work could be begun. The London Missionary Society, catching up Livingstone's watchword "Onward," were planning a mission at Linyanti, on the banks of the Zambesi. Mr. Moffat was about to pay a visit to the great Mosilikatse, with a view to the commencement of a mission to the Matebele. As for Livingstone himself, his heart was yearning after his friends the Makololo. He had been quite willing to go and be their missionary, but in the meantime other duty called him. Not being aware of any purpose to plant a mission among them, he made an arrangement with his brother-in-law, Mr. John Moffat, to become their missionary. Out of his private resources he promised him £500, for outfit, etc., and £150 a year for five years as salary, besides other sums, amounting in all to £1400. Nearly three years of his own salary as Consul (£500) were thus pledged and paid. In one word, Africa, which had long been a symbol of all that is dry and uninviting, suddenly became the most interesting part of the globe.

As the time of Dr. Livingstone's departure for Africa drew near, a strong desire arose among many of his friends, chiefly the geographers, to take leave of him in a way that should emphatically mark the strength of their admiration and the cordiality of their good wishes. It was accordingly resolved that he should be invited to a public dinner on the 13th February, 1858, and that Sir Roderick Murchison should occupy the chair. On the morning of that day he had the honor of an interview with Her Majesty the Queen. A Scottish correspondent of an American journal, whose letter at other points shows that he had good information,[1] after referring to the fact that Livingstone was not presented in the usual way, says:

"He was honored by the Queen with a private interview. . . . She sent for Livingstone, who attended Her Majesty at the palace, without ceremony, in his black coat and blue trousers, and his cap surrounded with a stripe of gold lace. This was his usual attire, and the cap had now become the appropriate distinction of one of Her Majesty's consuls, an official position to which the traveler attaches great importance, as giving him consequence in the eyes of the natives, and authority over the members of the expedition. The Queen conversed with him affably for half an hour on the subject of his travels. Dr. Livingstone told Her Majesty that he would now be able to say to the natives that he had seen his chief, his not having done so before having been a constant subject of surprise to the children of the African wilderness. He mentioned to Her Majesty also that the people were in the habit of inquiring whether his chief were wealthy; and that when he assured them she was very wealthy, they would ask how many cows she had got, a question at which the Queen laughed heartily."

In the only notice of this interview which we have found in Livingstone's own writing, he simply says that Her Majesty assured him of her good wishes in his journeys. It was the only interview with his Sovereign he ever had.

---

[1] We have ascertained that the correspondent was the late Mr. Keddie, of the Glasgow Free Church College, who got his information from Mr. James Young.

When he returned in 1864 he said that he would have been pleased to have another, but only if it came naturally, and without his seeking it. The Queen manifested the greatest interest in him, and showed great kindness to his family, when the rumor came of his death.

The banquet in Freemason's Tavern, which it had been intended to limit to 250 guests, overflowed the allotted bounds, and was attended by upward of 350, including the Ministers of Sweden and Norway, and of Denmark; Dukes of Argyll and Wellington; Earl of Shaftesbury and Earl Grey; Bishops of Oxford and St. David's; and hosts of other celebrities in almost every department of public life. The feeling was singularly cordial. Sir Roderick rehearsed the services of Livingstone, crowning them, as was his wont, with that memorable act—his keeping his promise to his black servants by returning with them from Loanda to the heart of Africa, in spite of all the perils of the way, and all the attractions of England, thereby "leaving for himself in that country a glorious name, and proving to the people of Africa what an English Christian is." Still more, perhaps, did Sir Roderick touch the heart of the audience when he said of Livingstone "that notwithstanding eighteen months of laudation, so justly bestowed on him by all classes of his countrymen, and after receiving all the honors which the Universities and cities of our country could shower upon him, he is still the same honest, true-hearted David Livingstone as when he issued from the wilds of Africa." It was natural for the Duke of Argyll to recall the fact that Livingstone's family was an Argyll-shire one, and it was a happy thought that as Ulva was close to Iona—"that illustrious island," as Dr. Samuel Johnson called it, "whence roving tribes and rude barbarians derived the benefits of knowledge and the blessings of religion,"—so might the son of Ulva carry the same blessings to Africa, and be remembered, perhaps, by millions of the human race as the first pioneer of civilization, and

22

the first harbinger of the gospel. It was graceful in the
Bishop of Oxford (Samuel Wilberforce) to advert to the
debt of unparalleled magnitude which England, founder
of the accursed slave-trade, owed to Africa, and to urge
the immediate prosecution of Livingstone's plans, inasmuch as the spots in Africa, where the so-called Christian
trader had come, were marked, more than any other, by
crime and distrust, and insecurity of life and property.
It was a good opportunity for Professor Owen to tell the
story of the spiral tusk, to rehearse some remarkable instances of Livingstone's accurate observations and happy
conjectures on the habits of animals, to rate him for destroying the moral character of the lion, and to claim
credit for having discovered, in the bone caves of England,
the remains of an animal of greater bulk than any living
species, that may have possessed all the qualities which the
most ardent admirer of the British lion could desire![1]

On no topic was the applause of the company more
enthusiastic than when mention was made of Mrs. Livingstone, who was then preparing to accompany her husband
on his journey. Livingstone's own words to the company
were simple and hearty, but they were the words of truth
and soberness. He was overwhelmed with the kindness he
had experienced. He did not expect any speedy result
from the Expedition, but he was sanguine as to its ultimate
benefit. He thought they would get in the thin end of the
wedge, and that it would be driven home by English
energy and spirit. For himself, with all eyes resting upon
him, he felt under an obligation to do better than he had
ever done. And as to Mrs. Livingstone:

"It is scarcely fair to ask a man to praise his own wife, but I can
only say that when I parted from her at the Cape, telling her that I

---

[1] Livingtone purposed to bequeath to Professor Owen a somewhat extraordinary legacy. Writing afterward to his friend Mr. Young, he said: "If
I die at home I would lie beside you. My left arm goes to Professor Owen,
mind. That is the will of David Livingstone."

should return in two years, and when it happened that I was absent four years and a half, I supposed that I should appear before her with a damaged character. I was, however, forgiven. My wife, who has always been the main spoke in my wheel, will accompany me in this expedition, and will be most useful to me. She is familiar with the languages of South Africa. She is able to work. She is willing to endure, and she well knows that in that country one must put one's hand to everything. In the country to which I am about to proceed she knows that at the missionary's station the wife must be the maid-of-all-work within, while the husband must be the jack-of-all-trades without, and glad am I indeed that I am to be accompanied by my guardian angel."

Of the many letters of adieu he received before setting out we have space for only two. The first came from the venerable Professor Sedgwick, of Cambridge, in the form of an apology for inability to attend the farewell banquet. It is a beautiful unfolding of the head and heart of the Christian philosopher, and must have been singularly welcome to Livingstone, whose views on some of the greatest subjects of thought were in thorough harmony with those of his friend:

"*Cambridge, February* 10, 1858.—MY DEAR SIR,—Your kind and very welcome letter came to me yesterday; and I take the first moment of leisure to thank you for it, and to send you a few more words of good-will, along with my prayers that God may, for many years, prolong your life and the lives of those who are most near and dear to you, and that he may support you in all coming trials, and crown with a success, far transcending your own hopes, your endeavors for the good of our poor humble fellow-creatures in Africa.

"There is but one God, the God who created all worlds and the natural laws whereby they are governed; and the God of revealed truth, who tells us of our destinies in an eternal world to come. All truth of whatever kind has therefore its creator in the will and essence of that great God who created all things, moral and natural. Great and good men have long upheld this grand conclusion. But, alas! such is too often our bigotry, or ignorance, or selfishness, that we try to divorce religious and moral from natural truth, as if they were inconsistent and in positive antagonism one to the other,—a true catholic spirit (oh that the word 'catholic' had not been so horribly abused by the foul deeds of men!) teaching us that all truths are linked together, and that all art and science, and all material discoveries (each held in its proper place and subordination), may be used to minister to the diffusion of Christian truth among men, with all its blessed fruits of peace and good-will. This is, I believe, your faith, as I see it shining out in your deeds, and set forth in the pages of your work on Southern Africa, which I have studied through from beginning to end with sentiments of reverence and honor for the past and good hopes for the future.

"What a glorious prospect is before you! the commencement of the civilization of Africa, the extension of our knowledge of all the kingdoms of nature, the production of great material benefits to the Old World, the gradual healing of that foul and fetid ulcer, the slave-trade, the one grand disgrace and weakness of Christendom, and that has defiled the hands of all those who have had any dealings with it; and last, but not least—nay, the greatest of all, and the true end of all— the lifting up of the poor African from the earth, the turning his face heavenward, and the glory of at length (after all his sufferings and all our sins) calling him a Christian brother. May our Lord and Saviour bless your labors, and may his Holy Spirit be with you to the end of your life upon this troubled world!

"I am an old man, and I shall (so far as I am permitted to look at the future) never see your face again. If I live till the 22d of March I shall have ended my 73d year, and not only from what we all know from the ordinary course of nature, but from what I myself know and feel from the experience of the two past years, I am assured that I have not long to live. How long, God only knows. It grieves me not to have seen you again in London, and I did hope that you might yourself introduce me to your wife and children. I hear that a farewell dinner is to be given you on Saturday, and greatly should I rejoice to be present on that occasion, and along with many other true-hearted friends wish you 'God-speed.' But it must not be. I am not a close prisoner to my room, as I was some weeks past, but I am still on the sick list, and dare not expose myself to any sudden change of temperature, or to the excitement of a public meeting. This is one of the frailties of old age and infirm health. I have gone on writing and writing more than I intended. Once for all, God bless you! and pray (though I do not personally know them) give my best and Christian love to your dear wife (Ma-Robert she was called, I think, in Africa) and children. Ever gratefully and affectionately yours,

"A. SEDGWICK."

Sir Roderick, too, had a kind parting word for his friend: "Accept my warmest acknowledgments for your last farewell note. Believe me, my dear friend, that no transaction in my somewhat long and very active life has so truly rewarded me as my intercourse with you, for, from the beginning to the end, it has been one continued bright gleam."

To this note Livingstone, as was his wont, made a hearty and Christian response: "Many blessings be on you and yours, and if we never meet again on earth, may we through infinite mercy meet in heaven!"

The last days in England were spent in arrangements for the expedition, settling family plans, and bidding farewell. Mrs. Livingstone accompanied her husband, along with Oswell, their youngest child. Dr. Livingstone's heart was deeply affected in parting with his other children. Amid all the hurry and bustle of leaving he snatches a few minutes almost daily for a note to one or more of them:

"*London, 2d February*, 1858.—MY DEAR TOM,—I am soon going off from this country, and will leave you to the care of Him who neither slumbers nor sleeps, and never disappointed any one who put his trust in Him. If you make him your friend He will be better to you than any companion can be. He is a friend that sticketh closer than a brother. May He grant you grace to seek Him and to serve Him. I have nothing better to say to you than to take God for your Father, Jesus for your Saviour, and the Holy Spirit for your sanctifier. Do this and you are safe for ever. No evil can then befall you. Hope you will learn quickly and well, so as to be fitted for God's service in the world."

" '*Pearl*,' in the *Mersey*, 10th *March*, 1858.—MY DEAR TOM,—We are off again, and we trust that He who rules the waves will watch over us and remain with you, to bless us and make us blessings to our fellowmen. The Lord be with you, and be very gracious to you! Avoid and hate sin, and cleave to Jesus as your Saviour from guilt. Tell grandma we are off again, and Janet will tell all about us."

In his letters to his children from first to last, the counsel most constantly and most earnestly pressed is to take Jesus for their friend. The personal Saviour is continually present to his heart, as the one inestimable treasure which he longs for them to secure. That treasure had been a source of unspeakable peace and joy to himself amid all the trials and troubles of his checkered life; if his children were only in friendship with Him, he could breathe freely in leaving them, and feel that they would indeed FARE WELL.

CHAPTER XII.

THE ZAMBESI, AND FIRST EXPLORATIONS OF THE SHIRÉ.

A.D. 1858–1859.

Dr. and Mrs. Livingstone sail in the "Pearl"—Characteristic instructions to members of Expedition—Dr. Livingstone conscious of difficult position— Letter to Robert—Sierra Leone—Effects of British Squadron and of Christian Missions—Dr. and Mrs. Moffat at Cape Town—Splendid reception there— Illness of Mrs. Livingstone—She remains behind—The five years of the Expedition—Letter to Mr. James Young—to Dr. Moffat—Kongone entrance to Zambesi—Collision with Naval Officer—Disturbed state of the country— Trip to Kebrabasa Rapids—Dr. Livingstone applies for new steamer— Willing to pay for one himself—Exploration of the Shiré—Murchison Cataracts—Extracts from private Journal—Discovery of Lake Shirwa—Correspondence—Letters to Agnes Livingstone—Trip to Tette—Kroomen and two members of Expedition dismissed—Livingstone's vindication—Discovery of Lake Nyassa—Bright hopes for the future—Idea of a colony—Generosity of Livingstone—Letters to Mr. Maclear, Mr. Young, and Sir Roderick Murchison—His sympathy with the "honest poor"—He hears of the birth of his youngest daughter.

ON the 10th March 1858, Dr. Livingstone, accompanied by Mrs. Livingstone, their youngest son, Oswell, and the members of his Expedition, sailed from Liverpool on board Her Majesty's colonial steamer, the "Pearl," which carried the sections of the "Ma-Robert," the steam launch with Mrs. Livingstone's African name, which was to be permanently used in the exploration of the Zambesi and its tributaries. At starting, the "Pearl" had fine weather and a favorable wind, and quickly ran down the Channel and across the Bay of Biscay. With that business-like precision which characterized him, Livingstone, as soon as sea-sickness was over, had the instructions of the Foreign

Office read in presence of all the members of the Expedition, and he afterward wrote out and delivered to each person a specific statement of the duties expected of him.

In these very characteristic papers, it is interesting to observe that his first business was to lay down to each man his specific work, this being done for the purpose of avoiding confusion and collision, acknowledging each man's gifts, and making him independent in his own sphere. While no pains were to be spared to make the Expedition successful in its scientific and commercial aims, and while, for this purpose, great stress was laid on the subsidiary instructions prepared by Professor Owen, Sir W. Hooker, and Sir R. Murchison, Dr. Livingstone showed still more earnestness in urging duties of a higher class, giving to all the same wise and most Christian counsel to maintain the *moral* of the Expedition at the highest point, especially in dealing with the natives:

"You will understand that Her Majesty's Government attach more importance to the moral influence which may be exerted on the minds of the natives by a well-regulated and orderly household of Europeans, setting an example of consistent moral conduct to all who may congregate around the settlement; treating the people with kindness, and relieving their wants; teaching them to make experiments in agriculture, explaining to them the more simple arts, imparting to them religious instruction, as far as they are capable of receiving it, and inculcating peace and good-will to each other.

"The expedition is well supplied with arms and ammunition, and it will be necessary to use these in order to obtain supplies of food, as well as to procure specimens for the purposes of Natural History. In many parts of the country which we hope to traverse, the larger animals exist in great numbers, and, being comparatively tame, may be easily shot. I would earnestly press on every member of the expedition a sacred regard to life, and never to destroy it unless some good end is to be answered by its extinction; the wanton waste of animal life which I have witnessed from night-hunting, and from the ferocious, but childlike, abuse of the instruments of destruction in the hands of Europeans, makes me anxious that this expedition should not be guilty of similar abominations.

"It is hoped that we may never have occasion to use our arms for protection from the natives, but the best security from attack consists in upright conduct, and the natives seeing that we are prepared to meet it. At the same time, you are strictly enjoined to exercise the greatest forbearance toward the people; and, while retaining proper firmness in the event of any misunderstanding, to conciliate, as far as possibly can be done with safety to our party.

"It is unnecessary for me to enjoin the strictest justice in dealing with the natives. This your own principles will lead you invariably to follow, but while doing so yourself, it is decidedly necessary to be careful not *to appear* to overreach or insult any one by the conduct of those under your command. . . .

"The chiefs of tribes and leading men of villages ought always to be treated with respect, and nothing should be done to weaken their authority. Any present of food should be accepted frankly, as it is impolitic to allow the ancient custom of feeding strangers to go into disuse. We come among them as members of a superior race, and servants of a Government that desires to elevate the more degraded portions of the human family. We are adherents of a benign, holy religion, and may, by consistent conduct, and wise, patient efforts, become the harbingers of peace to a hitherto distracted and trodden-down race. No great result is ever attained without patient, long-continued effort. In the enterprise in which we have the honor to be engaged, deeds of sympathy, consideration, and kindness, which, when viewed in detail, may seem thrown away, if steadily persisted in, are sure, ultimately, to exercise a commanding influence. Depend upon it, a kind word or deed is never lost."

Evidently, Dr. Livingstone felt himself in a difficult position at the head of this enterprise. He was aware of the trouble that had usually attended civil as contrasted with naval and military expeditions, from the absence of that habit of discipline and obedience which is so firmly established in the latter services. He had never served under Her Majesty's Government himself, nor had he been accustomed to command such men as were now under him, and there were some things in his antecedents that made the duty peculiarly difficult. On one thing only he was resolved: to do his own duty to the utmost, and to spare no pains to induce every member of the Expedition to do his. It was impossible for him not to be anxious as to how the team would pull together, especially as he knew

well the influence of a malarious atmosphere in causing intense irritability of temper. In some respects, though not
the most obvious, this was the most trying period of his
life. His letters and other written papers show one little
but not uninstructive effect of the pressure and distraction
that now came on him—in the great change which his
handwriting underwent—the neat, regular writing of his
youth giving place to a large and heavyish hand, as if he
had never had time to mend his pen, and his only thought
had been how to get on most quickly. Yet we see also,
very clearly, how nobly he strove after self-control and
conciliatory ways. The tone of courtesy, the recognition
of each man's independence in his own sphere, and the
appeal to his good sense and good feeling, apparent in the
instructions, show a studious desire, while he took and intended to keep his place as Commander, to conceal the
symbols of authority, and bind the members of the party
together as a band of brothers. And though in his
published book, *The Zambesi and its Tributaries*, which was
mainly a report of his doings to the Government and the
nation, he confined himself to the matters with which he
had been intrusted by them, there are many little proofs
of his seeking wisdom and strength from above with un-
diminished earnestness, and of his striving, as much as
ever, to do all to the glory of God.

As the swift motion of the ship bears him farther and
farther from home, he cannot but think of his orphan
children. As they near Sierra Leone, on the 25th March,
he sends a few lines to his eldest son:

"MY DEAR ROBERT,—We have been going at the rate of 200 miles
a day ever since we left Liverpool, and have been much favored by a
kind Providence in the weather. Poor Oswell was sorely sick while
rolling through the Bay of Biscay, and ate nothing for about three days;
but we soon got away from the ice and snow to beautiful summer
weather, and we are getting nicely thawed. We sleep with all our port-
holes open, and are glad of the awning by day. At night we see the
Southern Cross; and the Pole Star, which stands so high over you, is

here so low we cannot see it for the haze. We shall not see it again, but the same almighty gracious Father is over all, and is near to all who love Him. You are now alone in the world, and must seek his friendship and guidance, for if you do not lean on Him, you will go astray, and find that the way of transgressors is hard. The Lord be gracious to you, and accept you, though unworthy of his favor."

Sierra Leone was reached in a fortnight. Dr. Livingstone was gratified to learn that, during the last ten years, the health of the town had improved greatly—consequent on the abatement of the "whisky fever," and the draining and paving of the streets through the activity of Governor Hill. He found the Sunday as well kept as in Scotland, and was sure that posterity would acknowledge the great blessing which the operations of the English Squadron on the one hand and the various Christian missions on the other had effected. He was more than ever convinced, notwithstanding all that had been said against it, that the English Squadron had been a great blessing on the West Coast. The Christian missions, too, that had been planted under the protection of the Squadron, were an evidence of its beneficial influence. He used constantly to refer with intense gratitude to the work of Lord Palmerston in this cause, and to the very end of his life his Lordship was among the men whose memory he most highly honored. Often, when he wished to describe his aim briefly, in regard to slavery, commerce, and missions, he would say it was to do on the East Coast what had been done on the West. At Sierra Leone a crew of twelve Kroomen was engaged and taken on board for the navigation of the "Ma-Robert," after it should reach the Zambesi. On their leaving Sierra Leone, the weather became very rough, and from the state of Mrs. Livingstone's health, inclining very much to fever, it was deemed necessary that she, with Oswell, should be left at the Cape, go to Kuruman for a time, and after her coming confinement, join her husband on the Zambesi in 1860. "This," says Livingstone in his Journal, "is a great trial to me, for had she come on with us, she might have proved of essential service to the Expedition in case of sickness or otherwise; but it may all turn out for the best." It was the first disappointment, and it was but partially balanced by his learning from Dr. Moffat, who, with his wife, met them at the Cape, that he had made out his visit to Mosilikatse, and had learned that the men whom Livingstone had left at Tette had not returned home, so that they would still be waiting for him there. He knew of what value they would be to him in explaining his intentions to the natives. From Sir George Grey, the excellent Governor of the Cape, and the inhabitants of Cape Town generally, the Expedition met with an unusually cordial reception. At a great meeting at the Exchange, a silver box containing a testimonial of eight hundred guineas was presented to Livingstone by the Governor; and two days after, a grand dinner was given to the members of the Expedition, the Attorney-General being in the chair. Mr. Maclear was most enthusiastic in the reception of his friend, and at the public meeting had so much to say about him that he could hardly be brought to a close. It must have been highly amusing to Livingstone to contrast Cape Town in 1852 with Cape Town in 1858. In 1852 he was so suspected that he could hardly get a pound of gunpowder or a box of caps while preparing for his unprecedented journey, and he had to pay a heavy fine to get rid of a cantankerous post-master. Now he returns with the Queen's gold band round his cap, and with brighter decorations round his name than Sovereigns can give; and all Cape Town hastens to honor him. It was a great victory, as it was also a striking illustration of the world's ways.

It is not our object to follow Dr. Livingstone into all the details of his Expedition, but merely to note a few of the more salient points, in connection with the opportunities it afforded for the achievement of his object and the development of his character. It may be well to note here generally how the years were occupied. The remainder of 1858 was employed in exploring the mouths of the Zambesi, and the river itself up to Tette and the Kebrabasa Rapids, a few miles beyond. Next year—1859—was devoted mainly to three successive trips on the river Shiré, the third being signalized by the discovery of Lake Nyassa. In 1860 Livingstone went back with his Makololo up the Zambesi to the territories of Sekelétu. In 1861, after exploring the river Rovuma, and assisting Bishop Mackenzie to begin the Universities' Mission, he started for Lake Nyassa, returning to the ship toward the end of the year. In 1862 occurred the death of the Bishop and other missionaries, and also, during a detention at Shupanga, the death of Mrs. Livingstone : in the latter part of the year Livingstone again explored the Rovuma. In 1863 he was again exploring the Shiré Valley and Lake Nyassa, when an order came from Her Majesty's Government, recalling the Expedition. In 1864 he started in the "Lady Nyassa" for Bombay, and thence returned to England.

On the 1st May, 1858, the "Pearl" sailed from Simon's Bay, and on the 14th stood in for the entrance to the Zambesi, called the West Luabo, or Hoskins's Branch. Of their progress Dr. Livingstone gives his impressions in the following letter to his friend Mr. James Young :

" 'PEARL,' 10*th May,* 1858.

"Here we are, off Cape Corrientes ('Whaur's that, I wonner?'), and hope to be off the Luabo four days hence. We have been most remarkably favored in the weather, and it is well, for had our ship been in a gale with all this weight on her deck, it would have been perilous. Mrs. Livingstone was sea-sick all the way from Sierra Leone, and got as thin as a lath. As this was accompanied by fever, I was forced to run into Table Bay, and when I got ashore I found her father and mother down all the way from Kuruman to see us and help the young missionaries, whom the London Missionary Society has not yet sent. Glad, of course, to see the old couple again. We had a grand to-do at the Cape. Eight hundred guineas were presented in a silver box by the hand of the Governor, Sir George Grey, a fine fellow. Sure, no one might be more than ʾul to the Giver of all than myself. The Lord grant me grace to serve Him with heart and soul—the only return I can make! . . . It was a bitter parting with my wife, like tearing the heart out of one. It was so unexpected; and now we are screwing away up the coast. . . . We are all agreeable yet, and all looking forward with ardor to our enterprise. It is likely that I shall come down with the 'Pearl' through the Delta to doctor them if they become ill, and send them on to Ceylon with a blessing. All have behaved well, and I am really thankful to see it, and hope that God will graciously make some better use of us in promoting his glory. I met a Dr. King in Simon's Bay, of the 'Cambrian' frigate, one of our class-mates in the Andersonian. This frigate, by the way, saluted us handsomely when we sailed out. We have a man-of-war to help us (the 'Hermes'), but the lazy muff is far behind. He is, however, to carry our despatches to Quilimane. . . ."

A letter to Dr. Moffat lets us know in what manner he was preparing to teach the twelve Kroomen who were to navigate the "Ma-Robert," and his old Makololo men:

"First of all, supposing Mr. Skead should take this back by the 'Hermes' in time to catch you at the Cape, would you be kind enough to get a form of prayer printed for me? We have twelve Kroomen, who seem docile and willing to be taught; when we are parted from the 'Pearl' we shall have prayers with them every morning. . . . I think it will be an advantage to have the prayers in Sichuana when my men join us, and if we have a selection from the English Litany, with the Lord's Prayer in Sichuana, all may join. Will you translate it, beginning at 'Remember not, Lord, our offenses,' up to 'the right way'? Thence, petition for chiefs, and on to the end. . . . The Litany need not be literal. I suppose you are not a rabid nonconformist, or else I would not venture to ask this. . . ."

By the time they reached the mouth of the Zambesi, Livingstone was suffering from a severe attack of diarrhœa. On the 16th of May, being Sunday, while still suffering, he deemed it a work of necessity, in order to get as soon as possible out of the fever-breeding region of mangrove swamps where they had anchored, that they should at once remove the sections of the "Ma-Robert" from the

"Pearl"; accordingly, with the exception of the time occupied in the usual prayers, that day was spent in labor. His constant regard for the day of rest and great unwillingness to engage in labor then, is the best proof that on this occasion the necessity for working was to his mind absolutely irresistible. He had found that active exercise every day was one of the best preventives of fever; certainly it is very remarkable how thoroughly the men of the Expedition escaped it at this time. In his Journal he says: "After the experience gained by Dr. M'William, and communicated to the world in his admirable *Medical History of the Niger Expedition,* I should have considered myself personally guilty had any of the crew of the 'Pearl' or of the Expedition been cut off through delay in the mangrove swamps." Afterward, when Mrs. Livingstone died during a long but unavoidable delay at Shupanga, a little farther up, he was more than ever convinced that he had acted rightly. But some of his friends were troubled, and many reflections were thrown on him, especially by those who bore him no good-will.

The first important fact in the history of the Expedition was the discovery of the advantages of the Kongone entrance of the Zambesi, the best of all the mouths of the river for navigation. Soon after a site was fixed on as a depôt, and while the luggage and stores were being landed at it, there occurred an unfortunate collision with the naval officer, who tendered his resignation. At first Livingstone declined to accept of it, but on its being tendered a second time he allowed the officer to go. It vexed him to the last degree to have this difference so early, nor did he part with the officer without much forbearance and anxiety to ward off the breach. In his despatches to Government the whole circumstances were fully detailed. Letters to Mr. Maclear and other private friends give a still more detailed narrative. In a few quarters blame was cast upon him, and in the Cape newspapers the affair was much commented on. In due time there came a reply from Lord Malmesbury, then Foreign Secretary, dated 26th April, 1859, to the effect that after full inquiry by himself, and after consulting with the Admiralty, his opinion was that the officer had failed to clear himself, and that Dr. Livingstone's proceedings were fully approved. Livingstone had received authority to stop the pay of any member of the Expedition that should prove unsatisfactory; this, of course, subjected his conduct to the severer criticism.

When the officer left, Livingstone calmly took his place, adding the charge of the ship to his other duties. This step would appear alike rash and presumptuous, did we not know that he never undertook any work without full deliberation, and did we not remember that in the course of three sea-voyages which he had performed he had had opportunities of seeing how a ship was managed—opportunities of which, no doubt, with his great activity of mind, he had availed himself most thoroughly. The facility with which he could assume a new function, and do its duties as if he had been accustomed to it all his life, was one of the most remarkable things about him. His chief regret in taking the new burden was, that it would limit his intercourse with the natives, and prevent him from doing as much missionary work as he desired. Writing soon after to Miss Whately, of Dublin, he says: "It was imagined we could not help ourselves, but I took the task of navigating on myself, and have conducted the steamer over 1600 miles, though as far as my likings go, I would as soon drive a cab in November fogs in London as be 'skipper' in this hot sun; but I shall go through with it as a duty." To his friend Mr. Young he makes humorous reference to his awkwardness in nautical language: "My great difficulty is calling out 'starboard' when I mean 'port,' and feeling crusty when I see the helmsman putting the helm the wrong way."

Another difficulty arose from the state of the country north of the Zambesi, in consequence of the natives having rebelled against the Portuguese and being in a state of war. Livingstone was cautioned that he would be attacked if he ventured to penetrate into the country. He resolved to keep out of the quarrel, but to push on in spite of it. At one time his party, being mistaken for Portuguese, were on the point of being fired on, but on Livingstone shouting out that they were English the natives let them alone. On reaching Tette he found his old followers in ecstasies at seeing him; the Portuguese Government had done nothing for them, but Major Sicard, the excellent Governor of Tette, had helped them to find employment and maintain themselves. Thirty had died of small-pox; six had been killed by an unfriendly chief. When the survivors saw Dr. Livingstone, they said: "The Tette people often taunted us by saying, 'Your Englishman will never return;' but we trusted you, and now we shall sleep." It gave Livingstone a new hold on them and on the natives generally, that he had proved true to his promise, and had come back as he had said. As the men had found ways of living at Tette, Livingstone was not obliged to take them to their home immediately.

One of his first endeavors after reaching Tette was to ascertain how far the navigation of the Zambesi was impeded by the rapids at Kebrabasa, between twenty and thirty miles above Tette, which he had heard of but not seen on his journey from Linyanti to Quilimane. The distance was short and the enterprise apparently easy, but in reality it presented such difficulties as only his dogged perseverance could have overcome. After he had been twice at the rapids, and when he believed he had seen the whole, he accidentally learned, after a day's march on the way home, that there was another rapid which he had not yet seen. Determined to see all, he returned, with Dr. Kirk and four Makololo, and it was on this occasion that his followers, showing the blisters on their feet burst by the hot rocks, told him, when he urged them to make another effort, that hitherto they had always believed he had a heart, but now they saw he had none, and wondered if he were mad. Leaving them, he and Dr. Kirk pushed on alone; but their boots and clothes were destroyed; in three hours they made but a mile. Next day, however, they gained their point and saw the rapid. It was plain to Dr. Livingstone that had he taken this route in 1856, instead of through the level Shidina country, he must have perished. The party were of opinion that when the river was in full flood the rapids might be navigated, and this opinion was confirmed on a subsequent visit paid by Mr. Charles Livingstone and Mr. Baines during the rainy season. But the "Ma-Robert" with its single engine had not power to make way. It was resolved to apply to Her Majesty's Government for a more suitable vessel to carry them up the country, stores and all. Until the answer should come to this application, Dr. Livingstone could not return with his Makololo to their own country.

While making this application, he was preparing another string for his bow. He wrote to his friend Mr. James Young that if Government refused he would get a vessel at his own expense, and in a succession of letters authorized him to spend £2000 of his own money in the purchase of a suitable ship. Eventually, both suggestions were carried into effect. The Government gave the "Pioneer" for the navigation of the Zambesi and lower Shiré; Livingstone procured the "Lady Nyassa" for the Lake (where, however, she never floated), but the cost was more than £6000—the greater part, indeed, of the profits of his book.

The "Ma-Robert," which had promised so well at first, now turned out a great disappointment. Her consumption of fuel was enormous; her furnace had to be lighted hours before the steam was serviceable; she snorted so horribly that they called her " The Asthmatic," and after all she made so little progress that canoes could easily pass her. Having taken much interest in the purchase of the vessel, and thought he was getting a great bargain because its owner professed to do so much through " love of the cause," Livingstone was greatly mortified when he found he had got an inferior and unworthy article; and many a joke he made, as well as remarks of a more serious kind, in connection with the manner which the "eminent shipbuilder" had taken to show his love.

Eearly in 1859 the exploration of the Shiré was begun —a river hitherto absolutely unknown. The country around was rich and fertile, the natives not unfriendly, but suspicious. They had probably never been visited before but by man-stealers, and had never seen Europeans. The Shiré Valley was inhabited by the Manganja, a very warlike race. Some days' journey above the junction with the Zambesi, where the Shiré issues from the mountains, the progress of the party was stopped by rapids, to which they gave the name of the " Murchison Cataracts." It seemed in vain to penetrate among the people at that time without supplies, considering how suspicious they were. Crowds went along the banks watching them by day; they had guards over them all night, and these were always ready with their bows and poisoned arrows. Nevertheless, some progress was made in civilizing them, and at a future time it was hoped that further exploration might take place.

Some passages in Livingstone's private Journal give us a glimpse of the more serious thoughts that were passing through his mind at this time:

"*March* 3, 1859.—If we dedicate ourselves to God unreservedly He will make use of whatever peculiarities of constitution He has imparted for his own glory, and He will in answer to prayer give wisdom to guide. He will so guide as to make useful. O how far am I from that hearty devotion to God I read of in others! The Lord have mercy on me a sinner!"

"*March 5th.*—A woman left Tette yesterday with a cargo of slaves (20 men and 40 women) in irons to sell to St. Cruz [a trader], for exportation at Bourbon. Francisco at Shupanga is the great receiver for Cruz. This is carnival, and it is observed chiefly as a drinking feast."

"*March 6th.*—Teaching Makololo Lord's Prayer and Creed. Prayers as usual at 9½ A.M. When employed in active travel, my mind becomes inactive, and the heart cold and dead, but after remaining some time quiet, the heart revives and I become more spiritually-minded. This is a mercy which I have experienced before, and when I see a matter to be duty I go on regardless of my feelings. I do trust that the Lord is with me, though the mind is engaged in other matters than the spiritual. I want my whole life to be out and out for the Divine glory, and my earnest prayer is that God may accept what his own Spirit must have implanted—the desire to glorify Him. I have been more than usually drawn out in earnest prayer of late—for the Expedition—for my family—the fear lest ——'s misrepresentation may injure the cause of Christ—the hope that I may be permitted to open this dark land to the blessed gospel. I have cast all before my God. Good Lord, have mercy upon me. Leave me not, nor forsake me. He has guided well in time past. I commit my way to Him for the future. All I have received has come from Him. Will He be pleased in mercy to use me for his glory? I have prayed for this, and Jesus himself said, 'Ask, and ye shall receive,' and a host of statements to the same effect. There is a great deal of trifling frivolousness in not trusting in God. Not trusting in Him who is truth itself, faithfulness, the same yesterday, to-day, and for ever! It is presumption not to trust in Him implicitly, and yet this heart is sometimes fearfully guilty of distrust. I am ashamed to think of it. Ay; but He must put the trusting, loving, childlike spirit in by his grace. O Lord, I am Thine, truly I am Thine—take me—do what seemeth good in Thy sight with me, and give me complete resignation to Thy will in all things."

Two months later (May, 1859), a second ascent of the Shiré was performed, and friendly relations were established with a clever chief named Chibisa, "a jolly person, who laughs easily—which is always a good sign." Chibisa believed firmly in two things—the divine right of kings, and the impossibility that Chibisa should ever be in the wrong. He told them that his father had imparted an influence to him, which had come in by his head, whereby every person that had heard him speak respected him greatly. Livingstone evidently made a great impression on Chibisa; like other chiefs, he began to fall under th spell of his influence.

Making a détour to the east, the travelers now discovered Lake Shirwa, " a magnificent inland lake." This lake was absolutely unknown to the Portuguese, who, indeed, were never allowed by the natives to enter the Shiré. Livingstone had often to explain that he and his party were not Portuguese but British. After discovering this lake, the party returned to the ship, and then sailed to the Kongone harbor, in hopes of meeting a man-of-war and obtaining provisions. In this, however, they were disappointed.

Some idea of the voluminous correspondence carried on by Dr. Livingstone may be formed from the following enumeration of the friends to whom he addressed letters in May of this year: Lords Clarendon and Palmerston, Bishop of Oxford, Miss Burdett Coutts, Mr. Venn, Lord Kinnaird, Mr. James Wilson, Mr. Oswell, Colonel Steele, Dr. Newton of Philadelphia, his brother John in Canada, J. B. and C. Braithwaite, Dr. Andrew Smith, Admiral F. Grey, Sir R. Murchison, Captain Washington, Mr. Maclear, Professor Owen, Major Vardon, Mrs. Livingstone, Viscount Goderich.

Here is the account he gave of his proceedings to his little daughter Agnes:

" *River Shiré*, 1*st June*, 1859.—We have been down to the mouth of the river Zambesi in expectation of meeting a man-of-war with salt provisions, but, none appearing on the day appointed, we conclude that the Admiral has not received my letters in time to send her. We have no post-office here, so we buried a bottle containing a letter on an island in the entrance to Kongone harbor. This we told the Admiral we should do in case of not meeting the cruiser, and whoever comes will search for our bottle and see another appointment for 30th of July. This goes with despatches by way of Quilimane, and I hope some day to get from you a letter by the same route. We have got no news from home since we left Liverpool, and we long now to hear how all goes on in Europe and in India. I am now on my way to Tette, but we ran up the Shiré some forty miles to buy rice for our company. Uncle Charles is there. He has had some fever, but is better. We left him there about two months ago, and Dr. Kirk and I, with some fifteen Makololo, ascended this river one hundred miles in the 'Ma-Robert,' then left the vessel and proceeded beyond that on foot till we had discovered a magnificent lake called Shirwa (pronounced Shurwah). It was very grand, for we could not see the end of it, though some way up a mountain; and all around it are mountains much higher than any you see in Scotland. One mountain stands in the lake, and people live on it. Another, called Zomba, is more than six thousand feet high, and people live on it too, for we could see their gardens on its top, which is larger than from Glasgow to Hamilton, or about from fifteen to eighteen miles. The country is quite a Highland region, and many people live in it. Most of them were afraid of us. The women ran into their huts and shut the doors. The children screamed in terror, and even the hens would fly away and leave their chickens. I suppose you would be frightened, too, if you saw strange creatures, say a lot of Trundlemen, like those on ᵗhe Isle of Man pennies, come whirling up the street. No one was impudent to us except some slave-traders, but they became civil as soon as they learned we were English and not Portuguese. We saw the sticks they employ for training any one whom they have just bought. One is

is about eight feet long, the head, or neck rather, is put into the space between the dotted lines and shaft, and another slave carries the end. When they are considered tame they are allowed to go in chains.

"I am working in the hope that in the course of time this horrid system may cease. All the country we traveled through is capable of growing cotton and sugar, and the people now cultivate a good deal. They would grow much more if they could only sell it. At present we in England are the mainstay of slavery in America and elsewhere by buying slave-grown produce. Here there are hundreds of miles of land lying waste, and so rich that the grass towers far over one's head in walking. You cannot see where the narrow paths end, the grass is so tall and overhangs them so. If our countrymen were here they would soon render slave-buying unprofitable. Perhaps God may honor us to open up the way for this. My heart is sore when I think of so many of our countrymen in poverty and misery, while they might be doing so much good to themselves and others where our Heavenly Father has so abundantly provided fruitful hills and fertile valleys. If our people were out here they would not need to cultivate little snatches by the side of railways as they do. But all is in the hands of the all-wise Father. We must trust that He will bring all out right at last.

"My dear Agnes, you must take Him to be your Father and Guide.

Tell Him all that is in your heart, and make Him your confidant. His ear is ever open, and He despiseth not the humblest sigh. He is your best friend and loves at all times. It is not enough to be a servant, you must be a friend of Jesus. Love Him and surrender your entire being to Him. The more you trust Him, casting all your care upon Him, the more He is pleased, and He will so guide you that your life will be for his own glory. The Lord be with you. My kind love to Grandma and to all your friends. I hope your eyes are better, and that you are able to read books for yourself. Tell Tom that we caught a young elephant in coming down the Shiré, about the size of the largest dog he ever saw, but one of the Makololo, in a state of excitement, cut its trunk, so that it bled very much, and died in two days. Had it lived we should have sent it to the Queen, as no African elephant was ever seen in England. No news from mamma and Oswell.

Another evidence of the place of his children in his thoughts is found in the following lines in his Journal:

" 20th *June*, 1859.—I cannot and will not attribute any of the public attention which has been awakened to my own wisdom or ability. The great Power being my Helper, I shall always say that my success is all owing to his favor. I have been the channel of the Divine Power, and I pray that his gracious influence may penetrate me so that all may turn to the advancement of his gracious reign in this fallen world. "Oh, may the mild influence of the Eternal Spirit enter the bosoms of my children, penetrate their souls, and diffuse through their whole natures the everlasting love of God in Jesus Christ! Holy, gracious, almighty Power, I hide myself in Thee through Thy almighty Son. Take my children under Thy care. Purify them and fit them for Thy service. Let the beams of the Sun of Righteousness produce spring, summer, and harvest in them for Thee."

The short trip from Kongone to Tette and back was marked by some changes in the composition of the party. The Kroomen being found to be useless, were shipped on board a man-of-war. The services of two members of the Expedition were also dispensed with, as they were not found to be promoting its ends. Livingstone would not pay the public money to men who, he believed, were not thoroughly earning it. To these troubles was added the constantly increasing mortification arising from the state of the ship.

It has sometimes been represented, in view of such facts as have just been recorded, that Livingstone was imperious and despotic in the management of other men, otherwise he and his comrades would have got on better together. The accusation, even at first sight, has an air of improbability, for Livingstone's nature was most kindly, and it was the aim of his life to increase enjoyment. In explanation of the friction on board his ship it must be remembered that his party were a sort of scratch crew brought together without previous acquaintance or knowledge of each other's ways; that the heat and the mosquitoes, the delays, the stoppages on sandbanks, the perpetual struggle for fuel,[1] the monotony of existence, with so little to break it, and the irritating influence of the climate, did not tend to smooth their tempers or increase the amenities of life. The malarious climate had a most disturbing effect. No one, it is said, who has not experienced it, could imagine the sensation of misery connected with the feverish attacks so common in the low districts. And Livingstone had difficulties in managing his countrymen he had not in managing the natives. He was so conscientious, so deeply in earnest, so hard a worker himself, that he could endure nothing that seemed like playing or trifling with duty. Sometimes, too, things were harshly represented to him, on which a milder construction might have been put. One of those with whom he parted at this time afterward rejoined the Expedition, his pay being restored on Livingstone's intercession. Those who continued to enjoy his friendship were never weary of speaking of his delightful qualities as a companion in travel, and the warm sunshine which he had the knack of spreading around.

A third trip up the Shiré was made in August, and on the 16th of September Lake Nyassa was discovered.

---

[1] This was incredible. Livingstone wrote to his friend José Nunes that it took all hands a day and a half to cut one day's fuel.

Livingstone had no doubt that he and his party were the discoverers; Dr. Roscher, on whose behalf a claim was subsequently made, was two months later, and his unfortunate murder by the natives made it doubtful at what point he reached the lake. The discovery of Lake Nyassa, as well as Lake Shirwa, was of immense importance, because they were both parallel to the ocean, and the whole traffic of the regions beyond must pass by this line. The configuration of the Shiré Valley, too, was favorable to colonization. The valley occupied three different levels. First there was a plain on the level of the river, like that of the Nile, close and hot. Rising above this to the east there was another plain, 2000 feet high, three or four miles broad, salubrious and pleasant. Lastly, there was a third plain 3000 feet above the second, positively cold. To find such varieties of climate within a few miles of each other was most interesting.

In other respects the region opened up was remarkable. There was a great amount of fertile land, and the products were almost endless. The people were industrious; in the upper Shiré, notwithstanding a great love of beer, they lived usually to a great age. Cleanliness was not a universal virtue; the only way in which the Expedition could get rid of a troublesome follower was by threatening to wash him. The most disagreeable thing in the appearance of the women was their lip-ornament, consisting of a ring of ivory or tin, either hollow or made into a cup, inserted in the upper lip. Dr. Livingstone used to give full particulars of this fearful practice, having the idea that the taste of ladies at home in dress and ornament was not free from similar absurdity; or, as he wrote at this time to the Royal Geographical Society of Vienna, in acknowledging the honor of being made a corresponding member, "because our own ladies, who show so much virtuous perseverance with their waists, may wish to try lip-ornament too." In regard to the other sex, he in-

24

formed the same Society : " I could see nothing encouraging for the gentlemen who are anxious to prove that we are all descended from a race that wore tails."

In the highland regions of the Shiré Valley, the party were distinctly conscious of an increase of energy, from the more bracing climate. Dr. Livingstone was thoroughly convinced that these highlands of the Shiré Valley were the proper locality for commercial and missionary stations. Thus one great object of the Expedition was accomplished. In another point of view, this locality would be highly serviceable for stations. It was the great pathway for conveying slaves from the north and northwest to Zanzibar. Of this he had only too clear evidence in the gangs of slaves whom he saw marched along from time to time, and whom he would have been most eager to release had he known of any way of preventing them from falling again into the hands of the slave-sellers. In this region Englishmen " might enjoy good health, and also be of signal benefit, by leading the multitude of industrious inhabitants to cultivate cotton, maize, sugar, and other valuable produce, to exchange for goods of European manufacture, at the same time teaching them, by precept and example, the great truths of our holy religion." Water-carriage existed all the way from England, with the exception of the Murchison Cataracts, along which a road of forty miles might easily be made. A small steamer on the lake would do more good in suppressing the slave-trade than half-a-dozen men-of-war in the ocean. If the Zambesi could be opened to commerce the bright vision of the last ten years would be realized, and the Shiré Valley and banks of the Nyassa transformed into the garden of the Lord.

From the very first Livingstone saw the importance of the Shiré Valley and Lake Nyassa as the key to Central Africa. Ever since, it has become more and more evident that his surmise was correct. To make the occupation thoroughly effective, he thought much of the desirableness of a British colony, and was prepared to expend a great part of the remainder of his private means to carry it into effect. On August 4th, he says in his Journal:

"I have a very strong desire to commence a system of colonization of the honest poor; I would give £2000 or £3000 for the purpose. Intend to write my friend Young about it, and authorize him to draw if the project seems feasible. The Lord remember my desire, sanctify my motives, and purify all my desires. Wrote him.

"Colonization from a country such as ours ought to be one of hope, and not of despair. It ought not to be looked upon as the last and worst shift that a family can come to, but the performance of an imperative duty to our blood, our country, our religion, and to mankind. As soon as children begin to be felt an incumbrance, and what was properly in ancient times Old Testament blessings are no longer welcomed, parents ought to provide for removal to parts of this wide world where every accession is an addition of strength, and every member of the household feels in his inmost heart, 'the more the merrier.' It is a monstrous evil that all our healthy, handy, blooming daughters of England have not a fair chance at least to become the centres of domestic affections. The state of society, which precludes so many of them from occupying the position which Englishwomen are so well calculated to adorn, gives rise to enormous evils in the opposite sex,—evils and wrongs which we dare not even name,—and national colonization is almost the only remedy. Englishwomen are, in general, the most beautiful in the world, and yet our national emigration has often, by selecting the female emigrants from workhouses, sent forth the ugliest huzzies in creation to be the mothers—the model mothers— of new empires. Here, as in other cases, State necessities have led to the ill-formed and ill-informed being preferred to the well-formed and well-inclined honest poor, as if the worst as well as better qualities of mankind did not often run in the blood."

The idea of the colony quite fascinated Livingstone, and we find him writing on it fully to three of his most confidential business friends—Mr. Maclear, Mr. Young, and Sir Roderick Murchison. In all Livingstone's correspondence we find the tone of his letters modified by the character of his correspondents. While to Mr. Young and Sir Roderick he is somewhat cautious on the subject of the colony, knowing the keen practical eye they would direct on the proposal, to Mr. Maclear he is more gushing. He writes to him:

"I feel such a gush of emotion on thinking of the great work before us that I must unburden my mind. I am becoming every day more decidedly convinced that English colonization is an essential ingredient for our large success. . . . In this new region of Highlands no end of good could be effected in developing the trade in cotton and in discouraging that in slaves. . . . You know how I have been led on from one step to another by the overruling Providence of the great Parent, as I believe, in order to a great good for Africa. 'Commit thy way unto the Lord, trust also in Him, and He will bring it to pass.' I have tried to do this, and now see the prospect in front spreading out grandly. . . . But how is the land so promising to be occupied? . . . How many of our home poor are fighting hard to keep body and soul together! My heart yearns over our own poor when I see so much of God's fair earth unoccupied. Here it is really so; for the people have only a few sheep and goats, and no cattle. I wonder why we cannot have the old monastery system without the celibacy. In no other part where I have been does the prospect of self-support seem so inviting, and promising so much influence. Most of what is do‿ for the poor has especial reference to the blackguard poor."

In his letter to Mr. Young he expressed his conviction that a great desideratum in mission agency was missionary emigration by honest Christian poor to give living examples of Christian life that would insure permanency to the gospel once planted. He had always had a warm side to the English and Scottish poor—his own order, indeed. If twenty or thirty families would come out as an experiment, he was ready to give £2000 without saying from whom. He bids Mr. Young speak about the plan to Thom of Chorley, Turner of Manchester, Lord Shaftesbury, and the Duke of Argyll. "Now, my friend," he adds, "do your best, and God's blessing be with you. Much is done for the blackguard poor. Let us remember our own class, and do good while we have opportunity. I hereby authorize you to act in my behalf, and do whatever is to be done without hesitancy."

These letters, and their references to the honest poor, are characteristic. We have seen that among Dr. Livingstone's forefathers and connections were some very noble specimens of the honest poor. It touched him to think that, with all their worth, their life had been one protracted struggle. His sympathies were cordially with the class. He desired with all his heart to see them with a little less of the burden and more of the comfort of life. And he believed very thoroughly that, as Christian settlers in a heathen country, they might do more to promote Christianity among the natives than solitary missionaries could accomplish.

His parents and sisters were not forgotten. His letters to home are again somewhat in the apologetic vein. He feels that some explanation must be given of his own work, and some vindication of his coadjutors:

"We are working hard," he writes to his mother, "at what some can see at a glance the importance of, while to others we appear following after the glory of discovering lakes, mountains, jenny-nettles, and puddock-stools. In reference to these people I always remember a story told me by the late Dr. Philip with great glee. When a young minister in Aberdeen, he visited an old woman in affliction, and began to talk very fair to her on the duty of resignation, trusting, hoping, and all the rest of it, when the old woman looked up into his face, and said, 'Peer thing, ye ken naething aboot it.' This is what I say to those who set themselves up to judge another man's servant. We hope our good Master may permit us to do some good to our fellow-men."

His correspondence with Sir Roderick Murchison is likewise full of the idea of the colony. He is thoroughly persuaded that no good will ever be done by the Portuguese. They are a worn-out people—utterly worn out by disease—their stamina consumed. Fresh European blood must be poured into Africa. In consequence of recent discoveries, he now sees his way open, and all his hopes of benefit to England and Africa about to be realized. This must have been one of Livingstone's happiest times.

Visions of Christian colonies, of the spread of arts and civilization, of the progress of Christianity and the Christian graces, of the cultivation of cotton and the disappearance of the slave-trade, floated before him. Already the wilderness seemed to be blossoming. But the bright consummation was not so near as it seemed. One source of mischief was yet unchecked, and from it disastrous storms were preparing to break on the enterprise.

On his way home, Dr. Livingstone's health was not satisfactory, but this did not keep him from duty. "*14th October.*—Went on 17th part way up to Murchison's Cataracts, and yesterday reached it. Very ill with bleeding from the bowels and purging. Bled all night. Got up at one A.M. to take latitude."

At length, on 4th November, 1859, letters reached him from his family. "A letter from Mrs. L. says we were blessed with a little daughter on 16th November, 1858, at Kuruman. A fine healthy child. The Lord bless and make her his own child in heart and life!" She had been nearly a year in the world before he heard of her existence.

CHAPTER XIII.

GOING HOME WITH THE MAKOLOLO.

A.D. 1860.

Down to Kongone—State of the ship—Further delay—Letter to Secretary of Universities Mission—Letter to Mr. Braithwaite—At Tette—Miss Whately's sugar-mill—With his brother and Kirk at Kebrabasa—Mode of traveling—Reappearence of old friends—African warfare and its effects—Desolation— A European colony desirable—Escape from rhinoceros—Rumors of Moffat —The Portuguese local Governors oppose Livingstone—He becomes unpopular with them—Letter to Mr. Young—Wants of the country—The Makololo—Approach home—Some are disappointed—News of the death of the London missionaries, the Helmores and others—Letter to Dr. Moffat— The Victoria Falls re-examined—Sekelétu ill of leprosy—Treatment and recovery—His disappointment at not seeing Mrs. Livingstone—Efforts for the spiritual good of the Makololo—Careful observations in Natural History— The last of the "Ma-Robert"—Cheering prospect of the Universities Mission—Letter to Mr. Moore—to Mr. Young—He wishes another ship— Letter to Sir Roderick Murchison on the rumored journey of Silva Porto.

IT was necessary to go down to Kongone for the repair of the ship. Livingstone was greatly disappointed with it, and thought the greed of the vendor had supplied him with a very inferior article for the price of a good one. He thus pours forth his vexation in writing to a friend: "Very grievous it is to be standing here tinkering when we might be doing good service to the cause of African civilization, and that on account of insatiable greedineess. Burton may thank L. and B. that we are not at the other lakes before him. The loss of time greediness has inflicted on us has been frightful. My plan in this Expedition was excellent, but it did not include provisions against hypocrisy and fraud, which have sorely crippled us, and, indeed, ruined us, as a scientific Expedition."

Another delay was caused before they went inward, from their having to wait for a season suitable for hunting, as the party had to be kept in food. The mail from England had been lost, and they had the bitter disappointment of losing a year's correspondence from home. The following portions of a letter to the Secretary of the Committee for a Universities Mission gives a view of the situation at this time:

"RIVER ZAMBESI, *26th Jan.*, 1860.

"The defects we have unfortunately experienced in the 'Ma-Robert,' or rather the 'Asthmatic,' are so numerous that it would require a treatise as long as a lawyer's specification of any simple subject to give you any idea of them, and they have inflicted so much toil that a feeling of sickness comes over me when I advert to them.

"No one will ever believe the toil we have been put to in wood-cutting. The quantity consumed is enormous, and we cannot get sufficient for speed into the furnace. It was only a dogged determination not to be beaten that carried me through. . . . But all will come out right at last. We are not alone, though truly we deserve not his presence. He encourages the trust that is granted by the word, 'I am with you, even unto the end of the world.' . . .

"It is impossible for you to conceive how backward everything is here, and the Portuguese are not to be depended upon; their establishments are only small penal settlements, and as no women are sent out, the state of morals is frightful. The only chance of success is away from them; nothing would prosper in their vicinity. After all, I am convinced that were Christianity not divine, it would be trampled out by its professors. Dr. Kirk, Mr. C. Livingstone, and Mr. Rae, with two English seamen, do well. We are now on our way up the river to the Makololo country, but must go overland from Kebrabasa, or in a whaler. We should be better able to plan our course if our letters had not been lost. We have never been idle, and do not mean to be. We have been trying to get the Portuguese Government to acknowledge free-trade on this river, and but for long delay in our letters the negotiation might have been far advanced. I hope Lord John Russell will help in this matter, and then we must have a small colony or missionary and mercantile settlement. If this our desire is granted, it is probable we shall have no cause to lament our long toil and detention here. My wife's letters, too, were lost, so I don't know how or where she is. Our separation, and the work I have been engaged in, were not contemplated, but they have led to our opening a path into the fine cotton-field in the North. You will see that the discoveries of Burton and Speke confirm mine respecting the form of the continent and its fertility. It is an immense field. I crave the honor of establishing a focus of Christianity in it, but should it not be granted, I will submit as most unworthy. I have written Mr. Venn twice, and from yours I see something is contemplated in Cambridge. . . . If young men come to this country, they must lay their account with doing everything for themselves. They must not expect to find influence at once, and all the countries near to the Portuguese have been greatly depopulated. We are now ascending this river without vegetables, and living on salt beef and pork. The slave-trade has done its work, for formerly all kinds of provisions could be procured at every point, and at the cheapest rate. We cannot get anything for either love or money, in a country the fertility of which is truly astonishing.

A few more general topics are touched on in a letter to Mr. Braithwaite:

"I am sorry to hear of the death of Mr. Sturge. He wrote me a long letter on the 'Peace principle,' and before I could study it carefully, it was mislaid. I wrote him from Tette, as I did not wish him to suppose I neglected him, and mentioned the murder of the six Makololo and other things, as difficulties in the way of adopting his views, as they were perfectly unarmed, and there was no feud between the tribes. I fear that my letter may not have reached him alive. The departure of Sir Fowell Buxton and others is very unexpected. Sorry to see the loss of Dr. Bowen, of Sierra Leone—a good man and a true. But there is One who ever liveth to make intercession for us, and to carry on his own work. A terrible war that was in Italy, and the peace engenders more uneasy forebodings than any peace ever heard of. It is well that God and not the devil reigns, and will bring his own purposes to pass, right through the midst of the wars and passions of men. Have you any knowledge of a famous despatch written by Sir George Grey (late of the Cape), on the proper treatment of native tribes? I wish to study it.

"Tell your children that if I could get hold of a hippopotamus I would eat it rather than allow it to eat me. We see them often, but before we get near enough to get a shot they dive down, and remain hidden till we are past. As for lions, we never see them, sometimes hear a roar or two, but that is all, and I go on the plan put forth by a little girl in Scotland who saw a cow coming to her in a meadow, 'O boo! boo! you no hurt me, I no hurt you.'"

At Tette one of his occupations was to fit up a sugar-mill, the gift of Miss Whately, of Dublin, and some friends.

To that lady he writes a long letter of nineteen pages. He tells her he had just put up her beautiful sugar-mill, to show the natives what could be done by machinery. Then he adverts to the wonderful freedom from sickness that his party had enjoyed in the delta of the Zambesi, and proceeds to give an account of the Shiré Valley and its people. He finds ground for a favorable contrast between the Shiré natives and the Tette Portuguese:

"They (the natives) have fences made to guard the women from the alligators, all along the Shiré; at Tette they have none, and two women were taken past our vessel in the mouths of these horrid brutes. The number of women taken is so great as to make the Portuguese swear every time they speak of them, and yet, when I proposed to the priest to make a collection for a fence, and offered twenty dollars, he only smiled. You Protestants don't know all the good you do by keeping our friends of the only true and infallible Church up to their duty. Here, and in Angola, we see how it is, when they are not provoked— if not to love, to good works. . . .

"On telling the Makololo that the sugar-mill had been sent to Sekelétu by a lady, who collected a sum among other ladies to buy it, they replied, 'O na le pelu'—she has a heart. I was very proud of it, and so were they.

". . . With reference to the future, I am trying to do what I did before—obey the injunction, 'Commit thy way to the Lord, trust also in Him, and He shall bring it to pass.' And I hope that He will make some use of me. My attention is now directed specially to the fact that there is no country better adapted for producing the raw materials of English manufactures than this. . . .

"See to what a length I have run. I have become palaverist. I beg you to present my respectful salutation to the Archbishop and Mrs. Whately, and should you meet any of the kind contributors, say how thankful I am to them all."

From Tette he writes to Sir Roderick Murchison, 7th February, 1860, urging his plan for a steamer on Lake Nyassa: "If Government furnishes the means, all right; if not, I shall spend my book-money on it. I don't need to touch the children's fund, and mine could not be better spent. People who are born rich sometimes become miserable from a fear of becoming poor; but I have the advantage, you see, in not being afraid to die poor. If I live, I must succeed in what I have undertaken; death alone will put a stop to my efforts."

A month after he writes to the same friend, from Kongone, 10th March, 1860, that he is sending Rae home for a vessel:

"I tell Lord John Russell that he (Rae) may thereby do us more service than he can now do in a worn-out steamer, with 35 patches, covering at least 100 holes. I say to his Lordship, that after we have, by patient investigation and experiment, at the risk of life, rendered the fever not more formidable than a common cold; found access, from a good harbor on the coast, to the main stream; and discovered a pathway into the magnificent Highland lake region, which promises so fairly for our commerce in cotton, and for our policy in suppressing the trade in slaves, I earnestly hope that he will crown our efforts by securing our free passage through those parts of the Zambesi and Shiré of which the Portuguese make no use, and by enabling us to introduce civilization in a manner which will extend the honor and influence of the English name."

In his communications with the Government at home, Livingstone never failed to urge the importance of their securing the free navigation of the Zambesi. The Portuguese on the river were now beginning to get an inkling of his drift, and to feel indignant at any countenance he was receiving from their own Government.

Passing up the Zambesi with Charles Livingstone, Dr. Kirk, and such of the Makololo as were willing to go home, Dr. Livingstone took a new look at Kebrabasa, from a different point, still believing that in flood it would allow a steamer to pass. Of his mode of traveling we have some pleasant glimpses. He always tried to make progress more a pleasure than a toil, and found that kindly consideration for the feelings even of blacks, the pleasure of observing scenery and everything new, as one moves on at an ordinary pace, and the participation in the most delightful rest with his fellows, made traveling delightful. He was gratified to find that he was as able for the fatigue as the natives. Even the headman, who carried little more than he did himself, and never, like him, hunted in the afternoon, was not equal to him. The hunting was no small addition to the toil; the tired hunter was often tempted to give it up, after bringing what would have been only sufficient for the three whites, and leave the rest, thus sending "the idle, ungrateful poor" supperless to bed. But this was not his way. The blacks were thought of in hunting as well as the whites. "It is only by continuance in well-doing," he says, "even to the length of what the worldly-wise call weakness, that the conviction is produced anywhere, that our motives are high enough to secure sincere respect."

As they proceeded, some of his old acquaintances reappeared, notably Mpende, who had given him such a threatening reception, but had now learned that he belonged to a tribe "that loved the black man and did not make slaves." A chief named Pangola appeared, at first tipsy and talkative, demanding a rifle, and next morning, just as they were beginning divine service, reappeared sober to press his request. Among the Baenda-Pezi, or Go-Nakeds, whose only clothing is a coat of red ochre, a noble specimen of the race appeared in full dress, consisting of a long tobacco-pipe, and brought a handsome present.

The country bore the usual traces of the results of African warfare. At times a clever chief stands up, who brings large tracts under his dominion; at his death his empire dissolves, and a fresh series of desolating wars ensues. In one region which was once studded with villages, they walked a whole week without meeting any one. A European colony, he was sure, would be invaluable for constraining the tribes to live in peace. "Thousands of industrious natives would gladly settle round it, and engage in that peaceful pursuit of agriculture and trade of which they are so fond, and, undistracted by wars and rumors of wars, might listen to the purifying and ennobling truths of the gospel of Jesus Christ." At Zumbo, the most picturesque site in the country, they saw the ruins of Jesuit missions, reminding them that there men once met to utter the magnificent words, "Thou art the King of Glory, O Christ!" but without leaving one permanent trace of their labors in the belief and worship of the people.

Wherever they go, Dr. Livingstone has his eye on the trees and plants and fruits of the region, with a view to commerce; while he is no less interested to watch the treatment of fever, when cases occur, and greatly gratified that Dr. Kirk, who had been trying a variety of medicines on himself, made rapid recovery when he took Dr. Livingstone's pills. He used to say if he had followed Morison, and set up as pill-maker, he might have made his fortune. Passing through the Bazizulu he had an escape from a rhinoceros, as remarkable though not quite as romantic as his escape from the lion; the animal came dashing at him, and suddenly, for some unknown reason, stopped when close to him, and gave him time to escape, as if it had been struck by his color, and doubtful if hunting a white man would be good sport.

At a month's distance from Mosilikatse, they heard a report that the missionaries had been there, that they had told the chief that it was wrong to kill men, and that the chief had said he was born to kill people, but would drop the practice—an interesting testimony to the power of Mr. Moffat's words. Everywhere the Makololo proclaimed that they were the friends of peace, and their course was like a triumphal procession, the people of the villages loading them with presents.

But a new revelation came to Dr. Livingstone. Though the Portuguese Government had given public orders that he was to be aided in every possible way, it was evident that private instructions had come, which, unintentionally

25

perhaps, certainly produced the opposite effects. The Por‑ tuguese who were engaged in the slave-trade were far too much devoted to it ever to encourage an enterprise that aimed at extirpating it. Indeed, it became painfully apparent to Dr. Livingstone that the effect of his opening up the Zambesi had been to afford the Portuguese traders new facilities for conducting their unhallowed traffic; and had it not been for his promise to bring back the Makololo, he would now have abandoned the Zambesi and tried the Rovuma, as a way of reaching Nyassa. His future endeavors in connection with the Rovuma receive their explanation from this unwelcome discovery. The signifi‑ cance of the discovery in other respects cannot fail to be seen. Hitherto Livingstone had been on friendly terms with the Portuguese Government; he could be so no longer. The remarkable kindness he had so often received from Portuguese officers and traders made it a most pain‑ ful trial to break with the authorities. But there was no alternative. Livingstone's courage was equal to the occasion, though he could not but see that his new attitude to the Portuguese must give an altered aspect to his Expe‑ dition, and create difficulties that might bring it to an end.

A letter to Mr. James Young, dated 22d July, near Kalosi, gives a free and familiar account of "what he was about":

"This is July, 1860, and no letter from you except one written a few months after we sailed in the year of grace 1858. What you are doing I cannot divine. I am ready to believe any mortal thing except that Louis Napoleon has taken you away to make paraffin oil for the Tuile‑ ries. I don't believe that he is supreme ruler, or that he can go an inch beyond his tether. Well, as I cannot conceive what you are about, I must tell you what we are doing, and we are just trudging up the Zambesi as if there were no steam and no locomotive but shank's nag yet discovered. . . .

"We have heard of a mission for the Interior from the English Uni‑ versities, and this is the best news we have got since we came to Africa. I have recommended up Shiré as a proper sphere, and hasten back so as to be in the way if any assistance can be rendered. I rejoice at the prospect with all my heart, and am glad, too, that it is to be a Church of England Mission, for that Church has never put forth its strength, and I trust this may draw it forth. I am tired of discovery when no fruit follows. It was refreshing to be able to sit down every evening with the Makololo again, and tell them of Him who came down from heaven to save sinners. The unmerciful toil of the steamer prevented me from following my bent as I should have done. Poor fellows! they have learnt no good from their contact with slavery; many have imbibed the slave spirit; many had married slave-women and got children. These I did not expect to return, as they were captives of Sekelétu, and were not his own proper people. All professed a strong desire to return. To test them I proposed to burn their village, but to this they would not assent. We then went out a few miles and told them that any one wishing to remain might do so without guilt. A few returned, but though this was stated to them repeatedly afterward they preferred running away like slaves. I never saw any of the interior people so devoid of honor. Some complained of sickness, and all these I sent back, intrusting them with their burdens. About twenty-five returned in all to live at Tette. Some were drawn away by promises made to them as elephant-hunters. I had no objection to their trying to better their condition, but was annoyed at finding that they would not tell their intentions, but ran away as if I were using compulsion. I have learned more of the degrading nature of slavery of late than I ever conceived before. Our 20 millions were well spent in ridding ourselves of the incubus, and I think we ought to assist our countrymen in the West Indies to import free labor from India. . . . I cannot tell you how glad I am at a prospect of a better system being introduced into Eastern Africa than that which has prevailed for ages, the evils of which have only been intensified by Portuguese colonization, as it is called. Here we are passing through a well-peopled, fruitful region—a prolonged valley, for we have the highlands far on our right. I did not observe before that all the banks of the Zambesi are cotton-fields. I never intended to write a book and take no note of cotton, which I now see everywhere. On the Chongwe we found a species which is cultivated south of the Zambesi, which resembles some kinds from South America.

"All that is needed is religious and mercantile establishments to begin a better system and promote peaceful intercourse. Here we are among a people who go stark naked with no more sense of shame than we have with our clothes on. The women have more sense and go decently. You see great he-animals all about your camp carrying their indispensable tobacco-pipes and iron tongs to lift fire with, but the idea of a fig-leaf has never entered the mind. They cultivate largely, have had enormous crops of grain, work well in iron, and show taste in their dwellings, stools, baskets, and musical instruments. They are very hospitable, too, and appreciate our motives; but shame has been unaccountably left out of the question. They can give no reason for it except that all their ancestors went exactly as they do. Can you explain why Adam's first feeling has no trace of existence in his offspring?"

When the party reached the outskirts of Sekelétu's territory the news they heard was not encouraging. Some of the men heard that in their absence some of their wives had been variously disposed of. One had been killed for witchcraft, another had married again, while Masakasa was told that two years ago a kind of wild Irish wake had been celebrated in honor of his memory; the news made him resolve, when he presented himself among them, to declare himself an inhabitant from another world! One poor fellow's wail of anguish for his wife was most distressing to hear.

But far more tragical was the news of the missionaries who had gone from the London Missionary Society to Linyanti, to labor among Sekelétu's people. Mr. and Mrs. Helmore and several of his party had succumbed to fever, and the survivors had retired. Dr. Livingstone was greatly distressed, and not a little hurt, because he had not heard a word about the mission, nor been asked advice about any of the arrangements. If only the Helmores and their comrades had followed the treatment practiced by him so often, and in this very valley at this time by his brother Charles, they would probably have recovered. All spoke kindly of Mr. Helmore, who had quite won the hearts of the people. Knowing their language, he had at once begun to preach, and some of the young men at Seshéke were singing the hymns he had taught them. Rumors had gone abroad that some of the missionaries had been poisoned. In some quarters blame was cast on Livingstone for having misled the Society as to the character of Sekelétu and his disposition toward missionaries; but Livingstone satisfied himself that, though the missionaries had been neglected no foul play had taken place; fever alone had caused the deaths, and want of skill in managing the people had brought the remainder of the troubles. One piece of good news which he heard at Linyanti was that his old friend Sechéle was doing well. He had a Hanoverian missionary, nine tribes were under him, and the schools were numerously attended.

Writing to Dr. Moffat, 10th August, 1860, from Zambesi Falls, he says:

" With great sorrow we learned the death of our much-esteemed friends, Mr. and Mrs. Helmore, two days ago. We were too late to be of any service, for the younger missionaries had retired, probably dispirited by the loss of their leader. It is evident that the fever when untreated is as fatal now as it proved in the case of Commodore Owen's officers in this river, or in the great Niger Expedition. And yet what poor drivel was poured forth when I adopted energetic measures for speedily removing any Europeans out of the Delta. We were not then aware that the remedy which was first found efficacious in our own little Thomas on Lake 'Ngami, in 1850, and that cured myself and attendants during my solitary journeyings, was a certain cure for the disease, without loss of strength in Europeans generally. This we now know by ample experience to be the case. Warburg's drops, which have a great reputation in India, here cause profuse perspiration only, and the fever remains uncured. With our remedy, of which we make no secret, a man utterly prostrated is roused to resume his march next day. I have sent the prescription to John, as I doubt being able to go so far South as Mosilikatse's.

Again the grand Victoria Falls are reached, and Charles Livingstone, who has seen Niagara, gives the preference to Mosi-oa-tunya. By the route which they took, they would have passed the Falls at twenty miles' distance, but Dr. Livingstone could not resist the temptation to show them to his companions. All his former computations as to their size were found to be considerably within the mark; instead of a thousand yards broad they were more than eighteen hundred, and whereas he had said that the height of fall was about 100 feet, it turned out to be 310. His habit of keeping within the mark in all his statements of remarkable things was thus exemplified.

On coming among his old friends the Makololo, he found them in low spirits owing to protracted drought, and Sekelétu was ill of leprosy. He was in the hands of a native doctress, who was persuaded to suspend her treatment, and the lunar caustic applied by Drs. Livingstone and Kirk had excellent effects.[1] On going to Linyanti, Dr. Livingstone found the wagon and other articles which he had left there in 1853, safe and sound, except from the effects of weather and the white ants. The expressions of kindness and confidence toward him on the part of the natives greatly touched him. The people were much disappointed at not seeing Mrs. Livingstone and the children. But this confidence was the result of his way of dealing with them. "It ought never to be forgotten that influence among the heathen can be acquired only by patient continuance in well-doing, and that good manners are as necessary among barbarians as among the civilized." The Makololo were the most interesting tribe that Dr. Livingstone had ever seen. While now with them he was unwearied in his efforts for their spiritual good. In his Journal we find these entries:

"*September* 2, 1860.—On Sunday evening went over to the people, giving a general summary of Christian faith by the life of Christ. Asked them to speak about it afterward. Replied that these things were above them—they could not answer me. I said if I spoke of camels and buffaloes tamed, they understood, though they had never seen them; why not perceive the story of Christ, the witnesses to which refused to deny it, though killed for maintaining it? Went on to speak of the resurrection. All were listening eagerly to the statements about this, especially when they heard that they, too, must rise and be judged. Lerimo said, 'This I won't believe.' 'Well, the guilt lies between you and Jesus.' This always arrests attention. Spoke of blood shed by them; the conversation continued till they said, 'It was time for me to cross, for the river was dangerous at night.'"

[1] In 1864, while residing at Newstead Abbey, and writing his book, *The Zambesi and its Tributaries,* Dr. Livingstone heard of the death of Sekelétu.

"*September* 9.—Spoke to the people on the north side of the river—wind prevented evening service on the south."

The last subject on which he preached before leaving them on this occasion was the great resurrection. They told him they could not believe a reunion of the particles of the body possible. Dr. Livingstone gave them in reply a chemical illustration, and then referred to the authority of the Book that taught them the doctrine. And the poor people were more willing to give in to the authority of the Book than to the chemical illustration!

In *The Zambesi and its Tributaries* this journey to the Makololo country and back occupies one-third of the volume, though it did not lead to any very special results. But it enabled Dr. Livingstone to make great additions to his knowledge both of the people and the country. His observations are recorded with the utmost care, for though he might not be able to turn them to immediate use, it was likely, and even certain, that they would be useful some day. Indeed, the spirit of faith is apparent in the whole narrative, as if he could not pass over even the most insignificant details. The fish in the rivers, the wild animals in the woods, the fissures in the rocks, the course of the streams, the composition of the minerals and gravels, and a thousand other phenomena, are carefully observed and chronicled. The crowned cranes beginning to pair, the flocks of spurwinged geese, the habits of the ostrich, the nests of bee-eaters, pass under review in rapid succession. His sphere of observation ranges from the structure of the great continent itself to the serrated bone of the konokono, or the mandible of the ant.

Leaving Sesheke on the 17th September, they reached Tette on the 23d November, 1860, whence they started for Kongone with the unfortunate " Ma-Robert." But the days of that asthmatic old lady were numbered. On the 21st December she grounded on a sand-bank, and could not get off. A few days before this catastrophe Livingstone writes to Mr. Young:

"*Lupata, 4th Dec.*, 1860.—Many thanks for all you have been doing about the steamer and everything else. You seem to have gone about matters in a most business-like manner, and once for all I assure you I am deeply grateful.

"We are now on our way down to the sea, in hopes of meeting the new steamer for which you and other friends exerted yourselves so zealously. We are in the old 'Asthmatic,' though we gave her up before leaving in May last. Our engineer has been doctoring her bottom with fat and patches, and pronounced it safe to go down the river by dropping slowly. Every day a new leak bursts out, and he is in plastering and scoring, the pump going constantly. I would not have ventured again, but our whaler is as bad,—all eaten by the teredo,—so I thought it as well to take both, and stick to that which swims longest. You can put your thumb through either of them; they never can move again; I never expected to find either afloat, but the engineer had nothing else to do, and it saves us from buying dear canoes from the Portuguese.

"*20th Dec.*—One day, above Senna, the 'Ma-Robert' stuck on a sand-bank and filled, so we had to go ashore and leave her."

The correspondence of this year indicates a growing delight at the prospect of the Universities Mission. It was this, indeed, mainly that kept up his spirits under the depression caused by the failure of the "Ma-Robert," and other mishaps of the Expedition, the endless delays and worries that had resulted from that cause, and the manner in which both the Portuguese and the French were counter-working him by encouraging the slave-trade. While professedly encouraging emigration, the French were really extending slavery.

Here is his lively account of himself to his friend Mr. Moore:

"Tette, *28th November*, 1860.

"My dear Moore,—And why didn't you begin when you were so often on the point of writing, but didn't? This that you have accomplished is so far good, but very short. Hope you are not too old to learn. You have heard of our hindrances and annoyances, and, possibly, that we have done some work notwithstanding. Thanks to Providence, we have made some progress, and it is likely our operations will yet have a decided effect on slave-trading in Eastern Africa. I am greatly delighted with the prospect of a Church of England mission to

Central Africa. That is a good omen for those who are sitting in darkness, and I trust that in process of time great benefits will be conferred on our own overcrowded population at home. There is room enough and to spare in the fair world our Father has prepared for all his progeny. I pray to be made a harbinger of good to many, both white and black.

"I like to hear that some abuse me now, and say that I am *no* Christian. Many good things were said of me which I did not deserve, and I feared to read them. I shall read every word I can on the other side, and that will prove a sedative to what I was forced to hear of an opposite tendency. I pray that He who has lifted me up and guided me thus far, will not desert me now, but make me useful in my day and generation. 'I will never leave thee nor forsake thee.' So let it be.

"I saw poor Helmore's grave lately. Had my book been searched for excellencies, they might have seen a certain cure for African fever. We were curing it at a lower and worse part of the river at the very time that they were helplessly perishing, and so quickly, that more than a day was never lost after the operation of the remedy, though we were marching on foot. Our tramp was over 600 miles. We dropped down stream again in canoes from Sinamanero to Chicova—thence to this on shank's nag. We go down to the sea immediately, to meet our new steamer. Our punt was a sham and a snare.

"My love to Mary and all the children, with all our friends at Congleton."

In a letter to Mr. James Young, Dr. Livingstone gives good reasons for not wishing to push the colonization scheme at present, as he had recommended to the Universities Mission to add a similar enterprise to their undertaking:

"If you read all I have written you by this mail, you will deserve to be called a literary character. I find that I did not touch on the colonization scheme. I have not changed in respect to it, but the Oxford and Cambridge mission have taken the matter up, and as I shall do all I can to aid them, a little delay will, perhaps, be advisable.

"We are waiting for our steamer, and expect her every day; our first trip is a secret, and you will keep it so. We go to the Rovuma, a river exterior to the Portuguese claims, as soon as the vessel arrives. Captain Oldfield of the 'Lyra' is sent already, to explore, as far as he can, in that ship. The entrance is fine, and forty-five miles are known, but we keep our movements secret from the Portuguese—and so must you; they seize everything they see in the newspapers. Who are my imprudent friends that publish everything? I suspect Mr. ——, of ——, but no one gives me a name or a clue. Some expected me to feel sweet at being jewed by a false philanthropist, and bamboozled by a silly R. N. I did not, and could not, seem so; but I shall be more careful in future.

"Again back to the colony. It is not to sleep, but preparation must be made by collecting information, and maturing our plans. I shall be able to give definite instructions as soon as I see how the other mission works—at its beginning—and when we see if the new route we may discover has a better path to Nyassa than by Shiré—we shall choose the best, of course, and let you know as soon as possible. I think the Government will not hold back if we have a feasible plan to offer. I have recommended to the Universities Mission a little delay till we explore,—and for a working staff, two gardeners acquainted with farming; two country carpenters, capable of erecting sheds and any rough work; two traders to purchase and prepare cotton for exportation; one general steward of mission goods, his wife to be a good plain cook; one medical man, having knowledge of chemistry enough to regulate *indigo* and sugar-making. All the attendants to be married, and their wives to be employed in sewing, washing, attending the sick, etc., as need requires. The missionaries not to think themselves deserving a good English wife till they have erected a comfortable abode for her."

In the Royal Geographical Society this year (1860), certain communications were read which tended to call in question Livingstone's right to some of the discoveries he had claimed as his own. Mr. Macqueen, through whom these communications came, must have had peculiar notions of discovery, for some time before, there had appeared in the Cape papers a statement of his, that Lake 'Ngami of 1859 was no new discovery, as Dr. Livingstone had visited it seven years before; and Livingstone had to write to the papers in favor of the claims of Murray, Oswell, and Livingstone, against himself! It had been asserted to the Society by Mr. Macqueen, that Silva Porto, a Portuguese trader, had shown him a journal describing a journey of his from Benguela on the west to Ibo and Mozambique on the east, beginning November 26, 1852, and terminating August, 1854. Of that journal Mr.

Macqueen read a copious abstract to the Society (June 27, 1859), which is published in the Journal for 1860. In a letter to Sir Roderick Murchison (20th February, 1861), Livingstone, while exonerating Mr. Macqueen of all intention of misleading, gives his reasons for doubting whether the journey to the East Coast ever took place. He had met Porto at Linyanti in 1853, and subsequently at Naliele, the Barotse capital, and had been told by him that he had tried to go eastward, but had been obliged to turn, and was then going westward, and wished him to accompany him, which he declined, as he was a slave-trader; he had read his journal as it appeared in the Loanda "Boletim," but there was not a word in it of a journey to the East Coast; when the Portuguese minister had wished to find a rival to Dr. Livingstone, he had brought forward, not Porto, as he would naturally have done if this had been a genuine journey, but two black men who came to Tette in 1815; in the Boletim of Mozambique there was no word of the arrival of Porto there; in short, the part of the journal founded on could not have been authentic. Livingstone felt keenly on the subject of these rumors, not on his own account, but on account of the Geographical Society and of Sir Roderick who had introduced him to it; for nothing could have given him more pain than that either of these should have had any slur thrown on them through him, or even been placed for a time in an uncomfortable position.

CHAPTER XIV.

ROVUMA AND NYASSA—UNIVERSITIES MISSION.

A.D. 1861–1862.

Beginning of 1861—Arrival of the " Pioneer"—and of the agents of Universities Mission—Cordial welcome—Livingstone's catholic feelings—Ordered to explore the Rovuma—Bishop Mackenzie goes with him—Returns to the Shiré—Turning-point of prosperity past—Difficult navigation—The slave-sticks—Bishop settles at Magomero—Hostilities between Manganja and Ajawa—Attack of Mission party by Ajawa—Livingstone's advice to Bishop regarding them—Letter to his son Robert—Livingstone, Kirk, and Charles start for Lake Nyassa—Party robbed at north of Lake—Dismal activity of the slave-trade—Awful mortality in the process—Livingstone's fondness for *Punch*—Letter to Mr. Young—Joy at departure of new steamer "Lady Nyassa"—Colonization project—Letter against it from Sir R. Murchison—Hears of Dr. Stewart coming out from Free Church of Scotland—Visit at the ship from Bishop Mackenzie—News of defeat of Ajawa by missionaries—Anxiety of Livingstone—Arrangements for "Pioneer" to go to Kongone for new steamer and friends from home, then go to Ruo to meet Bishop—"Pioneer" detained—Dr. Livingstone's anxieties and depressions at New Year—"Pioneer" misses man-of-war "Gorgon"—At length "Gorgon" appears with brig from England and "Lady Nyassa"—Mrs. Livingstone and other ladies on board—Livingstone's meeting with his wife, and with Dr. Stewart—Stewart's recollections—Difficulties of navigation—Captain Wilson of "Gorgon" goes up river and hears of death of Bishop Mackenzie and Mr. Burrup—Great distress—Misrepresentations about Universities Mission—Miss Mackenzie and Mr. Burrup taken to "Gorgon"—Dr. and Mrs. Livingstone return to Shupanga—Illness and death of Mrs. Livingstone—Extracts from Livingstone's Journal and letters to the Moffats, Agnes, and the Murchisons.

THE beginning of 1861 brought some new features on the scene. The new steamer, the "Pioneer," at last arrived, and was a great improvement on the "Ma-Robert," though unfortunately she had too great draught of water. The agents of the Universities Missions also arrived, the first detachment consisting of Bishop Mackenzie and five other Englishmen, and five colored men from the Cape. Writing familiarly to his friend Moore, *àpropos* of his new comrades of the Church Mission, Livingstone says: " I have never felt anyway inclined to turn Churchman or dissenter either since I came out here. The feelings which we have toward different sects alter out here quite insensibly, till one looks upon all godly men as good and true brethren. I rejoiced when I heard that so many good and great men in the Universities had turned their thoughts toward Africa, and feeling sure that He who nad touched their hearts would lead them to promote his own glory, I welcomed the men they sent with a hearty, unfeigned welcome."

To his friend Mr. Maclear he wrote that he was very glad the Mission was to be under a bishop. He had seen so much idleness and folly result from missionaries being left to themselves, that it was a very great satisfaction to find that the new mission was to be superintended by one authorized and qualified to take the charge. Afterward when he came to know Bishop Mackenzie, he wrote of him to Mr. Maclear in the highest terms: " The Bishop is A 1, and in his readiness to put his hand to anything resembles much my good father-in-law Moffat."

It is not often that missions are over-manned, but in the first stage of such an undertaking as this, so large a body of men was an incumbrance, none of them knowing a word of the language or a bit of the way. It was Bishop Mackenzie's desire that Dr. Livingstone should accompany him at once to the scene of his future labors and help him to settle. But besides other reasons, the " Pioneer," as already stated, was under orders to explore the Rovuma, and, as the Portuguese put so many obstacles in the way on the Zambesi, to ascertain whether that river might not afford access to the Nyassa district. It was at last arranged that the Bishop should first go with the Doctor to the

26

Rovuma, and thereafter they should all go together to the
Shiré. In waiting for Bishop Mackenzie to accompany
him, Dr. Livingstone lost the most favorable part of the
season, and found that he could not get with the
"Pioneer" to the top of the Rovuma. He might have
left the ship and pushed forward on foot; but, not to delay
Bishop Mackenzie, he left the Rovuma in the meantime,
intending, after making arrangements with the Bishop, to
go to Nyassa, to find the point where the Rovuma left the
lake, if there were such a point, or, if not, get into its
headwaters and explore it downward.

Dr. Livingstone, as we have seen, welcomed the Mission
right cordially, for indeed it was what he had been most
eagerly praying for, and he believed that it would be the
beginning of all blessing to Eastern and Central Africa,
and help to assimilate the condition of the East Coast to
that of the West. The field for the cultivation of cotton
which he had discovered along the Shiré and Lake Nyassa
was immense, above 400 miles in length, and now it
seemed as if commerce and Christianity were going to take
possession of it. But it was found that the turning-point
of prosperity had been reached, and it was his lot to encounter dark reverses. The navigation of the Shiré was
difficult, for the "Pioneer" being deep in the water would
often run aground. On these occasions the Bishop, Mr.
Scudamore, and Mr. Waller, the best and the bravest of
the missionary party, were ever ready with their help in
hauling. Livingstone was sometimes scandalized to see
the Bishop toiling in the hot sun, while some of his subordinates were reading or writing in the cabin. As they
proceeded up the Shiré it was seen that the promises of
assistance from the Portuguese Government were worse
than fruitless. Evidently the Portuguese traders were
pushing the slave-trade with greater eagerness than ever.
Slave-hunting chiefs were marauding the country, driving
peaceful inhabitants before them, destroying their crops,

seizing on all the people they could lay hands on, and selling them as slaves. The contrast to what Livingstone had seen on his last journey was lamentable. All their prospects were overcast. How could commerce or Christianity flourish in countries desolated by war?

Every reader of *The Zambesi and its Tributaries* remembers the frightful picture of the slave-sticks, and the row of men, women, and children whom Livingstone and his companions set free. Nothing helped more than this picture to rouse in English bosoms an intense horror of the trade, and a burning sympathy with Livingstone and his friends. Livingstone and the Bishop, with his party, had gone up the Shiré to Chibisa's, and were halting at the village of Mbame, when a slave party came along. The flight of the drivers, the liberation of eighty-four men and women, and their reception by the good Bishop under his charge, speedily followed. The aggressors were the neighboring warlike tribe of Ajawa, and their victims were the Manganja, the inhabitants of the Shiré Valley. The Bishop accepted the invitation of Chigunda, a Manganja chief, to settle at Magomero. It was thought, however, desirable for the Bishop and Livingstone first to visit the Ajawa chief, and try to turn him from his murderous ways. The road was frightful—through burning villages resounding with the wailings of women and the shouts of the warriors. The Ajawa received the offered visit in a hostile spirit, and the shout being raised that Chibisa had come—a powerful chief with the reputation of being a sorcerer—they fired on the Bishop's party and compelled them, in self-defense, to fire in return. It was the first time that Livingstone had ever been so attacked by natives, often though they had threatened him. It was the first time he had had to repel an attack with violence; so little was he thinking of such a thing that he had not his rifle with him, and was obliged to borrow a revolver. The encounter was hot and serious, but it ended in the Ajawa being driven off without loss on the other side.

It now became a question for the Bishop in what relation he and his party were to stand to these murderous and marauding Ajawa—whether they should quietly witness their onslaughts or drive them from the country and rescue the captive Manganja. Livingstone's advice to them was to be patient, and to avoid taking part in the quarrels of the natives. He then left them at Magomero, and returned to his companions on the Shiré. For a time the Bishop's party followed Livingstone's advice, but circumstances afterward occurred which constrained them to take a different course, and led to very serious results in the history of the Mission.

Writing to his son Robert, Livingstone thus describes the atttack made by the Ajawa on him, the Bishop, and the missionaries:

" The slave-hunters had induced a number of another tribe to capture people for them. We came to this tribe while burning three villages, and though we told them that we came peaceably, and to talk with them, they saw that we were a small party, and might easily be overcome, rushed at us and shot their poisoned arrows. One fell between the Bishop and me, and another whizzed between another man and me. We had to drive them off, and they left that part of the country. Before going near them the Bishop engaged in prayer, and during the prayer we could hear the wail for the dead by some Manganja probably thought not worth killing, and the shouts of welcome home to these bloody murderers. It turned out that they were only some sixty or seventy robbers, and not the Ajawa tribe; so we had a narrow escape from being murdered.

'How are you doing? I fear from what I have observed of your temperament that you will have to strive against fickleness. Every one has his besetting fault—that is no disgrace to him, but it is a disgrace if he do not find it out, and by God's grace overcome it. I am not near to advise you what to do, but whatever line of life you choose, resolve to stick to it, and serve God therein to the last. Whatever failings you are conscious of, tell them to your heavenly Father; strive daily to master them and confess all to Him when conscious of having gone astray. And may the good Lord of all impart all the strength you need. Commit your way unto the Lord ; trust also in Him. Acknowledge Him in all your ways, and He will bless you."

Leaving the "Pioneer" at Chibisa's, on 6th August, 1861, Livingstone, accompanied by his brother and Dr. Kirk, started for Nyassa with a four-oared boat, which was carried by porters past the Murchison Cataracts. On 23d September they sailed into Lake Nyassa, naming the grand mountainous promontory at the end Cape Maclear, after Livingstone's great friend the Astronomer-Royal at the Cape. All about the lake was now examined with earnest eyes. The population was denser than he had seen anywhere else. The people were civil, and even friendly, but undoubtedly they were not handsome. At the north of the lake they were lawless, and at one point the party were robbed in the night—the first time such a thing had occurred in Livingstone's African life.[1] Of elephants there

---

[1] In *The Zambesi and its Tributaries,* Livingstone gives a grave account of the robbery. In his letters to his friends he makes fun of it, as he did of the raid of the Boers. To Mr. F. Fitch he writes: "You think I cannot get into a scrape. . . . For the first time in Africa we were robbed. Expert thieves crept into our sleeping-places, about four o'clock in the morning, and made off with what they could lay their hands on. Sheer over-modesty ruined me. It was Sunday, and such a black mass swarmed around our sail, which we used as a hut, that we could not hear prayers. I had before slipped away a quarter of a mile to dress for church, but seeing a crowd of women watching me through the reeds, I did not change my old 'unmentionables,'—they were so old, I had serious thoughts of converting them into—charity! Next morning early all our spare clothing was walked off with, and there I was left by my modesty nearly through at the knees, and no change of shirt, flannel, or stockings. After that, don't say that I can't get into a scrape!" The same letter thanks Mr. Fitch for sending him *Punch,* whom he deemed a sound divine! On the same subject he wrote at another time, regretting that *Punch* did not reach him, especially a number in which notice was taken of himself. "It never came. Who the miscreants are that steal them I cannot divine, I would not grudge them a reading if they would only send them on afterward. Perhaps binding the whole year's *Punches* would be the best plan; and then we need not label it 'Sermons in Lent,' or 'Tracts on Homœopathy,' but you may write inside, as Dr. Buckland did on his umbrella, 'Stolen from Dr. Livingstone.' We really enjoy them very much. They are good against fever. The 'Essence of Parliament,' for instance, is capital. One has to wade through an ocean of paper to get the same information, without any of the fun. And by the time the newspapers have reached us, most of the interest in public matters has evaporated.

was a great abundance,—indeed of all animal and vegetable life.

But the lake slave-trade was going on at a dismal rate.
An Arab dhow was seen on the lake, but it kept well out
of the way. Dr. Livingstone was informed by Colonel
Rigdy, late British Consul at Zanzibar, that 19,000 slaves
from this Nyassa region alone passed annually through
the custom-house there. This was besides those landed at
Portuguese slave ports. In addition to those captured,
thousands were killed or died of their wounds or of famine,
or perished in other ways, so that not one-fifth of the
victims became slaves—in the Nyassa district probably not
one-tenth. A small armed steamer on the lake might stop
nearly the whole of this wholesale robbery and murder.

Their stock of goods being exhausted, and no provisions
being procurable, the party had to return at the end of
October. They had to abandon the project of getting from
the lake to the Rovuma, and exploring eastward. They
reached the ship on 8th November, 1861, having suffered
more from hunger than on any previous trip.

In writing to his friend Young, 28th November, 1861,
Livingstone expresses his joy at the news of the departure
of the "Lady Nyassa;" gives him an account of the lake, and
of a terrific storm in which they were nearly lost; describes
the inhabitants, and the terrible slave-trade—the only
trade that was carried on in the district. It will take them
the best part of a year to put the ship on the lake, but it
will be such a blessing! He hopes the Government will
pay for it, once it is there.

The colonization project had not commended itself to
Sir R. Murchison. He had written of it sometime before:
"Your colonization scheme does not meet with supporters,
it being thought that you must have much more hold on
the country before you attract Scotch families to emigrate
and settle there, and then die off, or become a burden to
you and all concerned, like the settlers of old at Darien."

It was with much satisfaction that Livingstone now wrote to his friend (25th November, 1861): "A Dr. Stewart is sent out by the Free Church of Scotland to confer with me about a Scotch Colony. You will guess my answer. Dr. Kirk is with me in opinion, and if I could only get you out to take a trip up to the plateau of Zomba, and over the uplands which surround Lake Nyassa, you would give in too."

When the party returned to the ship they had a visit from Bishop Mackenzie, who was in good spirits and had excellent hopes of the Mission. The Ajawa had been defeated, and had professed a desire to be at peace with the English. But Dr. Livingstone was not without misgivings on this point. The details of the defeat of the Ajawa, in which the missionaries had taken an active part, troubled him, as we find from his private Journal. "The Bishop," he says (14th of November), "takes a totally different view of the affair from what I do." There were other points on which the utter inexperience of the missionaries, and want of skill in dealing with the natives, gave him serious anxiety. It is impossible not to see that even thus early, the Mission, in Livingstone's eyes, had lost something of its bloom.

It was arranged that the "Pioneer" should go down to the mouth of the Zambesi, to meet a man-of-war with provisions, and bring up the pieces of the new lake vessel, the "Lady Nyassa," which was eagerly expected, along with Mrs. Livingstone, Miss Mackenzie, the Bishop's sister, and other members of the Mission party. An appointment was made for January at the mouth of the river Ruo, a tributary of the Shiré, where the Bishop was to meet them. He and Mr. Burrup, who had just arrived, were meanwhile to explore the neighboring country.

The "Pioneer" was detained for five weeks on a shoal twenty miles below Chibisa's, and here the first death occurred—the carpenter's mate succumbed to fever. It was extremely irksome to suffer this long detention, to think of fuel and provisions wasting, and salaries running on, without one particle of progress. Livingstone was sensitive and anxious. He speaks in his Journal of the difficulty of feeling resigned to the Divine will in all things, and of believing that all things work together for good to those that love God. He seems to have been troubled at what had been said in some quarters of his treatment of members of the Expedition. In private letters, in the Cape papers, in the home papers, unfavorable representations of his conduct had been made. In one case, a prosecution at law had been threatened. On New Year's Day, 1862, he entered in his Journal an elaborate minute, as if for future use, bearing on the conduct of the Expedition. He refers to the difficulty to which civil expeditions are exposed, as compared with naval and military, in the matter of discipline, owing to the inferior authority and power of the chief. In the countries visited there is no enlightened public opinion to support the commander, and newspapers at home are but too ready to believe in his tyranny, and make themselves the champions of any dawdling fellow who would fain be counted a victim of his despotism. He enumerates the chief troubles to which his Expedition had been exposed from such causes. Then he explains how, at the beginning, to prevent collision, he had made every man independent in his own department, wishing only, for himself, to be the means of making known to the world what each man had done. His conclusion is a sad one, but it explains why in his last journeys he went alone: he is convinced that if he had been by himself he would have accomplished more, and undoubtedly he would have received more of the approbation of his countrymen.[1]

---

[1] Notwithstanding this expression of feeling, Dr. Livingstone was very sincere in his handsome acknowledgments, in the Introduction to *The Zambesi and its Tributaries*, of valuable services, especially from the members of the Expedition there named.

At length the "Pioneer" was got off the bank, and on the 11th January, 1862, they entered the Zambesi. They proceeded to the great Luabo mouth, as being more advantageous than the Kongone for a supply of wood. They were a month behind their appointment, and no ship was to be seen. The ship had been there, it turned out, on the 8th January, had looked eagerly for the "Pioneer," had fancied it saw the black funnel and its smoke in the river, and being disappointed had made for Mozambique, been caught in a gale, and was unable to return for three weeks. Livingstone's letters show him a little out of sorts at the manifold obstructions that had always been making him "too late"—"too late for Rovuma below, too late for Rovuma above, and now too late for our own appointment," but in greater trouble because the "Lady Nyassa" had not been sent by sea, as he had strongly urged, and as it afterward appeared might have been done quite well. To take out the pieces and fit them up would involve heavy expense and long delay, and perhaps the season would be lost again. But Livingstone had always a saving clause, in all his lamentations, and here it is: "I know that all was done for the best."

At length, on the last day of January, H.M.S. "Gorgon," with a brig in tow, hove in sight. When the "Pioneer" was seen, up went the signal from the "Gorgon"—"I have steamboat in the brig"; to which Livingstone replied— "Welcome news." Then "Wife aboard" was signaled from the ship. "Accept my best thanks" concluded what Livingstone called "the most interesting conversation he had engaged in for many a day." Next morning the "Pioneer" steamed out, and Dr. Livingstone found his wife "all right." In the same ship with Mrs. Livingstone, besides Miss Mackenzie and Mrs. Burrup, the Rev. E. Hawkins and others of the Universities Mission, had come the Rev. James Stewart, of the Free Church of Scotland (now Dr. Stewart, of Lovedale, South Africa), who had been sent out by a committee of that Church, "to meet with Dr. Livingstone, and obtain, by personal observation and otherwise, the information that might be necessary to enable a committee at home to form a correct judgment as to the possibility of founding a mission in that part of Africa." It happened that some time before Mr. Stewart had been tutor to Thomas Livingstone, while studying in Glasgow; this drew his sympathies to Livingstone and Africa, and was another link in that wonderful chain which Providence was making for the good of Africa. From Dr. Stewart's "Recollections of Dr. Livingstone and the Zambesi" in the *Sunday Magazine* (November, 1874), we get the picture from the other side. First, the sad disappointment of Mrs. Livingstone on the 8th January, when no "Pioneer" was to be found, with the anxious speculations raised in its absence as to the cause. Then a frightful tornado on the way to Mozambique, and the all but miraculous escape of the brig. Then the return to the Zambesi in company with H.M.S. "Gorgon," and on the 1st of February, in a lovely morning, the little cloud of smoke rising close to land, and afterward the white hull of a small paddle steamer making straight for the two ships outside.

"As the vessel approached," says Dr. Stewart, "I could make out with a glass a firmly built man of about the middle height, standing on the port paddle-box, and directing the ship's course. He was not exactly dressed as a naval officer, but he wore that gold-laced cap which has since become so well known both at home and in Africa. This was Dr. Livingstone, and I said to his wife, 'There he is at last." She looked brighter at this announcement than I had seen her do any day for seven months before."

Through the help of the men of the "Gorgon," the sections of the "Lady Nyassa" were speedily put on board the "Pioneer," and on the 10th February the vessel steamed off for the mouth of the Ruo, to meet the Bishop. But its progress through the river was miserable. Says Dr. Stewart:

" For ten days we were chiefly occupied in sailing or hauling the ship through sand-banks. The steamer was drawing between five and six feet of water, and though there were long reaches in the river with depth sufficient for a ship of larger draught, yet every now and then we found ourselves in shoal water of about three feet. No sooner was the boat got off one bank by might and main, and steady hauling on capstan and anchor laid out ahead, almost never astern, and we got a few miles of fair steering, than again we heard that sound, abhorred by all of us—a slight bump of the bow, and rush of sand along the ship's side, and we were again fast for a few hours, or a day or two, as the case might be."

The " Pioneer" was overladen, and the plan had to be changed. It was resolved to put the " Lady Nyassa" together at Shupanga, and tow her up to the Rapids.

"The detention," says Dr. Stewart, " was very trying to Dr. Living-stone, as it meant not a few weeks, but the loss of a year, inasmuch as by the time the ship was ready to be launched the river would be nearly at its lowest, and there would be no resource but to wait for the next rainy season. Yet, in the face of discouragement, he maintained his cheerfulness, and, after sunset, still enjoyed many an hour of pro-longed talk about current events at home, about his old College days in Glasgow, and about many of those who were unknown men then, but have since made their mark in life in the different paths they have taken. Amongst others his old friend Mr. Young, of Kelly, or Sir Paraffin, as he used subsequently to call him, came in for a large share of the conversation."

Meanwhile Captain Wilson (of the " Gorgon"), accom-panied by Dr. Kirk and others, had gone on in boats with Miss Mackenzie and Mrs. Burrup, and learned the sad fate of the Bishop and Mr. Burrup. It appeared that the Bishop, accompanied by the Makololo, had gone forth on an expedition to rescue the captive husbands of some of the Manganja women, and had been successful. But as the Bishop was trying to get to the mouth of the Ruo, his canoe was upset, his medicines and cordials were lost, and, being seized with fever, after languishing for some time, he died in distressing circumstances, on the 31st January. Mr. Burrup, who was with him, and who was also stricken, was carried back to Magomero, and died in a few days.

Captain Wilson, who had himself been prostrated by fever, and made a narrow escape, returned with this sad news, three weeks after he had left Shupanga, bringing the two broken-hearted ladies, who had expected to be welcomed, the one by her brother, the other by her husband. It was a great blow to Livingstone.

"It was difficult to say," writes Dr. Stewart, "whether he or the unhappy ladies, on whom the blow fell with the most personal weight, were most to be pitied. He felt the responsibility, and saw the widespread dismay which the news would occasion when it reached England, and at the very time when the Mission most needed support. 'This will hurt us all,' he said, as he sat resting his head on his hand, on the table of the dimly-lighted little cabin of the 'Pioneer.' His esteem for Bishop Mackenzie was afterward expressed in this way : 'For unselfish goodness of heart and earnest devotion to the work he had undertaken, it can safely be said that none of the commendations of his friends can exceed the reality.' He did what he could, I believe, to comfort those who were so unexpectedly bereaved ; but the night he spent must have been an uneasy one."

Livingstone says in his book that the unfavorable judgment which he had formed of the Bishop's conduct in fighting with the Ajawa was somewhat modified by a natural instinct, when he saw how keenly the Bishop was run down for it in England, and reflected more on the circumstances, and thought how excellent a man he was. Sometimes he even said that, had he been there, he would probably have done what the Bishop did.[1] Why, then, it may be asked, was Livingstone so ill-pleased when it was said that all that the Bishop had done was done by

---

[1] Writing to Mr. Waller, 12th February, 1863, Dr. Livingstone said: "I thought you wrong in attacking the Ajawa, till I looked on it as defense of your orphans. I thought that you had shut yourselves up to one tribe, and that, the Manganja; but I think differently now, and only wish they would send out Dr. Pusey here. He would learn a little sense, of which I suppose I have need myself."

his advice? No one will ask this question who reads the terms of a letter by Mr. Rowley, one of the Mission party, first published in the Cape papers, and copied into the *Times* in November, 1862. It was said there that "from the moment when Livingstone commenced the release of slaves, his course was one of aggression. He hunted for slaving parties in every direction, and when he heard of the Ajawa making slaves in order to sell to the slavers, he went designedly in search of them, and intended to take their captives from them by force if needful. It is true that when he came upon them he found them to be a more powerful body than he expected, and had they not fired first, he might have withdrawn. . . . His parting words to the chiefs just before he left . . . were to this effect: 'You have hitherto seen us only as fighting men, but it is not in such a character we wish you to know us.'"[1] How could Livingstone be otherwise than indignant to be spoken of as if the use of force had been his habit, while the whole tenor of his life had gone most wonderfully to show the efficacy of gentle and brotherly treatment? How could he but be vexed at having the odium of the whole proceedings thrown on him, when his last advice to the missionaries had been disregarded by them? Or how could he fail to be concerned at the discredit which the course ascribed to him must bring upon the Expedition under his command, which was entirely separate from the Mission? It was the unhandsome treatment of himself and reckless periling of the character and interests of his Expedition in order to shield others, that raised his indignation. "Good Bishop Mackenzie," he wrote to his friend Mr. Fitch, "would never have tried to screen himself by accusing me." In point of fact, a few years afterward the Portuguese Government, through

---

[1] Mr. Rowley afterward (February 22, 1865) expressed his regret that this letter was ever written, as it had produced an ill-effect. See *The Zambesi and its Tributaries*, p. 475 *note*.

27

Mr. Lacerda, when complaining bitterly of the statements of Livingstone in a speech at Bath, in 1865, referred to Mr. Rowley's letter as bearing out their complaint. It served admirably to give an unfavorable view of his aims and methods, *as from one of his own allies.* Dr. Livingstone never allowed himself to cherish any other feeling but that of high regard for the self-denial and Christian heroism of the Bishop, and many of his coadjutors; but he did feel that most of them were ill-adapted for their work and had a great deal to learn, and that the manner in which he had been turned aside from the direct objects of his own enterprise by having to look after so many inexperienced men, and then blamed for what he deprecated, and what was done in his absence, was rather more than it was reasonable for him to bear.[1]

Writing of the terrible loss of Mackenzie and Burrup to the Bishop of Cape Town, Livingstone says: "The blow is quite bewildering; the two strongest men so quickly cut down, and one of them, humanly speaking, indispensable to the success of the enterprise. We must bow to the will of Him who doeth all things well; but I cannot help feeling sadly disturbed in view of the effect the news may have at home. *I shall not swerve a hairbreadth from*

---

[1] It must not be supposed that the letter of Mr. Rowley expressed the mind of his brethren. Some of them were greatly annoyed at it, and used their influence to induce its author to write to the Cape papers that he had conveyed a wrong impression. In writing to Sir Thomas Maclear (20th November, 1862), after seeing Rowley's letter in the Cape papers, Dr. Livingstone said: " It is untrue that I ever on any one occasion adopted an aggressive policy against the Ajawa, or took slaves from them. Slaves were taken from Portuguese alone. I never hunted the Ajawa, or took the part of Manganja against Ajawa. In this I believe every member of the Mission will support my assertion." Livingstone declined to write a contradiction *to the public prints,* because he knew the harm that would be done by a charge against a clergyman. In this he showed the same magnanimity and high Christian self-denial which he had shown when he left Mabotsa. It was only when the Portuguese claimed the benefit of Rowley's testimony that he let the public see what its value was.

my work while life is spared, and I trust the supporters of the Mission may not shrink back from all that they have set their hearts to."

The next few weeks were employed in taking Miss Mackenzie and Mrs. Burrup to the "Gorgon" on their way home. It was a painful voyage to all—to Dr. and Mrs. Livingstone, to Miss Mackenzie and Mrs. Burrup, and last, not least, to Captain Wilson, who had been separated so long from his ship, and had risked life, position, and everything, to do service to a cause which in spite of all he left at a much lower ebb.

When the "Pioneer" arrived at the bar, it was found that owing to the weather the ship had been forced to leave the coast, and she did not return for a fortnight. There was thus another long waiting from 17th March to 2d April. Dr. and Mrs. Livingstone then returned to Shupanga. The long detention in the most unhealthy season of the year, and when fever was at its height, was a sad, sad calamity.

We are now arrived at the last illness and the death of Mrs. Livingstone. After she had parted from her husband at the Cape in the spring of 1858, she returned with her parents to Kuruman, and in November gave birth there to her youngest child, Anna Mary. Thereafter she returned to Scotland to be near her other children. Some of them were at school. No comfortable home for them all could be formed, and though many friends were kind, the time was not a happy one. Mrs. Livingstone's desire to be with her husband was intense; not only the longings of an affectionate heart, and the necessity of taking counsel with him about the family, but the feeling that when overshadowed by one whose faith was so strong her fluttering heart would regain its steady tone, and she would be better able to help both him and the children, gave vehemence to this desire. Her letters to her husband tell of much spiritual darkness; his replies were the very soul of tenderness and Christian earnestness. Providence seemed to favor her wish; the vessel in which she sailed was preserved from imminent destruction, and she had the great happiness of finding her husband alive and well.

On the 21st of April Mrs. Livingstone became ill. On the 25th the symptoms were alarming—vomitings every quarter of an hour, which prevented any medicine from remaining on her stomach. On the 26th she was worse and delirious. On the evening of Sunday the 27th Dr. Stewart got a message from her husband that the end was drawing near. " He was sitting by the side of a rude bed formed of boxes, but covered with a soft mattress, on which lay his dying wife. All consciousness had now departed, as she was in a state of deep coma, from which all efforts to rouse her had been unavailing. The strongest medical remedies and her husband's voice were both alike powerless to reach the spirit which was still there, but was now so rapidly sinking into the depths of slumber, and darkness and death. The fixedness of feature and the oppressed and heavy breathing only made it too plain that the end was near. And the man who had faced so many deaths, and braved so many dangers, was now utterly broken down and weeping like a child."

Dr. Livingstone asked Dr. Stewart to commend her spirit to God, and along with Dr. Kirk they kneeled in prayer beside her. In less than an hour, her spirit had returned to God. Half an hour after, Dr. Stewart was struck with her likeness to her father, Dr. Moffat. He was afraid to utter what struck him so much, but at last he said to Livingstone, " Do you notice any change?" "Yes," he replied, without raising his eyes from her face,—"the very features and expression of her father."

Every one is struck with the calmness of Dr. Livingstone's notice of his wife's death in *The Zambesi and its Tributaries.* Its matter-of-fact tone only shows that he regarded that book as a sort of official report to the nation, in which it would not be becoming for him to introduce personal feelings. A few extracts from his Journal and letters will show better the state of his heart.

"It is the first heavy stroke I have suffered, and quite takes away my strength. I wept over her who well deserved many tears. I loved her when I married her, and the longer I lived with her I loved her the more. God pity the poor children, who were all tenderly attached to her, and I am left alone in the world by one whom I felt to be a part of myself. I hope it may, by divine grace, lead me to realize heaven as my home, and that she has but preceded me in the journey. Oh my Mary, my Mary! how often we have longed for a quiet home, since you and I were cast adrift at Kolobeng; surely the removal by a kind Father who knoweth our frame means that He rewarded you by taking you to the best home, the eternal one in the heavens. The prayer was found in her papers —'Accept me, Lord, as I am, and make me such as Thou wouldst have me to be.' He who taught her to value this prayer would not leave his own work unfinished. On a letter she had written, 'Let others plead for pensions, I wrote to a friend I can be rich without money; I would give my services in the world from uninterested motives; I have motives for my own conduct I would not exchange for a hundred pensions.'

"She rests by the large baobab-tree at Shupanga, which is sixty feet in circumference, and is mentioned in the work of Commodore Owen. The men asked to be *allowed* to mount guard till we had got the grave built up, and we had it built with bricks dug from an old house.

"From her boxes we find evidence that she intended to make us all comfortable at Nyassa, though she seemed to have a presentiment of an early death,—she purposed to do more for me than ever.

"11*th May, Kongone.*—My dear, dear Mary has been this evening a fortnight in heaven,—absent from the body,

present with the Lord. To-day shalt thou be with Me in
Paradise. Angels carried her to Abraham's bosom—to be
with Christ is far better. Enoch, the seventh from Adam,
prophesied, ' Behold, the Lord cometh with ten thousand
of his saints'; ye also shall appear with Him in glory.
He comes with them; then they are now with Him. I go
to prepare a place for you; that where I am there ye may
be also, to behold his glory. Moses and Elias talked of the
decease He should accomplish at Jerusalem; then they
know what is going on here on certain occasions. They
had bodily organs to hear and speak. For the first time
in my life I feel willing to die.—D. L."

" *May* 19, 1862.—Vividly do I remember my first passage
down in 1856, passing Shupanga house without landing,
and looking at its red hills and white vales with the impression that it was a beautiful spot. No suspicion glanced
across my mind that there my loving wife would be called
to give up the ghost six years afterward. In some other
spot I may have looked at, my own resting-place may be
allotted. I have often wished that it might be in some far-
off still deep forest, where I may sleep sweetly till the
resurrection morn, when the trump of God will make all
start up into the glorious and active second existence.

" *25th May.*—Some of the histories of pious people in the
last century and previously tell of clouds of religious
gloom, or of paroxysms of opposition and fierce rebellion
against God, which found vent in terrible expressions.
These were followed by great elevations of faith, and
reactions of confiding love, the results of divine influence
which carried the soul far above the region of the intellect
into that of direct spiritual intuition. This seems to have
been the experience of my dear Mary. She had a strong
presentiment of death being near. She said that she
would never have a house in this country. Taking it to
be despondency alone, I only joked, and now my heart
smites me that I did not talk seriously on that and many
things besides.

"31*st* *May*, 1862.—The loss of my ever dear Mary lies like a heavy weight on my heart. In our intercourse in private there was more than what would be thought by some a decorous amount of merriment and play. I said to her a few days before her fatal illness: 'We old bodies ought now to be more sober, and not play so much.' 'Oh, no,' said she, 'you must always be as playful as you have always been; I would not like you to be as grave as some folks I have seen.' This, when I know her prayer was that she might be spared to be a help and comfort to me in my great work, led me to feel what I have always believed to be the true way, to let the head grow wise, but keep the heart always young and playful. She was ready and anxious to work, but has been called away to serve God in a higher sphere."

Livingstone could not be idle, even when his heart was broken; he occupied the days after the death in writing to her father and mother, to his children, and to many of the friends who would be interested in the sad news. Among these letters, that to Mrs. Moffat and her reply from Kuruman have a special interest. His letters went round by Europe, and the first news reached Kuruman by traders and newspapers. For a full month after her daughter's death, Mrs. Moffat was giving thanks for the mercy that had spared her to meet with her husband, and had made her lot so different from that of Miss Mackenzie and Mrs. Burrup. In a letter, dated 26th May, she writes to Mary a graphic account of the electrical thrill that passed through her when she saw David's handwriting—of the beating heart with which she tried to get the essence of his letter before she read the lines—of the overwhelming joy and gratitude with which she learned that they had met —and then the horror of great darkness that came over her when she read of the tragic death of the Bishop, to whom she had learned to feel as to a friend and brother. Then she pours out her tears over the "poor dear ladies,

Miss Mackenzie and Mrs. Burrup," and remembers the similar fate of the Helmores, who, like the Bishop and his friends, had had it in their hearts to build a temple to the Lord in Africa, but had not been permitted. Then comes some family news, especially about her son Robert, whose sudden death occurred a few days after, and was another bitter drop in the family cup. And then some motherly forecastings of her daughter's future, kindly counsel where she could offer any, and affectionate prayers for the guidance of God where the future was too dark for her to penetrate.

For a whole month before this letter was written, poor Mary had been sleeping under the baobab-tree at Shupanga!

In Livingstone's letter to Mrs. Moffat he gives the details of her illness, and pours his heart out in the same affectionate terms as in his Journal. He dwells on the many unhappy causes of delay which had detained them near the mouth of the river, contrary to all his wishes and arrangements. He is concerned that her deafness (through quinine) and comatose condition before her death prevented her from giving him the indications he would have desired respecting her state of mind in the view of eternity.

" I look," he says, " to her previous experience and life for comfort, and thank God for his mercy that we have it. . . . A good wife and mother was she. God have pity on the children—she was so much beloved by them. . . She was much respected by all the officers of the ' Gorgon,' —they would do anything for her. When they met this vessel at Mozambique, Captain Wilson offered his cabin in that fine large vessel, but she insisted rather that Miss Mackenzie and Mrs. Burrup should go. . . . I enjoyed her society during the three months we were together. It was the Lord who gave and He has taken away. I wish to say—Blessed be his name. I regret, as there always are regrets after our loved ones are gone, that the slander which, unfortunately, reached her ears from missionary gossips and others had an influence on me in allowing her to come, before we were fairly on Lake Nyassa. A doctor of divinity said, when her devotion to her family was praised: 'Oh, she is no good, she is here because her husband cannot live with her.' The last day will tell another tale."

To his daughter Agnes he writes, after the account of her death: " . . . Dear Nannie, she often thought of you, and when once, from the violence of the disease, she was delirious, she called out, 'See! Agnes is falling down a precipice.' May our Heavenly Saviour, who must be your Father and Guide, preserve you from falling into the gulf of sin over the precipice of temptation. . . . Dear Agnes, I feel alone in the world now, and what will the poor dear baby do without her mamma? She often spoke of her, and sometimes burst into a flood of tears, just as I now do in taking up and arranging the things left by my beloved partner of eighteen years. . . . I bow to the Divine hand that chastens me. God grant that I may learn the lesson He means to teach! All she told you to do she now enforces, ·as if beckoning from heaven. Nannie, dear, meet her there. Don't lose the crown of joy she now wears, and the Lord be gracious to you in all things. You will now need to act more and more from a feeling of responsibility to Jesus, seeing He has taken away one of your guardians. A right straightforward woman was she. No crooked way ever hers, and she could act with decision and energy when required. I pity you on receiving this, but it is the Lord.—Your sorrowing and lonely father."

Letters of the like tenor were written to every intimate friend. It was a relief to his heart to pour itself out in praise of her who was gone, and in some cases, when he had told all about the death, he returns to speak of her

life. A letter to Sir Roderick Murchison gives all the
particulars of the illness and its termination. Then he
thinks of the good and gentle Lady Murchison,—"la
spirituelle Lady Murchison," as Humboldt called her,—
and writes to her: "It will somewhat ease my aching
heart to tell you about my dear departed Mary Moffat,
the faithful companion of eighteen years." He tells of
her birth at Griqua Town in 1821, her education in
England, their marriage and their love. "At Kolobeng,
she managed all the household affairs by native servants
of her own training, made bread, butter, and all the
clothes of the family; taught her children most carefully;
kept also an infant and sewing school—by far the most
popular and best attended we had. It was a fine sight
to see her day by day walking a quarter of a mile to the
town, no matter how broiling hot the sun, to impart
instruction to the heathen Bakwains. Ma-Robert's name
is known through all that country, and 1800 miles beyond.
. . . A brave, good woman was she. All my hopes of
giving her one day a quiet home, for which we both had
many a sore longing, are now dashed to the ground.
She is, I trust, through divine mercy, in peace in the
home of the blest. . . . She spoke feelingly of your
kindness to her, and also of the kind reception she received from Miss Burdett Coutts. Please give that lady
and Mrs. Brown the sad intelligence of her death."

The reply of Mrs. Moffat to her son-in-law's letter was
touching and beautiful. "I do thank you for the detail
you have given us of the circumstances of the last days
and hours of our lamented and beloved Mary, our first-
born, over whom our fond hearts first beat with parental
affection!" She recounts the mercies that were mingled
with the trial—though Mary could not be called *eminently*
pious, she had the root of the matter in her, and though
the voyage of her life had been a trying and stormy one,
she had not become a wreck. God had remembered her;

had given her during her last year the counsels of faithful men—referring to her kind friend and valued counselor, the Rev. Professor Kirk, of Edinburgh, and the Rev. Dr. Stewart, of Lovedale—and, at last, the great privilege of dying in the arms of her husband. "As for the cruel scandal that seems to have hurt you both so much, those who said it did not know you *as a couple.* In all *our* intercourse with you, we never had a doubt as to your being comfortable together. I know there are some maudlin ladies who insinuate, when a man leaves his family fres quently, no matter how noble is his object, that he is not *comfortable* at home. But we can afford to smile at this, and say, 'The Day will declare it.' . . .

"Now, my dear Livingstone, I must conclude by assuring you of the tender interest we shall ever feel in your operations. It is not only as the husband of our departed Mary and the father of her children, but as one who has laid himself out for the emancipation of this poor wretched continent, and for opening new doors of entrance for the heralds of salvation (not that I would not have preferred your remaining in your former capacity). I nevertheless rejoice in what you are allowed to accomplish. We look anxiously for more news of you, and my heart bounded when I saw your letters the other day, thinking they were new. May our gracious God and Father comfort your sorrowful heart.—Believe me ever your affectionate mother,

"Maᴙʏ Moffat."

CHAPTER XV.

LAST TWO YEARS OF THE EXPEDITION.

A.D. 1862–1863.

Livingstone again buckles on his armor—Letter to Waller—Launch of "Lady Nyassa"—Too late for season—He explores the Rovuma—Fresh activity of the slave-trade—Letter to Governor of Mozambique about his discoveries—Letter to Sir Thomas Maclear—Generous offer of a party of Scotchmen—The Expedition proceeds up Zambesi with "Lady Nyassa" in tow—Appalling desolations of Marianno—Tidings of the Mission—Death of Scudamore—of Dickenson—of Thornton—Illness of Livingstone—Dr. Kirk and Charles Livingstone go home—He proceeds northward with Mr. Rae and Mr. E. D. Young of the "Gorgon"—Attempt to carry a boat over the rapids—Defeated—Recall of the Expedition—Livingstone's views—Letter to Mr. James Young—to Mr. Waller—Feeling of the Portuguese Government—Offer to the Rev. Dr. Stewart—Great discouragements—Why did he not go home?—Proceeds to explore Nyassa—Risks and sufferings—Occupation of his mind—Natural History—Obliged to turn back—More desolation—Report of his murder—Kindness of Chinsamba—Reaches the ship—Letter from Bishop Tozer, abandoning the Mission—Distress of Livingstone—Letter to Sir Thomas Maclear—Progress of Dr. Stewart—Livingstonia—Livingstone takes charge of the children of the Universities Mission—Letter to his daughter—Retrospect—The work of the Expedition—Livingstone's plans for the future.

IT could not have been easy for Livingstone to buckle on his armor anew. How he was able to do it at all may be inferred from some words of cheer written by him at the time to his friend Mr. Waller: "Thanks for your kind sympathy. In return, I say, Cherish exalted thoughts of the great work you have undertaken. It is a work which, if faithful, you will look back on with satisfaction while the eternal ages roll on their everlasting course. The devil will do all he can to hinder you by efforts from without and from within; but remember Him who *is* with you, and will be with you alway."

As soon as he was able to brace himself, he was again at his post, helping to put the "Lady Nyassa" together and launch her. This was achieved by the end of June, greatly to the wonder of the natives, who could not understand how iron should swim. The "Nyassa" was an excellent steamboat, and could she have been got to the lake would have done well. But, alas! the rainy season had passed, and until December this could not be done. Here was another great disappointment. Meanwhile, Dr. Livingstone resolved to renew the exploration of the Rovuma, in the hope of finding a way to Nyassa beyond the dominion of the Portuguese. This was the work in which he had been engaged at the time when he went with Bishop Mackenzie to help him to settle.

The voyage up the Rovuma did not lead to much. On one occasion they were attacked, fiercely and treacherously, by the natives. Cataracts occurred about 156 miles from the mouth, and the report was that farther up they were worse. The explorers did not venture beyond the banks of the rivers, but so far as they saw, the people were industrious, and the country fertile, and a steamer of light draft might carry on a very profitable trade among them. But there was no water-way to Nyassa. The Rovuma came from mountains to the west, having only a very minute connection with Nyassa. It seemed that it would be better in the meantime to reach the lake by the Zambesi and the Shiré, so the party returned. It was not till the beginning of 1863 that they were able to renew the ascent of these rivers. Livingstone writes touchingly to Sir Roderick, in reference to his returning to the Zambesi: "It may seem to some persons weak to feel a chord vibrating to the dust of her who rests on the banks of the Zambesi, and think that the path by that river is consecrated by her remains."

Meanwhile, Dr. Livingstone was busy with his pen. A

28

new energy had been imparted to him by the appalling
facts now fully apparent, that his discoveries had only
stimulated the activity of the slave-traders, that the Portuguese local authorities really promoted slave-trading, with
its inevitable concomitant slave-hunting, and that the
horror and desolation to which the country bore such
frightful testimony was the result. It seemed as if the
duel he had fought with the Boers when they determined
to close Africa, and he determined to open it, had now to
be repeated with the Portuguese. The attention of Dr.
Livingstone is more and more concentrated on this terrible
topic. Dr. Kirk writes to him that when at Tette he had
heard that the Portuguese Governor-General at Mozambique had instructed his brother, the Governor of that
town, to act on the principle that the slave-trade, though
prohibited on the ocean, was still lawful on the land,
and that any persons interfering with slave-traders, by
liberating their slaves, would be counted robbers. An
energetic despatch to Earl Russell, then Foreign Secretary,
calls attention to this outrage.

A few days after, a strong but polite letter is sent to the
Governor of Tette, calling attention to the forays of a man
named Belshore, in the Chibisa country, and entreating
him to stop them. About the same time he writes to the
Governor-General of Mozambique in reply to a paper by
the Viscount de Sa da Bandeira, published in the Almanac
by the Government press, in which the common charge
was made against him of arrogating to himself the glory
of discoveries which belonged to Senhor Candido and
other Portuguese. He affirms that before publishing his
book he examined all Portuguese books of travels he could
find; that he had actually shown Senhor Candido to have
been a discoverer before any Portuguese hinted that he
was such; that the lake which Candido spoke of as northwest of Tette could not be Nyassa, which was northeast of
it; that he did full justice to all the Portuguese explorers,

and that what he claimed as own discoveries were certainly not the discoveries of the Portuguese. A few days after, he writes to Mr. Layard, then our Portuguese Minister, and comments on the map published by the Viscount as representing Portuguese geography,—pointing out such blunders as that which made the Zambesi enter the sea at Quilimane, proving that by their map the Portuguese claimed territory that was certainly not theirs; adverting to their utter ignorance of the Victoria Falls, the most remarkable phenomenon in Africa; affirming that many so-called discoveries were mere vague rumors, heard by travelers; and showing the use that had been made of his own maps, the names being changed to suit the Portuguese orthography.

Livingstone had the satisfaction of knowing that his account of the trip to Lake Nyassa had excited much interest in the Cabinet at home, and that a strong remonstrance had been addressed to the Portuguese Government against slave-hunting. But it does not appear that this led to any improvement at the time.

While stung into more than ordinary energy by the atrocious deeds he witnessed around him, Livingstone was living near the borders of the unseen world. He writes to Sir Thomas Maclear on the 27th October, 1862:

"I suppose that I shall die in these uplands, and somebody will carry out the plan I have longed to put into practice. I have been thinking a great deal since the departure of my beloved one about the regions whither she has gone, and imagine from the manner the Bible describes it we have got too much monkery in our ideas. There will be work there as well as here, and possibly not such a vast difference in our being as is expected. But a short time there will give more insight than a thousand musings. We shall see Him by whose inexpressible love and mercy we get there, and all whom we loved, and all the lovable. I can sympathize with you now more fully than I did before. I work with as much vigor as I can, and mean to do so till the change comes; but ㅤprospect of a home is all dispelled."

In one of his despatches to Lord Russell, Livingstone

reports an offer that had been made by a party consisting
of an Englishman and five Scotch working men at the
Cape, which must have been extremely gratifying to him,
and served to deepen his conviction that sooner or later
his plan of colonization would certainly be carried into
effect. The leader of the party, John Jehan, formerly of
the London City Mission, in reading Dr. Livingstone's
book, became convinced that if a few mechanics could be
induced to take a journey of exploration it would prove
very useful. His views being communicated to five other
young men (two masons, two carpenters, one smith), they
formed themselves into a company in July, 1861, and had
been working together, throwing their earnings into a
common fund, and now they had arms, two wagons, two
spans of oxen, and means of procuring outfits. In September, 1862, they were ready to start from Aliwal in
South Africa.[1]

After going to Johanna for provisions, and to discharge
the crew of Johanna men whose term of service had
expired, the Expedition returned to Tette. On the 10th
January, 1863, they steamed off with the " Lady Nyassa "
in tow. The desolation that had been caused by Marianno,
the Portuguese slave-agent, was heart-breaking. Corpses
floated past them. In the morning the paddles had to be
cleared of corpses caught by the floats during the night.
Livingstone summed up his impressions in one terrible
sentence :

" Wherever we took a walk, human skeletons were seen
in every direction, and it was painfully interesting to

---

[1] The recall of Livingstone's Expedition and the removal of the Universities
Mission seem to have knocked this most promising scheme on the head.
Writing of it to Sir Roderick Murchison on the 14th December, 1862, he
says : " I like the Scotchmen, and think them much better adapted for our
plans than those on whom the Universities Mission has lighted. If employed
as I shall wish them to be in trade, and setting an example of industry in
cotton or coffee planting, I think they are just the men I need brought to my
hand. Don't you think this sensible ? "

observe the different postures in which the poor wretches had breathed their last. A whole heap had been thrown down a slope behind a village, where the fugitives often crossed the river from the east; and in one hut of the same village no fewer than twenty drums had been collected, probably the ferryman's fees. Many had ended their misery under shady trees, others under projecting crags in the hills, while others lay in their huts with closed doors, which when opened disclosed the mouldering corpse with the poor rags round the loins, the skull fallen off the pillow, the little skeleton of the child, that had perished first, rolled up in a mat between two large skeletons. The sight of this desert, but eighteen months ago a well-peopled valley, now literally strewn with human bones, forced the conviction upon us that the destruction of human life in the middle passage, however great, constitutes but a small portion of the waste, and made us feel that unless the slave-trade—that monster iniquity which has so long brooded over Africa—is put down, lawful commerce cannot be established."

In passing up, Livingstone's heart was saddened as he visited the Bishop's grave, and still more by the tidings which he got of the Mission, which had now removed from Magomero to the low lands of Chibisa. Some time before, Mr. Scudamore, a man greatly beloved, had succumbed, and now Mr. Dickenson was added to the number of victims. Mr. Thornton, too, who left the Expedition in 1859, but returned to it, died under an attack of fever, consequent on too violent exertion undertaken in order to be of service to the Mission party. Dr. Kirk and Mr. C. Livingstone were so much reduced by illness that it was deemed necessary for them to return to England. Livingstone himself had a most serious attack of fever, which lasted all the month of May, Dr. Kirk remaining with him till he got over it. When his brother and Dr. Kirk left, the only Europeans remaining with him were Mr.

Rae, the ship's engineer, and Mr. Edward D. Young, formerly of the "Gorgon," who had volunteered to join the Expedition, and whose after services, both in the search for Livingstone and in establishing the mission of Livingstonia, were so valuable. On the noble spirit shown by Livingstone in remaining in the country after all his early companions had left, and amid such appalling scenes as everywhere met him, we do not need to dwell.

Here are glimpses of the inner heart of Livingstone about this time:

"*1st March*, 1863.—I feel very often that I have not long to live, and say, ' My dear children, I leave you. Be manly Christians, and never do a mean thing. Be honest to men, and to the Almighty One.' "

"*10th April.*—Reached the Cataracts. Very thankful indeed after our three months' toil from Shupanga."

"*27th April.*—On this day twelvemonths my beloved Mary Moffat was removed from me by death.

"' If I can, I'll come again, mother, from out my resting-place;
Though you'll not see me, mother, I shall look upon your face;
Though I cannot speak a word, I shall hearken what you say,
And be often, often with you when you think I'm far away.'
"Tennyson."

The "Lady Nyassa" being taken to pieces, the party began to construct a road over the thirty-five or forty miles of the rapids, in order to convey the steamer to the lake. After a few miles of the road had been completed, it was thought desirable to ascertain whether the boat left near the lake two years before was fit for service, so as to avoid the necessity of carrying another boat past the rapids. On reaching it the boat was found to have been burnt. The party therefore returned to carry up another. They had got to the very last rapid, and had placed the boat for a short space in the water, when, through the carelessness of five Zambesi men, she was overturned, and away she went like an arrow down the rapids. To keep calm under such a crowning disappointment must have taxed Livingstone's self-control to the very utmost.

It was now that he received a despatch from Earl Russell intimating that the Expedition was recalled. This, though a great disappointment, was not altogether a surprise. On the 24th April he had written to Mr. Waller, "I should not wonder in the least to be recalled, for should the Portuguese persist in keeping the rivers shut, there would be no use in trying to develop trade." He states his views on the recall calmly in a letter to Mr. James Young:

"*Murchison Cataracts, 3d July,* 1863.— . . . Got instructions for our recall yesterday, at which I do not wonder. The Government has behaved well to us throughout, and I feel abundantly thankful to H.M.'s ministers for enabling me so far to carry on the experiment of turning the industrial and trading propensities of the natives to good account, with a view of thereby eradicating the trade in slaves. But the Portuguese dogged our footsteps, and, as is generally understood, with the approbation of their Home Government, neutralized our labors. Not that the Portuguese statesmen approved of slaving, but being enormously jealous lest their pretended dominion from sea to sea and elsewhere should in the least degree, now or any future time, become aught else than a slave 'preserve,' the Governors have been instructed, and have carried out their instructions further than their employers intended. Major Sicard was removed from Tette as too friendly, and his successor had emmissaries in the Ajawa camp. Well, we saw their policy, and regretted that they should be allowed to follow us into perfectly new regions. The regret was the more poignant, inasmuch as but for our entering in by gentleness, they durst not have gone. No Portuguese dared, for instance, to come up this Shiré Valley; but after our dispelling the fear of the natives by fair treatment, they came in calling themselves our 'children.' The whole thing culminated when this quarter was inundated with Tette slavers, whose operations, with a marauding tribe of Ajawas, and a drought, completely depopulated the country. The sight of this made me conclude that unless something could be done to prevent these raids, and take off their foolish obstructions on the rivers, which they never use, our work in this region was at an end. . . . Please the Supreme, I shall work some other point yet. In leaving, it is bitter to see some 900 miles of coast abandoned to those who were the first to begin the slave-trade, and seem determined to be the last to abandon it."

Writing to Mr. Waller at this time he said: "I don't know whether I am to go on the shelf or not. If I do, I

make Africa the shelf. If the "Lady Nyassa" is well sold.
I shall manage. There is a Ruler above, and his providence guides all things. He is our Friend, and has plenty
of work for all his people to do. Don't fear of being left
idle, if willing to work for Him. I am glad to her of
Alington. If the work is of God it will come out all right
at last. To Him shall be given of the gold of Sheba, and
daily shall He be praised. I always think it was such a
blessing and privilege to be led into his work instead of
into the service of the hard taskmasters—the Devil and
Sin."

The reason assigned by Earl Russell for the recall of the
Expedition were, that, not through any fault of Dr. Livingstone's, it had not accomplished the objects for which it
had been designed, and that it had proved much more
costly than was originally expected. Probably the Government felt likewise that their remonstrances with the
Portuguese Government were unavailing, and that their
relations were becoming too uncomfortable. Even among
those most friendly to Dr. Livingstone's great aim, and
most opposed to the slave-trade, and to the Portuguese
policy in Africa, there were some who doubted whether his
proposed methods of procedure were quite consistent with
the rights of the Portuguese Government. His Royal
Highness the Prince-Consort indicated some feeling of this
kind in his interview with Livingstone in 1857. He expressed the feeling more strongly when he declined the
request, made to him through Professor Sedgwick of Cambridge, that he would allow himself to be Patron of the Universities Mission. Dr. Livingstone knew well that from that
exalted quarter his plans would receive no active support.
That he should have obtained the support he did from
successive Governments and successive Foreign Secretaries,
Liberal and Conservative, was a great gratification, if not
something of a surprise. Hence the calmness with which
he received the intelligence of the recall. Toward the

Portuguese Government his feelings were not very sweet. On them lay the guilt of arresting a work that would have conferred untold blessing on Africa. He determined to make this known very clearly when he should return to England. At a future period of his life, he purposed, if spared, to go more fully into the reasons of his recall. Meanwhile, his course was simply to acquiesce in the resolution of the British Government.

It was unfortunate that the recall took place before he had been able to carry into effect his favorite scheme of placing a steamer on Lake Nyassa; nor could he do this now, although the vessel on which he had spent half his fortune lay at the Murchison Cataracts. He had always cherished the hope that the Government would repay him at least a part of the outlay, which, instead of £3000, as he had intended, had mounted up to £6000. He had very generously told Dr. Stewart that if this should be done, and if he should be willing to return from Scotland to labor on the shores of Nyassa, he would pay him his expenses out, and £150 yearly, so anxious was he that he should begin the work. On the recall of the Expedition, without any allowance for the ship, or even mention of it, all these expectations and intentions came abruptly to an end.

At no previous time had Dr. Livingstone been under greater discouragements than now. The Expedition had been recalled; his heart had not recovered from the desolation caused by the death of the Bishop and his brethren, as well as the Helmores in the Makololo country, and still more by the removal of Mrs. Livingstone, and the thought of his motherless children; the most heart-rending scenes had been witnessed everywhere in regions that a short time ago had been so bright; all his efforts to do good had been turned to evil, every new path he had opened having been seized as it were by the devil and turned to the most diabolical ends; his countrymen were

nearly all away from him; the most depressing of diseases
had produced its natural effect; he had had worries, delays, and disappointments about ships and boats of the
most harrassing kind; and now the "Lady Nyassa" could
not be floated in the waters of which he had fondly hoped
to see her the angel and the queen. It is hardly possible
to exaggerate the noble quality of the heart that, unde-
terred by all these troubles, resolved to take this last
chance of exploring the banks of Nyassa, although it
could only be by the weary process of trudge, trudge,
trudging; although hunger, if not starvation, blocked the
path, and fever and dysentery flitted around it like imps
of darkness; although tribes, demoralized by the slave-
trade, might at any moment put an end to him and his
enterprise;—not to speak of the ordinary risks of travel,
the difficulty of finding guides, the liability to bodily hurt,
the scarcity of food, the perils from wild beasts by night
and by day,—risks which no ordinary traveler could think
of lightly, but which in Livingstone's journeys drop out
of sight, because they are so overtopped and dwarfed by
risks that ordinary travelers never know.

Why did not Livingstone go home? A single sentence
in a letter to Mr. Waller, while the recall was only in con-
templation, explains: "In my case, duty would not lead
me home, and home therefore I would not go." Away
then goes Livingstone, accompanied by the steward of the
"Pioneer" and a handful of native servants (Mr. Young
being left in charge of the vessel), to get to the northern
end of the lake, and ascertain whether any large river
flowed into it from the west, and if possible to visit Lake
Moero, of which he had heard, lying a considerable way
to the west. For the first time in his travels he carried
some bottles of wine,—a present from the missionaries
Waller and Alington; for water had hitherto been his only
drink, with a little hot coffee in the mornings to warm the
stomach and ward off the feeling of sinking. At one time

the two white men are lost three days in the woods, without food or the means of purchasing it; but some poor natives out of their poverty show them kindness. At another they can procure no guides, though the country is difficult and the way intersected by deep gullies that can only be scaled at certain known parts; anon they are taken for slave-dealers, and make a narrow escape of a night attack. Another time, the cries of children remind Livingstone of his own home and family, where the very same tones of sorrow had often been heard; the thought brought its own pang, only he could feel thankful that in the case of his children the woes of the slave-trade would never be added to the ordinary sorrows of childhood. Then he would enjoy the joyous laugh of some Manganja women, and think of the good influence of a merry heart, and remember that whenever he had observed a chief with a joyous twinkle of the eye accompanying his laugh, he had always set him down as a good fellow, and had never been disappointed in him afterward. Then he would cheer his monotony by making some researches into the origin of civilization, coming to the clear conclusion that born savages must die out, because they could devise no means of living through disease. By and by he would examine the Arab character, and find Mahometanism as it now is in Africa worse than African heathenism, and remark on the callousness of the Mahometans to the welfare of one another, and on the especial glory of Christianity, the only religion that seeks to propagate itself, and through the influence of love share its blessings with others. Anon he would dwell on the primitive African faith; its recognition of one Almighty Creator, its moral code, so like our own, save in the one article of polygamy; its pious recognition of a future life, though the element of punishment is not very conspicuous; its mild character generally, notwithstanding the bloodthirstiness sometimes ascribed to it, which, however, Livingstone held to be, at Dahomey for example, purely exceptional.

Another subject that occupied him was the natural history of the country. He would account for desert tracts like Kalahari by the fact that the east and south-east winds, laden with moisture from the Indian Ocean, get cooled over the coast ranges of mountains, and having discharged their vapor there had no spare moisture to deposit over the regions that for want of it became deserts. The geology of Southern Africa was peculiar; the geographical series described in books was not to be found here, for, as Sir Roderick Murchison had shown, the great submarine depressions and elevations that had so greatly affected the other continents during the secondary, tertiary, and more recent periods, had not affected Africa. It had preserved its terrestrial conditions during a long period, unaffected by any changes save those dependent on atmospheric influences. There was also a peculiarity in pre-historic Africa—it had no stone period; at least no flint weapons had been found, and the familiarity and skill of the natives with the manufacture of iron seemed to indicate that they had used iron weapons from the first.

The travelers had got as far as the river Loangwa (of Nyassa), when a halt had to be called. Some of the natives had been ill, and indeed one had died in the comparatively cold climate of the highlands. But nothing would have hindered Livingstone from working his way round the head of the lake if only time had been on his side. But time was inexorably against him; the orders from Government were strict. He must get the "Pioneer" down to the sea while the river was in flood. A month or six weeks would have enabled him to finish his researches, but he could not run the risk. It would have been otherwise had he foreseen that when he got to the ship he would be detained two months waiting for the rising of the river. On their way back, they took a nearer cut, but found the villages all deserted. The reeds along the banks of the lake were crowded with fugitives. "In passing mile after mile, marked with the sad proofs that 'man's inhumanity to man makes countless thousands mourn,' one experiences an overpowering sense of helplessness to alleviate human woe, and breathes a silent prayer to the Almighty to hasten the good time coming when 'man to man, the world o'er, shall brothers be for all that.'" Near a village called Bangwé they were pursued by a body of Mazitu, who retired when they came within ear-shot. This little adventure seemed to give rise to the report that Dr. Livingstone had been murdered by the Makololo, which reached England, and created no small alarm. Referring to the report in his jocular way, in a letter to his friend Mr. Fitch, he says, " A report of my having been murdered at the lake has been very industriously circulated by the Portuguese. Don't become so pale on getting a letter from a dead man."

Reaching the stockade of Chinsamba in Mosapo, they were much pleased with that chief's kindness. Dr. Livingstone followed his usual method, and gained his usual influence. " When a chief has made any inquiries of us, we have found that we gave most satisfaction in our answers when we tried to fancy ourselves in the position of the interrogator, and him that of a poor uneducated fellow-countryman in England. The polite, respectful way of speaking, and behavior of what we call 'a thorough gentleman,' almost always secures the friendship and good-will of the Africans."

On 1st November, 1863, the party reached the ship, and found all well. Here, as has been said, two months had to be spent waiting for the flood, to Dr. Livingstone's intense chagrin.

While waiting here he received a letter from Bishop Tozer, the successor of Bishop Mackenzie, informing him that he had resolved to abandon the Mission on the continent and transfer operations to Zanzibar. Dr. Livingstone had very sincerely welcomed the new Bishop, and

29

at first liked him, and thought that his caution would lead to good results. Indeed, when he saw that his own scheme was destroyed by the Portuguese, he had great hopes that what he had been defeated in, the Mission would accomplish. Some time before, his hopes had begun to wane, and now the news conveyed in Bishop Tozer's letter was their death-blow. In his reply he implored the Bishop to reconsider the matter. After urging strongly some considerations bearing on the duty of missionaries, the reputation of Englishmen, and the impression likely to be made on the native mind, he concluded thus: " I hope, dear Bishop, you will not deem me guilty of impertinence in thus writing to you with a sore heart. I see that if you go, the last ray of hope for this wretched, trodden-down people disappears, and I again from the bottom of my heart entreat you to reconsider the matter, and may the All-wise One guide to that decision which will be most for his glory."

And thus, for Livingstone's life-time, ended the Universities Mission to Central Africa, with all the hopes which its bright dawn had inspired, that the great Church of England would bend its strength against the curse of Africa, and sweep it from the face of the earth. Writing to Sir Thomas Maclear, he said that he felt this much more than his own recall. He could hardly write of it; he was more inclined " to sit down and cry." No mission had ever had such bright prospects; notwithstanding all that had been said against it, he stood by the climate as firmly as ever, and if he were only young, he would go himself and plant the gospel there. It would be done one day without fail, though he might not live to see it.

As usual, Livingstone found himself blamed for the removal of the Mission. The Makololo had behaved badly, and they were Livingstone's people. " Isn't it interesting," he writes to Mr. Moore, " to get blamed for everything? But I must be thankful in feeling that I would rather perish than blame another for my misdeeds and deficiencies."

We have lost sight of Dr. Stewart and the projected mission of the Free Church of Scotland. As Dr. Livingstone's arrangements did not admit of his accompanying Dr. Stewart up the Shiré, he set out alone, falling in afterward with the Rev. Mr. Scudamore, a member, and as we have already said ultimately a martyr, of the Universities Mission. The report which Dr. Stewart made of the prospects of a mission was that, owing to the disturbed state of the country, no immediate action could be taken. Livingstone seemed to think him hasty in this conclusion. The scheme continued to be ardently cherished, and some ten or twelve years after—in 1874—in the formation of the "Livingstonia" mission and colony, a most promising and practical step was taken toward the fulfillment of Dr. Livingstone's views. Dr. Stewart has proved one of the best friends and noblest workers for African regeneration both at Lovedale and Livingstonia—a strong man on whom other men may lean, with his whole heart in the cause of Africa.

In the breaking up of the Universities Mission, it was necessary that some arrangement should be made on behalf of about thirty boys and a few helpless old persons and others, a portion of the rescued slaves, who had been taken under the charge of the Mission, and could not be abandoned. The fear of the Portuguese seemed likely to lead to their being left behind. But Livingstone could not bear the idea. He thought it would be highly discreditable to the good name of England, and an affront to the memory of Bishop Mackenzie, to "repudiate" his act in taking them under his protection. Therefore, when Bishop Tozer would not accept the charge, he himself took them in hand, giving orders to Mr. E. D. Young (as he says in his Journal), "in the event of any Portuguese interfering with them in his absence, to pitch him overboard!" Through his influence arrangements were made, as we shall see, for conveying them to the Cape. Mr. R. M. Ballantyne, in his *Six Months at the Cape*, tells us that he found, some years afterward, among the most efficient teachers in St. George's Orphanage, Cape Town, one of these black girls, named Dauma, whom Bishop Mackenzie had personally rescued and carried on his shoulders, and whom Livingstone now rescued a second time.

Livingstone's plan for himself was to sail to Bombay in the "Lady Nyassa," and endeavor to sell her there, before returning home. The Portuguese would have liked to get her, to employ her as a slaver—"But," he wrote to his daughter (10th August, 1863), "I would rather see her go down to the depths of the Indian Ocean than that. We have not been able to do all that we intended for this country, owing to the jealousy and slave-hunting of the Portuguese. They have hindered us effectually by sweeping away the population into slavery. Thousands have perished, and wherever we go human skeletons appear. I suppose that our Government could not prevail on the Portuguese to put a stop to this; so we are recalled. I am only sorry that we ever began near these slavers, but the great men of Portugal professed so loudly their eager desire to help us (and in the case of the late King I think there was sincerity), that I believed them, and now find out that it was all for show in Europe. . . . If missions were established as we hoped, I should still hope for good being done to this land, but the new Bishop had to pay fourpence for every pound weight of calico he bought, and calico is as much currency here as money is in Glasgow. It looks as if they wished to prohibit any one else coming, and, unfortunately, Bishop Tozer, a good man enough, lacks courage. . . . What a mission it would be if there were no difficulties—nothing but walking about in slippers made by admiring young ladies! Hey! that would not suit me. It would give me the doldrums; but there are many tastes in the world."

Looking back on the work of the last six years, while deeply grieved that the great object of the Expedition had not been achieved, Dr. Livingstone was able to point to some important results:

1. The discovery of the Kongone harbor, and the ascertaining of the condition of the Zambesi River, and its fitness for navigation.

2. The ascertaining of the capacity of the soil. It was found to be admirably adapted for indigo and cotton, as well as tobacco, castor-oil, and sugar. Its great fertility was shown by its gigantic grasses, and abundant crops of corn and maize. The highlands were free from tsetse and mosquitoes. The drawback to all this was the occurrence of periodical droughts, once every few years.

But every fine feature of the country was bathed in gloom by the slave-trade. The image left in Dr. Livingstone's mind was not that of the rich, sunny, luxuriant country, but that of the woe and wretchedness of the people. The real service of the Expedition was, that it had exposed slavery at its fountain-head, and in all its phases. First, there was the internal slave-trade between hostile native tribes. Then, there were the slave-traders from the coast, Arabs, or half-caste Portuguese, for whom natives were encouraged to collect slaves by all the horrible means of marauding and murder. And further, there were the parties sent out from Portuguese and Arab coast towns, with cloth and beads, muskets and ammunition. The destructive and murderous effects of the last were the climax of the system.

Dr. Livingstone had seen nothing to make him regard the African as of a different species from the rest of the human family. Nor was he the lowest of the species. He had a strong frame and a wonderfully persistent vitality, was free from many European diseases, and could withstand privations with wonderful light-heartedness.

He did not deem it necessary formally to answer a

question sometimes put, whether the African had enough of intellect to receive Christianity. The reception of Christianity did not depend on intellect. It depended, as Sir James Stephen had remarked, on a spiritual intuition, which was not the fruit of intellectual culture. But, in fact, the success of missions on the West Coast showed that not only could the African be converted to Christianity, but that Christianity might take root and be cordially supported by the African race.

It was the accursed slave-trade, promoted by the Portuguese, that had frustrated everything. For some time to come his efforts and his prayers must be directed to getting influential men to see to this, so that one way or other the trade might be abolished forever. The hope of obtaining access to the heart of Africa by another route than that through the Portuguese settlements was still in Livingstone's heart. He would go home, but only for a few months; at the earliest possible moment he would return to look for a new route to the interior.

CHAPTER XVI.

QUILIMANE TO BOMBAY AND ENGLAND.

A.D. 1864.

Livingstone returns the "Pioneer" to the Navy, and is to sail in the "Nyassa" to Bombay—Terrific circular storm—Imminent peril of the "Nyassa"—He reaches Mozambique—Letter to his daughter—Proceeds to Zanzibar—His engineer leaves him—Scanty crew of "Nyassa"—Livingstone captain and engineer—Peril of the voyage of 2500 miles—Risk of the monsoons—The "Nyassa" becalmed—Illness of the men—Remarks on African travel— Flying-fish—Dolphins—Curiosities of his Journal—Idea of a colony—Furious squall—Two sea-serpents seen—More squalls—The "Nyassa" enters Bombay harbor—Is unnoticed—First visit from officers with Custom-house schedules—How filled up—Attention of Sir Bartle Frere and others—Livingstone goes with the Governor to Dapuri—His feelings on landing in India— Letter to Sir Thomas Maclear—He visits mission-schools, etc., at Poonah— Slaving in Persian Gulf—Returns to Bombay—Leaves two boys with Dr. Wilson—Borrows passage-money and sails for England—At Aden—At Alexandria—Reaches Charing Cross—Encouragement derived from his Bombay visit—Two projects contemplated on his way home.

On reaching the mouth of the Zambesi, Dr. Livingstone was fortunate in falling in, on the 13th February, with H.M.S. "Orestes," which was joined on the 14th by the "Ariel." The "Orestes" took the "Pioneer" in tow, and the "Ariel" the "Lady Nyassa," and brought them to Mozambique. The day after they set out, a circular storm passed over them, raging with the utmost fury, and creating the greatest danger. Often as Dr. Livingstone had been near the gates of death, he was never nearer than now. He had been offered a passage on board the "Ariel," but while there was danger he would not leave the "Lady Nyassa." Had the latter not been an excellent sea-ship she could not have survived the tempest; all the greater was Dr. Livingstone's grief that she ha l never reached the lake for which she was adapted so wel l.

Writing to his daughter Agnes from Mozambique, he gives a very graphic account of the storm, after telling her the manner of their leaving the Zambesi:

"*Mozambique, 24th Feb.,* 1864.—When our patience had been well nigh exhausted the river rose and we steamed gladly down the Shiré on the 19th of last month. An accident detained us some time, but on the 1st February we were close by Morumbala, where the Bishop [Tozer] passed a short time before bolting out of the country. I took two members of the Mission away in the 'Pioneer,' and thirteen women and children, whom having liberated we did not like to leave to become the certain prey of slavers again. The Bishop left twenty-five boys, too, and these also I took with me, hoping to get them conveyed to the Cape, where I trust they may become acquainted with our holy religion. We had thus quite a swarm on board, all very glad to get away from a land of slaves. There were many more liberated, but we took only the helpless and those very anxious to be free and with English people. Those who could cultivate the soil we encouraged to do so, and left up the river. Only one boy was unwilling to go, and he was taken by the Bishop. It is a great pity that the Bishop withdrew the Mission, for he had a noble chance of doing great things. The captives would have formed a fine school, and as they had no parents he could have educated them as he liked.

"When we reached the sea-coast at Luabo we met a man-of-war, H.M.S. 'Orestes.' I went to her with 'Pioneer,' and sent 'Lady Nyassa' round by inland canal to Kongone. Next day I went into Kongone in 'Pioneer'; took our things out of her, and handed her over to the officers of the 'Orestes.' Then H.M.S 'Ariel' came and took 'Nyassa' in tow, 'Orestes' having 'Pioneer.' Captain Chapman of 'Ariel' very kindly invited me on board to save me from the knocking about of the "Lady Nyassa,' but I did not like to leave so long as there was any danger, and accepted his invitation for Mr. Waller, who was dreadfully sea-sick. On 15th we were caught by a hurricane which whirled the 'Ariel' right round. Her sails, quickly put to rights, were again backed so that the ship was driven backward and a hawser wound itself round her screw, so as to stop the engines. By this time she was turned so as to be looking right across 'Lady Nyassa,' and the wind alone propelling her as if to go over the little vessel. I saw no hope of escape except by catching a rope's-end of the big ship as she passed over us, but by God's goodness she glided past, and we felt free to breathe. That night it blew a furious gale. The captain

offered to lower a boat if I would come to the 'Ariel,' but it would
have endangered all in the boat: the waves dashed so hard against
the sides of the vessel, it might have been swamped, and my going
away would have taken heart out of those that remained. We then
passed a terrible night, but the 'Lady Nyassa' did wonderfully well,
rising like a little duck over the foaming billows. She took in spray
alone, and no green water. The man-of-war's people expected that
she would go down, and it was wonderful to see how well she did
when the big man-of-war, only about 200 feet off, plunged so as to
show a large portion of copper on her bottom, then down behind
so as to have the sea level with the top of her bulwarks. A boat
hung at that level was smashed. If we had gone down we could not
have been helped in the least—pitch dark, and wind whistling above;
the black folks, 'ane bocking here, anither there,' and wanting us to
go to the 'bank.' On 18th the weather moderated, and, the captain
repeating his very kind offer, I went on board with a good conscience,
and even then the boat got damaged. I was hoisted up in it, and got
rested in what was quite a steady ship as compared with the 'Lady
Nyassa.' The 'Ariel' was three days cutting off the hawser, though
nine feet under water, the men diving and cutting it with immensely
long chisels. On the 19th we spoke to a Liverpool ship, requesting
the captain to report me alive, a silly report having been circulated by
the Portuguese that I had been killed at Lake Nyassa, and on the
24th we entered Mozambique harbor, very thankful for our kind and
merciful preservation. The 'Orestes' has not arrived with the 'Pioneer,'
though she is a much more powerful vessel than the 'Ariel.' Here
we have a fort, built in 1500, and said to be of stones brought from
Lisbon. It is a square massive-looking structure. The town adjacent
is Arab in appearance. The houses flat-roofed and colored white, pink,
and yellow; streets narrow, with plenty of slaves on them. It is on
an island, the mainland on the north being about a mile off."

The "Pioneer" was delivered over to the Navy, being
Her Majesty's property, and proceeded to the Cape with
the "Valorous," Mr. Waller being on board with a portion
of the mission flock. Of Mr. Waller (subsequently editor
of the *Last Journals*) Dr. Livingstone remarked that "he
continued his generous services to all connected with the
Mission, whether white or black, till they were no longer
needed; his conduct to them throughout was truly noble,
and worthy of the highest praise."

After remaining some weeks at Mozambique for thorough

repairs, the " Lady Nyassa" left on 16th April for Johanna and Zanzibar. She was unable to touch at the former place, and reached Zanzibar on the 24th. Offers were made for her there, which might have led to her being sold, but her owner did not think them sufficient, and in point of fact, he could not make up his mind to part with her. He clung to the hope that she might yet be useful, and to sell her seemed equivalent to abandon all hope of carrying out his philanthropic schemes. At all events, till he should consult Mr. Young he would not sell her at such a sacrifice. At Zanzibar he found that a naval gentleman, who had been lately there, had not spoken of him in the most complimentary terms. But it had not hurt him with his best friends. " Indeed, I find that evil-speaking against me has, by the good providence of my God, turned rather to my benefit. I got two of my best friends by being spoken ill of, for they found me so different from what they had been led to expect that they befriended me more than they otherwise would have done. It is the good hand of Him who has all in his power that influences other hearts to show me kindness."

The only available plan now was to cross the Indian Ocean for Bombay, or possibly Aden, in the " Nyassa" and leave the ship there till he should make a run home, consult with his friends as to the future, and find means for the prosecution of his work. At Zanzibar a new difficulty arose. Mr. Rae, the engineer, who had now been with him for many years, and with whom, despite his peculiarities, he got on very well, signified his intention of leaving him. He had the offer of a good situation, and wished to accept of it. He was not without compunctions at leaving his friend in the lurch, and told Livingstone that if he had had no offer for the ship he would have gone with him, but as he had declined the offer made to him, he did not feel under obligation to do so. Livingstone was too generous to press him to remain. It was impossible to supply Mr. Rae's place, and if anything should go wrong with the engines, what was to be done? The entire crew of the vessel consisted of four Europeans, namely, Dr. Livingstone—"skipper," one stoker, one carpenter, and one sailor; seven native Zambesians, who, till they volunteered, had never seen the sea, and two boys, one of whom was Chuma, afterward his attendant on the last journey. With this somewhat sorry complement, and fourteen tons of coal, Dr. Livingstone set out on 30th April, on a voyage of 2500 miles, over an ocean which he had never crossed.

It was a very perilous enterprise, for he was informed that the breaking of the monsoon occurred at the end of May or the beginning of June. This, as he came to think, was too early; but in any case, he would come very near the dangerous time. As he wrote to one of his friends, he felt jammed into a corner, and what could he do? He believed from the best information he could get that he would reach Bombay in eighteen days. Had any one told him that he would be forty-five days at sea, and that for twenty-five of these his ship would be becalmed, and even when she had a favorable wind would not sail fast, even he would have looked pale at the thought of what was before him. The voyage was certainly a memorable one, and has only escaped fame by the still greater wonders performed by Livingstone on land.

On the first day of the voyage, he made considerable way, but Collyer, one of his white men, was prostrated by a bilious attack. However, one of the black men speedily learned to steer, and took Dr. Livingstone's place at the wheel. Hardly was Collyer better when Pennell, another of his men, was seized. The chief foes of the ship were currents and calms. Owing to the illness of the men they could not steam, and the sails were almost useless. Even steam, when they got it up, enabled them only to creep. On 20th May, Livingstone, after recording but sixteen knots in the last twenty-four hours, says in his Journal:

"This very unusual weather has a very depressing influence on my mind. I often feel as if I am to die on this
voyage, and wish I had sent the accounts to the Government, as also my chart to the Zambesi. I often wish that
I may be permitted to do something for the benighted of
Africa. I shall have nothing to do at home; by the failure
of the Universities Mission my work seems vain. No fruit
likely to come from J. Moffat's mission either. Have I not
labored in vain? Am I to be cut off before I do anything
to effect permanent improvement in Africa? I have been
unprofitable enough, but may do something yet, in giving
information. If spared, God grant that I may be more
faithful than I have been, and may He open up the way
for me!"

Next day the weather was as still as ever; the sea a
glassy calm, with a hot glaring sun, and sharks stalking
about. "All ill-natured," says honest Livingstone, "and
in this I am sorry to feel compelled to join."

There is no sign of ill-nature, however, in the following remarks on African travel, in his Journal for 23d
May:

"In traveling in Africa, with the specific object in view of ameliora-
ting the benighted condition of the country, every act is ennobled. In
obtaining shelter for the night, and exchanging the customary civilities,
purchasing food for one's party and asking the news of the country, and
answering in their own polite way any inquiries made respecting the
object of the journey, we begin to spread information respecting that
people by whose agency their land will yet be made free from the evils
that now oppress it. The mere animal pleasure of traveling is very
great. The elastic muscles have been exercised. Fresh and healthy
blood circulates in the veins, the eye is clear, the step firm, but the day's
exertion has been enough to make rest thoroughly enjoyable. There is
always the influence of the remote chances of danger on the mind,
either from men or wild beasts, and there is the fellow-feeling drawn out
to one's humble, hardy companions, with whom a community of interests
and perils renders one friends indeed. The effect of travel on my mind
has been to make it more self-reliant, confident of resources and presence
of mind. On the body the limbs become well-knit, the muscles after

six months' tramping are as hard as a board, the countenance bronzed
as was Adam's, and no dyspepsia.

"In remaining at any spot, it is to work. The sweat of the brow is
no longer a curse when one works for God; it is converted into a blessing. It is a tonic to the system. The charms of repose cannot be
known without the excitement of exertion. Most travelers seem taken
up with the difficulties of the way, the pleasures of roaming free in the
most picturesque localities seem forgotten."

Toward the end of May a breeze at last springs up;
many flying-fish come on board, and Livingstone is as
usual intent on observation. He observes them fly with
great ease a hundred yards, the dolphin pursuing them
swiftly, but not so swiftly as they can fly. He notices that
the dolphin's bright colors afford a warning to his enemies,
and give them a chance of escape. Incessant activity is a
law in obtaining food. If the prey could be caught with
ease, and no warning were given, the balance would be
turned against the feebler animals, and carnivora alone
would prevail. The cat shows her shortened tail, and the
rattlesnake shakes his tail, to give warning to the prey.
The flying-fish has large eyes in proportion to other fish,
yet leaps on board very often at night, and kills himself
by the concussion.

Livingstone is in great perplexity what to do. At the
rate at which his ship is going it would take him fifteen
days to reach Bombay, being one day before the breaking
of the monsoon, which would be running it too close to
danger. He thinks of going to Aden, but that would
require him to go first to Maculla for water and provisions.
When he tries Aden the wind is against him; so he turns
the ship's head to Bombay, though he has water enough
for but ten or twelve days on short allowance. "May the
Almighty be gracious to us all, and help us!"

His Journal is a curious combination of nautical observations and reflections on Africa and his work. We
seem to hear him pacing his little deck, and thinking
aloud:

"The idea of a colony in Africa, as the term colony is usually under-stood, cannot be entertained. English races cannot compete in manual labor of any kind with the natives, but they can take a leading part in managing the land, improving the quality, in creating the quantity and extending the varieties of the productions of the soil; and by taking a lead, too, in trade, and in all public matters, the Englishman would be an unmixed advantage to every one below and around him, for he would fill a place which is now practically vacant.

"It is difficult to convey an idea of the country; it is so different from all preconceived notions. The country in many parts rises up to plateaus, slopes up to which are diversified by valleys lined with trees; or here and there rocky bluffs jut out; the plateaus themselves are open prairies covered with grass dotted over with trees, and watered by numerous streams. Nor are they absolutely flat, their surface is varied by picturesque undulations. Deep gorges and ravines leading down to the lower levels offer special beauties, and landscapes from the edges of the higher plateaus are in their way unequaled. Thence the winding of the Shiré may be followed like a silver thread or broad lake with its dark mountain mass behind.

"I think that the Oxford and Cambridge missionaries have treated me badly in trying to make me the scapegoat of their own blunders and inefficiency. . . . But I shall try equitably and gently to make allowances for human weakness, though that weakness has caused me much suffering."

On 28th May they had something like a foretaste of the breaking of the monsoon, though happily that event did not yet take place. "At noon a dense cloud came down on us from E. and N.E., and blew a furious gale; tore sails; the ship, as is her wont, rolled broadside into it, and nearly rolled quite over. Everything was hurled hither and thither. It lasted half an hour, then passed with a little rain. It was terrible while it lasted. We had calm after it, and sky brightened up. Thank God for his goodness."

In June there was more wind, but a peculiarity in the construction of the ship impeded her progress through the water. It was still very tedious and trying. Livingstone seems to have been reading books that would take his attention off the very trying weather.

"Lord Ravensworth has been trying for twenty years to render the lines in Horace—

' Dulce ridentem Lalagen amabo
Dulce loquentem.'

And after every conceivable variety of form this is the
best:

'The softly speaking Lalage,
The softly smiling still for me.'

Pity he had nothing better to engage his powers, for
instance the translating of the Bible into one of the
languages of the world."

The 10th of June was introduced by a furious squall
which tore the fore square-sail to ribbons. A curious
sight is seen at sea: "two serpents—said to be often seen
on the coast. One dark olive, with light yellow rings
round it, and flattened tail; the other lighter in color.
They seem to be salt-water animals."

Next day, a wet scowling morning. Frequent rains,
and thunder in the distance. "A poor weak creature.
Permit me to lean on an all-powerful arm."

"The squalls usually come up right against the wind,
and cast all our sails aback. This makes them so dangerous, active men are required to trim them to the other
side. We sighted land a little before 12, the high land
of Rutnagerry. I thought of going in, but finding that
we have twenty-eight hours' steam, I changed my mind,
and pushed on for Bombay, 115 miles distant. We are
nearer the land down here than we like, but our N.W.
wind has prevented us from making northing. We hope
for a little change, and possibly may get in nicely. The
good Lord of all help us!

"At 3 P.M. wind and sea high; very hazy. Raining,
with a strong head wind; at 8 P.M. a heavy squall came
off the land on our east. Wind whistled through the
rigging loudly, and we made but little progress steaming.
At 11 P.M. a nice breeze sprang up from east and helped
us. About 12 a white patch reported seemed a shoal, but

none is marked on the chart. Steered a point more out from land; another white patch marked in middle watch. Sea and wind lower at 3 A.M. At daylight we found ourselves abreast high land at least 500 feet above sea-level. Wind light, and from east, which enables us to use fore and aft try-sails. A groundswell on, but we are getting along, and feel very thankful to Him who has favored us. Hills not so beautifully colored as those in Africa. . . .

"At 7 P.M. a furious squall came off the land; could scarcely keep the bonnets on our heads. Pitchy dark, except the white curl on the waves, which was phosphorescent. Seeing that we could not enter the harbor, though we had been near, I stopped the steaming and got up the try-sails, and let Pennell, who has been up thirty hours, get a sleep.

"13*th June*, 1864.—We found that we had come north only about ten miles. We had calms after the squall, and this morning the sea is as smooth as glass, and a thick haze over the land. A scum as of dust on face of water. We are, as near as I can guess by the chart, about twenty-five miles from the port of Bombay. Came to Choul Rock at mid-day, and, latitude agreeing thereto, pushed on N. by w. till we came to light-ship. It was so hazy inland we could see nothing whatever, then took the direction by chart, and steered right into Bombay most thankfully. I mention God's good providence over me, and beg that He may accept my spared life for his service."

Between the fog and the small size of the Nyassa, her entrance into the harbor was not observed. Among Livingstone's first acts on anchoring was to give handsome gratuities to those who had shared his danger and helped him in his straits. Going ashore, he called on the Governor and the police magistrate, but the one was absent and the other busy, and so he returned to the ship unrecognized. The schedules of the custom-house sent to be filled up were his first recognition by the authorities of Bombay. He replied that except a few bales of calico and a box of beads he had no merchandise; he was consigned to no one; the seamen had only their clothes, and he did not know a single soul in Bombay. As soon as his arrival was known every attention was showered on him by Sir Bartle Frere, the Governor, and others. They had been looking out for him, but he had eluded their notice. The Governor was residing at Dapuri, and on his invitation Livingstone went there. Stopping at Poona, he called on the missionaries, and riding on an elephant he saw some of the "lions" of the place. Colonel Stewart, who accompanied him, threw some light on the sea-serpent. "He told us that the yellow sea-serpent which we had seen before reaching Bombay is poisonous; there are two kinds —one dark olive, the other pale lemon color; both have rings of brighter yellow on their tails."

Landing in India was a strange experience, as he tells Sir Thomas Maclear. "To walk among the teeming thousands of all classes of population, and see so many things that reading and pictures had made familiar to the mind, was very interesting. The herds of the buffaloes, kept I believe for their milk, invariably made the question glance across the mind, 'Where's your rifle?' Nor could I look at the elephants either without something of the same feeling. Hundreds of bales of cotton were lying on the wharves."

"20*th June,* 1864.—Went with Captain Leith to Poona to visit the Free Church Mission Schools there, under the Rev. Mr. Mitchell, Gardner, etc. A very fine school of 500 boys and young men answered questions very well. . . . All collected together, and a few ladies and gentlemen for whom I answered questions about Africa. We then went to a girls' school; the girls sang very nicely, then acted a little play. There were different castes in all the schools, and quite mixed. After this we went to College, where young men are preparing for degrees of the University under Dr. Haug and Mr. Wordsworth; then to the Roman Catholic Orphanage, where 200 girls are assembled, clothed, and fed under a French Lady Superior—dormitory clean and well aired, but many had scrofulous-looking sore eyes; then home to see some friends whom Lady Frere had invited, to save me the trouble of calling on them. Saw Mr. Cowan's daughter."

"21*st June*, 1864.— . . . Had a conversation with the Governor after breakfast about the slaving going on toward the Persian Gulf. His idea is that they are now only beginning to put a stop to slavery—they did not know of it previously. . . . The merchants of Bombay have got the whole of the trade of East Africa thrown on their hands, and would, it is thought, engage in an effort to establish commerce on the coast. The present Sultan is, for an Arab, likely to do a good deal. He asked if I would undertake to be consul at a settlement, but I think I have not experience enough for a position of that kind among Europeans."

On returning to Bombay, he saw the missionary institutions of the Scotch Established and Free Churches, and arranged with Dr. Wilson of the latter mission to take his two boys, Chuma and Wikatani. He arranged also that the "Lady Nyassa," which he had not yet sold, should be taken care of, and borrowing £133, 10s. for the passage-money of himself and John Reid, one of his men, embarked for old England.

At Aden considerable rain had fallen lately; he observed that there was much more vegetation than when he was there before, and it occurred to him that at the time of the Exodus the same effects probably followed the storms of rain, lightning, and hail in Egypt. Egypt was very far from green, so that Dr. Stanley must have visited it at another part of the year. At Alexandria, when he went on board the "Ripon," he found the Maharaja Dhuleep

Singh and his young Princess—the girl he had fancied and married from an English Egyptian school. Paris is reached on the 21st July; a day is spent in resting; and on the evening of the 23d he reaches Charing Cross, and is regaled with what, after nearly eight years' absence, must have been true music—the roar of the mighty Babylon.

The desponding views of his work which we find in such entries in his Journal as that of 20th May must not be held to express his deliberate mind. It must not be thought that he had thrown aside the motto which had helped him as much as it had helped his royal countryman, Robert Bruce—"Try again." He had still some arrows in his quiver. And his short visit to Bombay was a source of considerable encouragement. The merchants there, who had the East African trade in their hands, encouraged him to hope that a settlement for honest traffic might be established to the north of the region over which the Portuguese claimed authority. As Livingstone moved homeward he was revolving two projects. The first was to expose the atrocious slave-trading of the Portuguese, which had not only made all his labor fruitless, but had used his very discoveries as channels for spreading fresh misery over Africa. The thought warmed his blood, and he felt like a Highlander with his hand on his claymore. The second project was to find means for a new settlement at the head of the Rovuma, or somewhere else beyond the Portuguese lines, which he would return in the end of the year to establish. Writing a short book might help to accomplish both these projects. As yet, the idea of finding the sources of the Nile was not in his mind. It was at the earnest request of others that he undertook the work that cost him so many years of suffering, and at last his life.

CHAPTER XVII.

SECOND VISIT HOME.

A.D. 1864-65.

Dr. Livingstone and Sir R. Murchison—At Lady Palmerston's reception—at other places in London—Sad news of his son Robert—His early death—Dr. Livingstone goes to Scotland—Pays visits—Consultation with Professor Syme as to operation—Visit to Duke of Argyll—to Ulva—He meets Dr. Duff—At launch of a Turkish frigate—At Hamilton—Goes to Bath to British Association—Delivers an address—Dr. Colenso—At funeral of Captain Speke—Bath speech offends the Portuguese—Charges of Lacerda—He visits Mr. and Mrs. Webb at Newstead—Their great hospitality—The Livingstone room—He spends eight months there writing his book—He regains elasticity and playfulness—His book—Charles Livingstone's share—He uses his influence for Dr. Kirk—Delivers a lecture at Mansfield—Proposal made to him by Sir R. Murchison to return to Africa—Letter from Sir Roderick—His reply—He will not cease to be a missionary—Letter to Mr. James Young—Overtures from Foreign Office—Livingstone displeased—At dinner of Royal Academy—His speech not reported—President Lincoln's assassination—Examination by Committee of House of Commons—His opinion on the capacity of the negro—He goes down to Scotland—*Tom Brown's School Days*—His mother very ill—She rallies—He goes to Oxford —Hears of his mother's death—Returns—He attends examination of Oswell's school—His speech—Goes to London, preparing to leave—Parts from Mr. and Mrs. Webb—Stays with Dr. and Mrs. Hamilton—Last days in England.

ON reaching London, Dr. Livingstone took up his quarters at the Tavistock Hotel; but he had hardly swallowed dinner, when he was off to call on Sir Roderick and Lady Murchison.

"Sir Roderick took me off with him, just as I was, to Lady Palmerston's reception. My lady very gracious— gave me tea herself. Lord Palmerston looking well. Had two conversations with him about slave-trade. Sir Roderick says that he is more intent on maintaining his policy on that than on any other thing. And so is she—a wonderfully fine, matronly lady. Her daughters are grown up. Lady Shaftesbury like her mother in beauty and grace. Saw and spoke to Sir Charles Wood about India, 'his Eastern Empire,' as he laughingly called it. Spoke to Duke and Duchess of Somerset. All say very polite things, and all wonderfully considerate."

An invitation to dine with Lord Palmerston on the 29th detained him for a few days from going down to Scotland. "*Monday*, 25*th July.*—Went to Foreign Office. . . . Got a dress suit at Nicol & Co.'s, and dined with Lord and Lady Dunmore. Very clever and intelligent man, and lady very sprightly. Thence to Duchess of Wellington's reception. A grand company—magnificent rooms. Met Lord and Lady Colchester, Mrs. F. Peel, Lady Emily Peel, Lady de Redcliffe, Lord Broughton, Lord Houghton, and many more whose names escaped me. Ladies wonderfully beautiful—rich and rare were the gems they wore.

"26*th July.*—Go to Wimbledon with Mr. Murray, and see Sir Bartle Frere's children. . . . See Lord Russell —his manner is very cold, as all the Russells are. Saw Mr. Layard too; he is warm and frank. Received an invitation from the Lord Mayor to dine with Her Majesty's Ministers.

"27*th July.*—Hear the sad news that Robert is in the American army. . . . Went to Lord Mayor Lawrence's to dinner. . . ."

With reference to the "sad news" of Robert, which made his father very heavy-hearted during the first part of his visit home, it is right to state a few particulars, as the painful subject found its way into print, and was not always recorded accurately. Robert had some promising qualities, and those who knew and understood him had good hopes of his turning out well. But he was extremely restless, as if, to use Livingstone's phrase, he had got " a

deal of the vagabond nature from his father;" and school-
life was very irksome to him. With the view of joining
his father, he was sent to Natal, but he found no opportunity of getting thence to the Zambesi. Leaving Natal, he
found his way to America, and at Boston he enlisted in
the Federal army. The service was as hot as could be.
In one battle, two men were killed close to him by shrap-
nel shell, a rifle bullet passed close to his head, and killed
a man behind him; other two were wounded close by him.
His letters to his sister expressed his regret at the course
of his life, and confessed that his troubles were due to his
disobedience. So far was he from desiring to trade on his
father's name, that in enlisting he assumed another, nor
did any one in the army know whose son it was that was
fighting for the freedom of the slave. Meeting the risks
of battle with dauntless courage, he purposely abstained,
even in the heat of a charge, from destroying life. Not
long after, Dr. Livingstone learned that in one of his
battles he was wounded and taken prisoner; then came a
letter from a hospital, in which he again expressed his
intense desire to travel. But his career had come to its
close. He died in his nineteenth year. His body lies in
the great national cemetery of Gettysburg, in Pennsyl-
vania, in opening which Lincoln uttered one of those
speeches that made his name dear to Livingstone. Whatever degree of comfort or hope his father might derive
from Robert's last letters, he felt saddened by his unsatisfactory career. Writing to his friend Moore (5th August)
he says: "I hope your eldest son will do well in the distant land to which he has gone. My son is in the Federal
army in America, and no comfort. The secret ballast is
often applied by a kind hand above, when to outsiders we
appear to be sailing gloriously with the wind."

"*29th July.*—Called on Mr. Gladstone; he was very affable—spoke
about the Mission, and asked if I had told Lord Russell about it. . . .
Visited Lady Franklin and Miss Cracroft, her niece. . . . Dined

with Lord and Lady Palmerston, Lady Shaftesbury, and Lady Victoria Ashley, the Portuguese Minister, Count d'Azeglio (Sardinian Minister), Mr. Calcraft—a very agreeable party. Mr. Calcraft and I walked home after retiring. He is cousin to Colonel Steele; the colonel has gone abroad with his daughter, who is delicate."

"*Saturday, 31st July*, 1864.—Came down by the morning train to Harburn, and met my old friend Mr. Young, who took me to Limefield, and introduced me to a nice family."

Dr. Livingstone's relation to Mr. Young's family was very close and cordial. Hardly one of the many notes and letters he wrote to his friend fails to send greetings to "Ma-James," as he liked to call Mrs. Young, after the African fashion. It is not only the playful ease of his letters that shows how much he felt at home with Mr. Young,—the same thing appears from the frequency with which he sought his counsel in matters of business. and the value which he set upon it.

"*Sunday, 1st August.*—Went to the U.P. church, and heard excellent sermons. Was colder this time than on my former visit to Scotland.

"*2d August.*—Reached Hamilton. Mother did not know me at first. Anna Mary, a nice sprightly child, told me that she preferred Garibaldi buttons on her dress, as I walked down to Dr. Loudon to thank him for his kindness to my mother.

"*3d August.*—Agnes, Oswell, and Thomas came. I did not recognize Tom, he has grown so much. Has been poorly a long while; congestion of the kidney, it is said. Agnes quite tall, and Anna Mary a nice little girl."

The next few days were spent with his family, and in visits to the neighborhood. He had a consultation with Professor Syme as to a surgical operation recommended for an ailment that had troubled him ever since his first great journey; he was strongly urged to have the operation performed, and probably it would have been better if he had; but he finally declined, partly because an old medical friend was against it, but chiefly, as he told Sir Roderick, because the matter would get into the newspapers, and he did not like the public to be speaking of his infirmities. On the 17th he went to Inveraray to visit the Duke of Argyll. He was greatly pleased with his reception, and his Journal records the most trifling details. What especially charmed him was the considerate forethought in making him feel at his ease. "On Monday morning I had the honor of planting two trees beside those planted by Sir John Lawrence and the Marquis of Lansdowne, and by the Princess of Prussia and the Crown Prince. The coach came at twelve o'clock, and I finished the most delightful visit I ever made."

Next day he went to Oban, and the day after by steamer to Iona and Staffa, and thereafter to Aros, in Mull. Next day Captain Greenhill took him in his yacht to Ulva.

"In 1848 the kelp and potatoes failed, and the proprietor, a writer from Stirling, reduced the population from six hundred to one hundred. None of my family remain. The minister, Mr. Fraser, had made inquiries some years ago, and found an old woman who remembered my grandfather living at Uamh, or the Cave. It is a sheltered spot, with basaltic rocks jutting out of the ground below the cave; the walls of the house remain, and the corn and potato patches are green, but no one lives there. . . ."

Returning to Oban on the 24th August, ". . . I then came to the Crinan Canal, and at Glasgow end thereof met that famous missionary, Dr. Duff, from India. A fine, tall, noble-looking man, with a white beard and a twitch in his muscles which shows that the Indian climate has done its work on him. . . . Home to Hamilton."

The Highlanders everywhere claimed him; "they cheered me," he writes to Sir Roderick, "as a man and a brother."

The British Association was to meet at Bath this autumn, and Livingstone was to give a lecture on Africa. It was a dreadful thought. "Worked at my Bath speech. A cold shiver comes over me when I think of it. Ugh!"

Then he went with his daughter Agnes to see a beautiful sight, the launching of a Turkish frigate from Mr. Napier's yard—"8000 tons weight plunged into the Clyde, and sent a wave of its dirty water over to the other side." The Turkish Ambassador, Musurus Pasha, was one of the party at Shandon, and he and Livingstone traveled in the same carriage. At one of the stations they were greatly cheered by the Volunteers. "The cheers are for you," Livingstone said to the Ambassador, with a smile. "No," said the Turk, "I am only what my master made me; you are what you made yourself." When the party reached the Queen's Hotel, a working man rushed across the road, seized Livingstone's hand, saying, "I must shake your hand," clapped him on the back, and rushed back again. "You'll not deny now," said the Ambassador, "that that's for you."

Returning to Hamilton, he notes, on 4th September: "Church in the forenoon to hear a stranger, in the afternoon to hear Mr. Buchan give an excellent sermon." On 5th, 6th, 7th, he is at the speech. On 8th he receives a most kind invitation from Mr. and Mrs. Webb of Newstead Abbey, to make their house his home. Mr. Webb was a very old friend, a great hunter, who had seen Livingstone at Kolobeng, and formed an attachment to him which continued as warm as ever to the last day of Livingstone's life. Livingstone and his daughter Agnes reach Bath on the 15th, and become the guests of Dr. and Miss Watson, of both of whom he writes in the highest terms.

"On Sunday, heard a good sermon from Mr. Fleming. Bishop Colenso called on me. He was very much cheered by many people; it is evident that they admire his pluck, and consider him a persecuted man. Went to the theatre on Monday, 19th, to deliver my address. When in the green-room, a loud cheering was made for Bishop Colenso, and some hisses. It was a pity that he came to the British Association, as it looks like taking sides. Sir Charles Lyell cheered and clapped his hands in a most vigorous

way. Got over the address nicely. People very kind and
indulgent—2500 persons present, but it is a place easily
spoken in."

When Bishop Colenso moved the vote of thanks to Dr.
Livingstone for his address, occasion was taken by some
narrow and not very scrupulous journals to raise a prejudice against him. He was represented as sharing the
Bishop's theological views. For this charge there was no
foundation, and the preceding extract from his Journal
will show that he felt the Bishop's presence to be somewhat
embarrassing. Dr. Livingstone was eminently capable of
appreciating Dr. Colenso's chivalrous backing of native
races in Africa, while he differed *toto cœlo* from his theological views. In an entry in his Journal a few days later
he refers to an African traveler who had got a high reputation without deserving it, for " he sank to the low estate
of the natives, and rather admired *Essays and Reviews.*"

The next passage we give from his Journal refers to the
melancholy end of another brother-traveler, of whom he
always spoke with respect:

" *23d Sept.*—Went to the funeral of poor Captain Speke,
who, when out shooting on the 15th, the day I arrived at
Bath, was killed by the accidental discharge of his gun. It
was a sad shock to me, for, having corresponded with him,
I anticipated the pleasure of meeting him, and the first
news Dr. Watson gave me was that of his death. He was
buried at Dowlish, a village where his family have a vault.
Captain Grant, a fine fellow, put a wreath or immortelle
upon the coffin as it passed us in church. It was composed
of mignonette and wild violets."

The Bath speech gave desperate offense to the Portuguese. Livingstone thought it a good sign, wrote playfully to Mr. Webb that they were " cussin' and swearin'
dreadful," and wondered if they would keep their senses
when the book came out. In a postscript to the preface
to *The Zambesi and its Tributaries*, he says, " Senhor Lacerda

has endeavored to extinguish the facts adduced by me at Bath by a series of papers in the Portuguese official journal; and their Minister for Foreign Affairs has since devoted some of the funds of his Government to the translation and circulation of Senhor Lacerda's articles in the form of an English tract." He replies to the allegations of the pamphlet on the main points. But he was too magnanimous to make allusion to the shameless indecency of the personal charges against himself. "It is manifest," said Lacerda, "without the least reason to doubt, that Dr. Livingstone, under the pretext of propagating the Word of God (this being the least in which he employed himself) and the advancement of geographical and natural science, made all his steps and exertions subservient to the idea of . . . eventually causing the loss to Portugal of the advantages of the rich commerce of the interior, and in the end, when a favorable occasion arose that of the very territory itself." Lacerda then quoted the bitter letter of Mr. Rowley in illustration of Livingstone's plans and methods, and urged remonstrance as a duty of the Portuguese Government. "Nor," he continued, "ought the Government of Portugal to stop here. It ought, as we have said, to go further; because from what his countrymen say of Livingstone—and to which he only answers by a mere vain negation,—from what he unhesitatingly declares of himself and his intentions, and from what must be known to the Government by private information from their delegates, it is obvious that such men as Livingstone may become extremely prejudicial to the interests of Portugal, especially when resident in a public capacity in our African possessions, if not efficiently watched, if their audacious and mischievous actions are not restrained. If steps are not taken in a proper and effective manner, so that they may be permitted only to do good, if indeed good can come from such," etc.

"*26th Sept.*—Agnes and I go to-day to Newstead Abbey, Notts.
Reach it about 9 P.M., and find Mr. and Mrs. Webb all I anticipated
and more. A splendid old mansion with a wonderful number of
curiosities in it, and magnificent scenery around. It was the residence
of Lord Byron, and his furniture is kept" [in his private rooms] "just
as he left it. His character does not shine. It appears to have been
horrid. . . . He made a drinking cup of a monk's skull found under
the high altar, with profane verses on the silver setting, and kept his
wine in the stone coffin. These Mrs. Webb buried, and all the bones
she could find that had been desecrated by the poet."

In a letter to Sir Thomas Maclear he speaks of the poet
as one of those who, like many others—some of them
travelers who abused missionaries,—considered it a fine
thing to be thought awfully bad fellows.

"*27th.*—Went through the whole house with our kind hosts, and
saw all the wonders, which would require many days properly to
examine. . . .

"*2d October.*—Took Communion in the chapel of the Abbey. God
grant me to be and always to act as a true Christian.

"*3d.*—Mr. and Mrs. Webb kindness itself personified. A blessing be
on them and their children from the Almighty!"

When first invited to reside at Newstead Abbey, Dr.
Livingstone declined, on the ground that he was to be
busy writing a book, and that he wished to have some of
his children with him, and in the case of Agnes, to let
her have music lessons. His kind friends, however, were
resolved that these reasons should not stand in the way,
and arrangements were made by them accordingly. Dr.
Livingstone continued to be their guest for eight months,
and received from them all manner of assistance. Sometimes Mr. and Mrs. Webb, Mrs. Goodlake (Mrs. Webb's
mother), and his daughter Agnes would all be busy copy-
ing his journals. The "Livingstone room," as it is called,
in the Sussex tower, is likely to be associated with his
name while the building lasts. It was his habit to rise
early and work at his book, to return to his task after
breakfast and continue till luncheon, and in the afternoon

have a long walk with Mr. Webb. It is only when the book is approaching its close that we find him working "till two in the morning." One of his chief recreations was in the field of natural history, watching experiments with the spawning of trout. He endeared himself to all, high and low; was a special favorite with the children, and did not lose opportunities to commend, in the way he thought best, those high views of life and duty which had been so signally exemplified in his own career. The playfulness of his nature found full and constant scope at Newstead; he regained an almost boyish flow of animal spirits, reveled in fun and frolic in his short notes to friends like Mr. Young, or Mr. Webb when he happened to be absent; wrote in the style of Mr. Punch, and called his opponents by ludicrous names; yet never forgot the stern duty that loomed before him, or allowed the enjoyment and *abandon* of the moment to divert him from the death-struggle on behalf of Africa in which he had yet to engage.

The book was at first to be a little one,—a blast of the trumpet against the monstrous slave-trade of the Portuguese; but it swelled to a goodly octavo, and embraced the history of the Zambesi Expedition. Charles Livingstone had written a full diary, and in order that his name might be on the title-page, and he might have the profits of the American edition, his journal was made use of in the writing of the book; but the arrangement was awkward; sometimes Livingstone forgot the understanding of joint-authorship, and he found that he could more easily have written the whole from the foundation. At first it was designed that the book should appear early in the summer of 1865, but when the printing was finished the map was not ready; and the publication had to be delayed till the usual season in autumn.

The entries in his Journal are brief, and of little general interest during the time the book was getting ready. Most of them have reference to the affairs of other people. As he finds that Dr. Kirk is unable to undertake a work on the botany and natural history of the Expedition, unless he should hold some permanent situation, he exerts himself to procure a Government appointment for him, recommending him strongly to Sir R. Murchison and others, and is particularly gratified by a reply to his application from the Earl of Dalhousie, who wrote that he regarded his request as a command. He is pleased to learn that, through the kind efforts of Sir Roderick, his brother Charles has been appointed Consul at Fernando Po. He sees the American Minister, who promises to do all he can for Robert, but almost immediately after, the report comes that poor Robert has died in a hospital in Salisbury, North Carolina. He delivers a lecture at the Mechanics' Institute at Mansfield, but the very idea of a speech always makes him ill, and in this case it brings on an attack of hæmorrhoids, with which he had not been troubled for long. He goes to London to a meeting of the Geographical Society, and hears a paper of Burton's—a gentleman from whose geographical views he dissents, as he does from his views on subjects more important. In regard to his book he says very little; four days, he tells us, were spent in writing the description of the Victoria Falls; and on the 15th April, 1865, he summons his daughter Agnes to take his pen and write FINIS at the end of his manuscript. On leaving Newstead on the 25th, he writes, "Parted with our good friends the Webbs. And may God Almighty bless and reward them and their family!"

Some time before this, a proposal was made to him by Sir Roderick Murchison which in the end gave a new direction to the remaining part of his life. It was brought before him in the following letter:

"*Jan.* 5, 1865.

"MY DEAR LIVINGSTONE:—As to *your future,* I am anxious to know what *your own wish is* as respects a renewal of African exploration.

"Quite irrespective of missionaries or political affairs, there is at this moment a question of intense geographical interest to be settled; namely, the watershed, or watersheds, of South Africa.

"Now, if you would really like to be the person to finish off your remarkable career by completing such a survey, unshackled by other avocations than those of the geographical explorer, I should be delighted to consult my friends of the Society, and take the best steps to promote such an enterprise.

"For example, you might take your little steamer to the Rovuma, and, getting up by water as far as possible in the rainy season, then try to reach the south end of the Tanganyika. Thither you might transport a light boat, or build one there, and so get to the end of that sheet of water.

"Various questions might be decided by the way, and if you could get to the west, and come out on that coast, or should be able to reach the White Nile (!), you would bring back an unrivaled reputation, and would have settled all the great disputes now pending.

"If you do not like to undertake *the purely geographical work*, I am of opinion that no one, after yourself, is so fitted to carry it out as Dr. Kirk. I know that he thinks of settling down now at home. But if he could delay this home-settlement for a couple of years, he would not only make a large sum of money by his book of travels, but would have a renown that would give him an excellent introduction as a medical man.

"I have heard you so often talk of the enjoyment you feel when in Africa, that I cannot believe you now think of anchoring for the rest of your life on the mud and sand-banks of England.

"Let me know your mind on the subject. When is the book to appear? Kind love to your daughter.—Yours sincerely,

"Rod^{CK} I. Murchison."

Livingstone begins his answer by assuring Sir Roderick that he never contemplated settling down quietly in England; it would be time enough for that when he was in his dotage. "I should like the exploration you propose very much, and had already made up my mind to go up the Rovuma, pass by the head of Lake Nyassa, and away west or northwest as might be found practicable." He would have been at this ere now, but his book chained him, and he feared that he could not take back the "Lady Nyassa" to Africa, with the monsoon against him, so that he must get a boat to explore the Rovuma.

"What my inclination leads me to prefer is to have intercourse with the people, and do what I can by talking, to enlighten them on the slave-trade, and give them some idea of our religion. It may not be much that I can do, but I feel when doing that I am not living in vain. You remember that when, to prevent our coming to a standstill, I had to turn skipper myself, the task was endurable only because I was determined that no fellow should prove himself indispensable to our further progress. To be debarred from spending most of my time in traveling, in exploration, and continual intercourse with the natives, I always felt to be a severe privation, and if I can get a few hearty native companions, I shall enjoy myself, and feel that I am doing my duty. As soon as my book is out, I shall start."

In Livingstone's Journal, 7th January, 1865, we find, this entry: "Answered Sir Roderick about going out. Said I could only feel in the way of duty by working as a missionary." The answer is very noteworthy in the view of what has so often been said against Livingstone— that he dropped the missionary to become an explorer. To understand the precise bearing of the proposal, and of Livingstone's reply, it is necessary to say that Sir Roderick had a conviction, which he never concealed, that the missionary enterprise encumbered and impeded the geographical. He had a special objection to an Episcopal mission, holding that the planting of a Bishop and staff on territory dominated by the Portuguese was an additional irritant, rousing ecclesiastical jealousy, and bringing it to the aid of commercial and political apprehensions as to the tendency of the English enterprise. Neither mission nor colony could succeed in the present state of the country; they could only be a trouble to the geographical explorer. On this point Livingstone held his own views. He could only feel in the line of duty as a missionary. Whatever he might or might not be able to do in that capacity, he would never abandon it, and, in particular, he would never come under an obligation to the Geographical Society that he would serve them "unshackled by other avocations than those of the geographical explorer."

A letter to Mr. James Young throws light on the feelings with which he regarded Sir Roderick's proposal:

"*20th January*, 1865.—I am not sure but I told you already that Sir Roderick and I have been writing about going out, and my fears that I must sell 'Lady Nyassa,' because the monsoon will be blowing from Africa to India before I get out, and it won't do for me to keep her idle. I must go down to the Seychelles Islands (tak' yer speks and keek at the map or gougrafy), then run my chance to get over by a dhow or man-of-war to the Rovuma, going up that river in a boat, till we get to the cataracts, and then tramp. I must take Belochees from India, and may go down the lake to get Makololo, if the Indians don't answer. I would not consent to go simply as a geographer, but as a missionary, and do geography by the way, because I feel I am in the way of duty when trying either to enlighten these poor people, or open their land to lawful commerce."

It was at this time that Mr. Hayward, Q.C., while on a visit to Newstead, brought an informal message from Lord Palmerston, who wished to know what he could do for Livingstone. Had Livingstone been a vain man, wishing a handle to his name, or had he even been bent on getting what would be reasonable in the way of salary for himself, or of allowance for his children, now was his chance of accomplishing his object. But so single-hearted was he in his philanthropy that such thoughts did not so much as enter his mind; there was one thing, and one only, which he wished Lord Palmerston to secure—free access to the highlands, by the Zambesi and Shiré, to be made good by a treaty with Portugal. It is satisfactory to record that the Foreign Office has at last made arrangements to this effect.

While the proposal on the part of the President of the Geographical Society was undergoing consideration, certain overtures were made to Dr. Livingstone by the Foreign Office. On the 11th of March he called at the office, at the request of Mr. Layard, who propounded a scheme that he should have a commission giving him authority over the chiefs, from the Portuguese boundary

to Abyssinia and Egypt; the office to carry no salary.
When a formal proposal to this effect was submitted to
him, with the additional proviso that he was to be entitled
to no pension, he could not conceal his irritation. For
himself he was just as willing as ever to work as before,
without hope of earthly recompense, and to depend on
the petition, "Give us this day our daily bread;" but he
thought it ungenerous to take advantage of his well-
known interest in Africa to deprive him of the honora-
rium which the most insignificant servant of Her Majesty
enjoyed. He did not like to be treated like a charwoman.
As for the pension, he had never asked it, and counted it
offensive to be treated as if he had shown a greed which
required to be repressed. It came out, subsequently, that
the letter had been written by an underling, but when
Earl Russell was appealed to, he would only promise a
salary when Dr. Livingstone should have settled somewhere! The whole transaction had a very ungracious
aspect.

Before publishing his book, Dr. Livingstone had asked
Sir Roderick Murchison's advice as to the wisdom of
speaking his mind on two somewhat delicate points. In
reply, Sir Roderick wrote: "If you think you have been
too hard as to the Bishop or the Portuguese, you can
modify the phrases. But I think that the truth ought to
be known, if only in vindication of your own conduct, and
to account for the little success attending your last mission."
We continue our extracts from his Journal:

"*26th April,* 1865.—In London. Horrified by news of President
Lincoln's assassination, and the attempt to murder Seward."

"*29th April.*—Went down to Crystal Palace, with Agnes, to a Saturday Concert. The music very fine. Met Waller, and lost a train.
Came up in hot haste to the dinner of the Royal Academy. . . .
Sir Charles Eastlake, President; Archbishops of Canterbury and York
on each side of the chair; all the Ministers present, except Lord Palmerston, who is ill of gout in the hand. Lord Russell, Lord Granville,
and Duke of Somerset sat on other side of table from Sir Henry Hol-

land, Sir Roderick, and myself. Lord Clarendon was close enough to lean back and clap me on the shoulder, and ask me when I was going out. Duke of Argyll, Bishops of Oxford and London, were within earshot; Sir J. Romilly, the Master of the Rolls, was directly in front, on the other side of our table. He said that he watched all my movements with great interest. . . . Lord Derby made a good speech. The speeches were much above the average. I was not told that I was expected to speak till I got in, and this prevented my eating. When Lord John Manners complimented me after my speech, I mentioned the effect the anticipation had on me. To comfort me he said that the late Sir Robert Peel never enjoyed a dinner in these circumstances, but sat crumbling up his bread till it became quite a heap on the table. . . . My speech was not reported."

"*2d May.*—Met Mr. Elwin, formerly editor of the *Quarterly*. He said that Forster, one of our first-class writers, had told him that the most characteristic speech was not reported, and mentioned the heads —as, the slave-trade being of the same nature as thuggee, garrotting; the tribute I paid to our statesmen; and the way that Africans have been drawn, pointing to a picture of a woman spinning. This non-reporting was much commented on, which might, if I needed it, prove a solace to my wounded vanity. But I did not feel offended. Everything good for me will be given, and I take all as a little child from its father.

"Heard a capital sermon from Dr. Hamilton [Regent Square Church], on President Lincoln's assassination. 'It is impossible but that offenses will come,' etc. He read part of the President's address at second inauguration. In the light of subsequent events it is grand. If every drop of blood shed by the lash must be atoned for by an equal number of white men's vital fluid,—righteous, O Lord, are Thy judgments! The assassination has awakened universal sympathy and indignation, and will lead to more cordiality between the countries. The Queen has written an autograph letter to Mrs. Lincoln, and Lords and Commons have presented addresses to Her Majesty, praying her to convey their sentiments of horror at the fearful crime."

"*18th May*, 1865.—Was examined by the Committee [of the House of Commons] on the West Coast; was rather nervous and confused, but let them know pretty plainly that I did not agree with the aspersions cast on missions."

In a letter to Mr. Webb, he writes *à propos* of this examination:

"The monstrous mistake of the Burton school is this : they ignore the point-blank fact that the men that do the most for the mean whites are the same that do the most for the mean blacks, and you never hear one mother's son of them say, You do wrong to give to the whites. I told the Committee I had heard people say that Christianity made the blacks worse, but did not agree with them. I might have said it was 'rot,' and truly. I can stand a good deal of bosh, but to tell me that Christianity makes people worse—ugh! Tell that to the young trouts. You know on what side I am, and I shall stand to my side, Old Pam fashion, through thick and thin. I don't agree with all my side say and do. I won't justify many things, but for the great cause of human progress I am heart and soul, *and so are you.*"

Dr. Livingstone was asked at this time to attend a public meeting on behalf of American freedom. It was not in his power to go, but, in apologizing, he was at pains to express his opinion on the capacity of the negro, in connection with what was going on in the United States:

"Our kinsmen across the Atlantic deserve our warmest sympathy. They have passed, and are passing, through trials, and are encompassed with difficulties which completely dwarf those of our Irish famine, and not the least of them is the question, what to do with those freedmen for whose existence as slaves in America our own forefathers have so much to answer. The introduction of a degraded race from a barbarous country was a gigantic evil, and if the race cannot be elevated, an evil beyond remedy. Millions can neither be amalgamated nor transported, and the presence of degradation is a contagion which propagates itself among the more civilized. But I have no fears as to the mental and moral capacity of the Africans for civilization and upward progress. We who suppose ourselves to have vaulted at one bound to the extreme of civilization, and smack our lips so loudly over our high elevation, may find it difficult to realize the debasement to which slavery has sunk those men, or to appreciate what, in the discipline of the sad school of bondage, is in a state of freedom real and substantial progress. But I, who have been intimate with Africans who have never been defiled by the slave-trade, believe them to be capable of holding an honorable rank in the family of man."

Wherever slavery prevailed, or the effects of slavery were found, Dr. Livingstone's testimony against it was clear and emphatic. Neither personal friendship nor any other consideration under the sun could repress it. When his friends Sir Roderick and Mr. Webb afterward expressed their sympathy with Governor Eyre, of Jamaica, he did not scruple to tell them how different an estimate he had formed of the Governor's conduct.

We continue our extracts from his Journal and letters:

"*24th May.*—Came down to Scotland by last night's train; found mother very poorly; and, being now eighty-two, I fear she may not have long to live among us."

27th May (to Mr. Webb).—"I have been reading *Tom Brown's School Days*—a capital book. Dr. Arnold was a man worth his weight in something better than gold. You know Oswell" [his early friend] "was one of his Rugby boys. One could see his training in always doing what was brave and true and right."

"*2d June.*—Tom better, but kept back in his education by his complaint. Oswell getting on well at school at Hamilton. Anna Mary well. Mother gradually becoming weaker. Robert we shall never hear of again in this world, I fear; but the Lord is merciful and just and right in all his ways. He would hear the cry for mercy in the hospital at Salisbury. I have lost my part in that gigantic struggle which the Highest guided to a consummation never contemplated by the Southerners when they began; and many others have borne more numerous losses."

"*5th June.*—Went about a tombstone to my dear Mary. Got a good one of cast-iron to be sent out to the Cape.

"Mother very low. . . . Has been a good affectionate mother to us all. The Lord be with her. . . . Whatever is good for me and mine the Lord will give.

"To-morrow, Communion in kirk. The Lord strip off all imperfections, wash away all guilt, breathe love and goodness through all my nature, and make his image shine out from my soul.

"Mother continued very low, and her mind ran on poor Robert. Thought I was his brother, and asked me frequently, 'Where is your brother? where is that puir laddie?' . . . Sisters most attentive. . . . Contrary to expectation she revived, and I went to Oxford. The Vice-Chancellor offered me the theatre to lecture in, but I expected a telegram if any change took place on mother. Gave an address to a number of friends in Dr. Daubeny's chemical class-room."

"*Monday, 19th June.*—A telegram came, saying that mother had died the day before. I started at once for Scotland. No change was observed till within an hour and a half of her departure. . . . Seeing the end was near, sister Agnes said, 'The Saviour has come for you, mother. You can "lippen" yourself to him?' She replied, 'Oh yes.' Little Anna Mary was held up to her. She gave her the last

32

look, and said 'Bonnie wee lassie,' gave a few long inspirations, and all was still, with a look of reverence on her countenance. She had wished William Logan, a good Christian man, to lay her head in the grave, if I were not there. When going away in 1858, she said to me that she would have liked one of her laddies to lay her head in the grave. It so happened that I was there to pay the last tribute to a dear good mother."

The last thing we find him doing in Scotland is attending the examination of Oswell's school, with Anna Mary, and seeing him receive prizes. Dr. Loudon, of Hamilton, the medical attendant and much-valued friend of the Livingstones, furnishes us with a reminiscence of this occasion. He had great difficulty in persuading Livingstone to go. The awful bugbear was that he would be asked to make a speech. Being assured that it would be thought strange if, in a gathering of the children's parents, he were absent, he agreed to go. And of course he had to speak. What he said was pointed and practical, and in winding up, he said he had just two things to say to them— "FEAR GOD, AND WORK HARD." These appear to have been Livingstone's last public words in his native Scotland.

His Journal is continued in London:

"*8th August.*—Went to Zoological Gardens with Mr. Webb and Dr. Kirk; then to lunch with Miss Coutts" [Baroness Burdett Coutts]. "Queen Emma of Honolulu is to be there. It is not fair for High Church people to ignore the labors of the Americans, for [the present state of Christianity] is the fruit of their labors, and not of the present Bishop. Dined at Lady Franklin's with Queen Emma; a nice, sensible person the Queen seems to be.

"*9th August.*—Parted with my friends Mr. and Mrs. Webb at King's Cross station to-day. He gracefully said that he wished I had been coming rather than going away, and she shook me very cordially with both hands, and said, 'You will come back again to us, won't you?' and shed a womanly tear. The good Lord bless and save them both, and have mercy on their whole household!"

"*11th August.*—Went down to say good-bye to the Duchess-Dowager of Sutherland, at Maidenhead. Garibaldi's rooms are shown: a good man he was, but followed by a crowd of harpies who tried to use him for their own purposes. . . . He was so utterly worn out by shaking hands, that a detective policeman who was with him in the carriage, . . . and a . . . ceremony for him.

"Took leave at Foreign Office. Mr. Layard very kind in his expressions at parting, and so was Mr. Wylde.

" *12th August.*—Went down to Wimbledon to dine with Mr. Murray,
and take leave. Mr. and Mrs. Oswell came up to say farewell. He
offers to go over to Paris at any time to bring Agnes" [who was going
to school there] " home, or do anything that a father would. [" I love
him," Livingstone writes to Mr. Webb, " with true affection, and I
believe he does the same to me ; and yet we never show it."]

" We have been with Dr. and Mrs. Hamilton for some time—good,
gracious people. The Lord bless them and their household ! Dr. Kirk
and Mr. Waller go down to Folkestone to-morrow. and take leave of us
there. This is very kind. The Lord puts it into their hearts to show
kindness, and blessed be his name."

Dr. Livingstone's last weeks in England were passed
under the roof of the late Rev. Dr. Hamilton, author of
Life in Earnest, and could hardly have been passed in a
more congenial home. Natives of the same part of Scotland, nearly of an age, and resembling each other much
in taste and character, the two men drew greatly to each
other. The same Puritan faith lay at the basis of their
religious character, with all its stability and firmness. But
above all, they had put on charity, which is the bond of
perfectness. In Natural History, too, they had an equal
enthusiasm. In Dr. Hamilton, Livingstone found what
he missed in many orthodox men. On the evening of his
last Sunday, he was prevailed on to give an address in
Dr. Hamilton's church, after having in the morning
received the Communion with the congregation. In his
address he vindicated his character as a missionary, and
declared that it was as much as ever his great object
to proclaim the love of Christ, which they had been com-
memorating that day. His prayers made a deep impression ; they were like the communings of a child with his
father. At the railway station, the last Scotch hands
grasped by him were those of Dr. and Mrs. Hamilton.
The news of Dr. Hamilton's death was received by Livingstone a few years after, in the heart of Africa, with no
small emotion. Their next meeting was in the better land.

CHAPTER XVIII.

FROM ENGLAND TO BOMBAY AND ZANZIBAR.

A.D. 1865-1866.

Object of new journey—Double scheme—He goes to Paris with Agnes—Baron Hausmann—Anecdote at Marseilles—He reaches Bombay—Letter to Agnes—Reminiscences of Dr. Livingstone at Bombay by Rev. D. C. Boyd—by Alex. Brown, Esq.—Livingstone's dress—He visits the caves of Kenhari—Rumors of murder of Baron van der Decken—He delivers a lecture at Bombay—Great success—He sells the " Lady Nyassa"—Letter to Mr. Young—Letter to Anna Mary—Hears that Dr. Kirk has got an appointment—Sets out for Zanzibar in " Thule"—Letter to Mr. Young—His experience at sea—Letter to Agnes—He reaches Zanzibar—Calls on Sultan—Presents the "Thule" to him from Bombay Government—Monotony of Zanzibar life—Leaves in " Penguin" for the continent.

THE object for which Dr. Livingstone set out on his third and last great African journey is thus stated in the preface to *The Zambesi and its Tributaries:* " Our Government have supported the proposal of the Royal Geographical Society made by my friend Sir Roderick Murchison, and have united with that body to aid me in another attempt to open Africa to civilizing influences, and a valued private friend has given a thousand pounds for the same object. I propose to go inland, north of the territory which the Portuguese in Europe claim, and endeavor to commence that system on the East which has been so eminently successful on the West Coast: a system combining the repressive efforts of Her Majesty's cruisers with lawful trade and Christian missions—the moral and material results of which have been so gratifying. I hope to ascend the Rovuma, or some other river north of Cape Delgado, and, in addition to my other work, shall strive, by passing along the northern end of Lake Nyassa, and round the southern end of Lake Tanganyika, to ascertain the watershed of that part of Africa."

The first part of the scheme was his own, the second he had been urged to undertake by the Geographical Society. The sums in aid contributed by Government and the Geographical society were only £500 each; but it was not thought that the work would occupy a long time. The Geographical Society coupled their contribution with some instructions as to observations and reports which seemed to Dr. Livingstone needlessly stringent, and which certainly ruffled his relation to the Society. The honorary position of Consul at large he was willing to accept for the sake of the influence which it gave him, though still retaining his opinion of the shabbiness which had so explicitly bargained that he was to have no salary and to expect no pension.

The truth is, if Livingstone had not been the most single-minded and trustful of men, he would never have returned to Africa on such terms. The whole sum placed at his disposal was utterly inadequate to defray the cost of the Expedition, and support his family at home. Had it not been for promises that were never fulfilled, he would not have left his family at this time as he did. But in nothing is the purity of his character seen more beautifully than in his bearing toward some of those who had gained not a little consideration by their connection with him, and had made him fair promises, but left him to work on as best he might. No trace of bitter feeling disturbed him, or abated the strength of his love and confidence.

Dr. Livingston went first to Paris with his daughter, and left her there for education. Passing on he reached Marseilles on the 19th August, and wrote her a few lines, in which he informed her that the man who was now transforming Paris [Baron Hausmann] was a Protestant, and had once taught a Sunday-school in the south of

France; and that probably he had greater pleasure in the first than in the second work. The remark had a certain applicability to his own case, and probably let out a little of his own feeling; it showed at least his estimate of the relative place of temporal and spiritual philanthropy. The prayer that followed was expressive of his deepest feelings toward his best-beloved on earth : "May the Almighty qualify you to be a blessing to those around you, wherever your lot is cast. I know that you hate all that is mean and false. May God make you good, and to delight in doing good to others. If you ask He will give abundantly. The Lord bless you!"

From a Bombay gentleman who was his fellow-traveler to India a little anecdote has casually come to our knowledge illustrating the unobtrusiveness of Livingstone—his dislike to be made a lion of. At the *table-d'hôte* of the hotel in Marseilles, where some Bombay merchants were sitting, the conversation turned on Africa in connection with ivory—an extensive article of trade in Bombay. One friend dropped the remark, "I wonder where that old chap Livingstone is now." To his surprise and discomfiture, a voice replied, "Here he is." They were fast friends all through the voyage that followed. Little of much interest happened during that voyage. Livingstone writes that Palgrave was in Cairo when he passed through, but he did not see him. Of Baker he could hear nothing. Miss Tinné, the Dutch lady, of whom he thought highly as a traveler, had not been very satisfactory to the religious part of the English community at Cairo. Miss Whately was going home for six weeks, but was to be back to her Egyptian Ragged School. He saw the end of the Lesseps Canal, about the partial opening of which they were making a great noise. Many thought it would succeed, though an Egyptian Commodore had said to him, "It is hombog." The Red Sea was fearfully hot and steamy. The "Lady Nyassa" hung like a millstone around his neck, and he was prepared to sell her for whatever she might bring. Bombay was reached on 11th September.

"*Bombay,* 20*th Sept.,* 1865.— . . . By advice of the Governor, I went up to Nassick to see if the Africans there under Government instruction would suit my purpose as members of the Expedition. I was present at the examination of a large school under Mr. Price by the Bishop of Bombay. It is partly supported by Government. The pupils (108) are not exclusively African, but all showed very great proficiency. They excelled in music. I found some of the Africans to have come from parts I know—one from Ndonde on the Rovuma— and all had learned some handicraft, besides reading, writing, etc., and it is probable that some of them will go back to their own country with me. Eight have since volunteered to go. Besides these I am to get some men from the 'Marine Battalion,' who have been accustomed to rough it in various ways, and their pensions will be given to their widows if they should die. The Governor (Sir Bartle Frere) is going to do what he can for my success.

"After going back to Bombay I came up to near Poonah, and am now at Government House, the guest of the Governor.

"Society here consists mainly of officers and their wives. . . . Miss Frere, in the absence of Lady Frere, does the honors of the establishment, and very nicely she does it. She is very clever, and quite unaffected—very like her father. . . .

"Christianity is gradually diffusing itself, leavening as it were in various ways the whole mass. When a man becomes a professor of Christianity, he is at present cast out, abandoned by all his relations, even by wife and children. This state of things makes some who don't care about Christian progress say that all Christian servants are useless. They are degraded by their own countrymen, and despised by others, but time will work changes. Mr. Maine, who came out here with us, intends to introduce a law whereby a convert deserted by his wife may marry again. It is in accordance with the text in Corinthians—If an unbelieving wife depart, let her depart. People will gradually show more sympathy with the poor fellows who come out of heathenism, and discriminate between the worthy and unworthy. You should read Lady Duff Gordon's *Letters from Egypt.* They show a nice sympathizing heart, and are otherwise very interesting. She saw the people as they are. Most people see only the outsides of things. . . . Avoid all nasty French novels. They are very injurious, and effect a lasting injury on the mind and heart. I go up to Government House again three days hence, and am to deliver two lectures,—one at Poonah and one at Bombay."

Some slight reminiscences of Livingstone at Bombay, derived from admiring countrymen of his own, will not be out of place, considering that the three or four months spent there was the last period of his life passed in any part of the dominions of Great Britain.

The Rev. Dugald C. Boyd, of Bombay (now of Portsoy, Banffshire), an intimate friend of Dr. Stewart, of Lovedale, writing to a correspondent on 10th October, 1865, says:

"Yesterday evening I had the pleasure of meeting Livingstone at dinner in a very quiet way. . . . It was an exceedingly pleasant evening. Dr. Wilson was in great 'fig,' and Livingstone was, though quiet, very communicative, and greatly disposed to talk about Africa. . . . I had known Mrs. Livingstone, and I had known Robert and Agnes, his son and daughter, and I had known Stewart. He spoke very kindly of Stewart, and seems to hope that he may yet join him in Central Africa. . . . He is much stouter, better, and healthier-looking than he was last year. . . .

"12*th October.*—Livingstone was at the *tamasha* yesterday. He was dressed very unlike a minister—more like a post-captain or admiral. He wore a blue dress-coat, trimmed with lace, and bearing a Government gilt button. In his hand he carried a cocked hat. At the Communion on Sunday (he sat on Dr. Wilson's right hand, who sat on my right) he wore a blue surtout, with Government gilt buttons, and shepherd-tartan trousers; and he had a gold band round his cap.[1] I

---

[1] Dr. Livingstone's habit of dressing as a layman, and accepting the designation of David Livingstone, Esquire, as readily as that of the Rev. Dr. Livingstone, probably helped to propagate the idea that he had sunk the missionary in the explorer. The truth, however, is, that from the first he wished to be a lay missionary, not under any Society, and it was only at the instigation of his friends that he accepted ordination. He had an intense dislike of what was merely professional and conventional, and he thought that as a free-lance he would have more influence. Whether in this he sufficiently appreciated the position and office of one set aside by the Church for the service of the gospel may be a question: but there can be no question that he had the same view of the matter from first to last. He would have worn a blue dress and gilt buttons, if it had been suitable, as readily as any other, at the most ardent period of his missionary life. His heart was as truly that of a missionary under the Consul's dress as it had ever been when he wore black, or whatever else he could get, in the wilds of Africa. At the time of his encounter with the lion he wore a coat of tartan, and he thought that that material might have had some effect in preventing the usual irritating results of a lion's bite.

spent two hours in his society last evening at Dr. Wilson's. He was not very complimentary to Burton. He is to lecture in public this evening."

Another friend, Mr. Alexander Brown, now of Liverpool, sends a brief note of a very delightful excursion given by him, in honor of Livingstone, to the caves of Kennery or Kenhari, in the island of Salsette. There was a pretty large party. After leaving the railway station, they rode on ponies to the caves.

"We spent a most charming day in the caves, and the wild jungle around them. Dr. Wilson, you may believe, was in his element, pouring forth volumes of Oriental lore in connection with the Buddhist faith and the Kenhari caves, which are among the most striking and interesting monuments of it in India. They are of great extent, and the main temple is in good preservation. Doctor Livingstone's almost boyish enjoyment of the whole thing impressed me greatly. The stern, almost impassive, man seemed to unbend, and enter most thoroughly into the spirit of a day in which pleasure and instruction, under circumstances of no little interest, were so delightfully combined."

At Bombay he heard disquieting tidings of the Hanoverian traveler, Baron van der Decken. In his Journal he says:

"*29th December*, 1865.—The expedition of the Baron van der Decken has met with a disaster up the Juba. He had gone up 300 miles, and met only with the loss of his steam launch. He then ran his steamer on two rocks and made two large holes in her bottom. The Baron and Dr. Link got out in order to go to the chief to conciliate him. He had been led to suspect war. Then a large party came and attacked them, killing the artist Trenn and the chief engineer. They were beaten off, and Lieutenant von Schift with four survivors left in the boat, and in four days came down the stream. Thence they came in a dhow to Zanzibar. It is feared that the Baron may be murdered, but possibly not. It looks ill that the attack was made after he landed.

"My times are in thy hand, O Lord! Go Thou with me and I am safe. And above all, make me useful in promoting Thy cause of peace and good-will among men."

The rumor of the Baron's death was subsequently confirmed. His mode of treating the natives was the very opposite of Livingstone's, who regarded the manner of his death as another proof that it was not safe to disregard the manhood of the African people.

The Bombay lecture was a great success. Dr. Wilson, Free Church Missionary, was in the chair, and after the lecture tried to rouse the Bombay merchants, and especially the Scotch ones, to help the enterprise. Referring to the driblets that had been contributed by Government and the Geographical Society, he proposed that in Bombay they should raise as much as both. In his next letter to his daughter, Livingstone tells of the success of the lecture, of the subscription, which promised to amount to £1000 (it did not quite do so), and of his wish that the Bombay merchants should use the money for setting up a trading establishment in Africa. "I must first of all find a suitable spot; then send back here to let it be known. I shall then be off in my work for the Geographical Society, and when that is done, if I am well, I shall come back to the first station." He goes on to speak of the facilities he had received for transporting Indian buffaloes and other animals to Africa, and of the extraordinary kindness and interest of Sir Bartle Frere, and the pains he had taken to commend him to the good graces of the Sultan of Zanzibar, then in Bombay. He speaks pleasantly of his sojourn with Dr. Wilson and other friends. He is particularly pleased with the management and *menu* of a house kept by four bachelors—and then he adds: "Your mamma was an excellent manager of the house, and made everything comfortable. I suppose it is the habit of attending to little things that makes such a difference in different houses. As I am to be away from all luxuries soon, I may as well live comfortably with the bachelors while I can."

To Mr. James Young he writes about the "Lady Nyassa," which he had sold, after several advertisements, but only for £2300: "The whole of the money given for her

I dedicated to the great object for which she was built. I am satisfied at having made the effort; would of course have preferred to have succeeded, but we are not responsible for results." In reference to the investment of the money, it was intended ultimately to be sunk in Government or railway securities; but meanwhile he had been recommended to invest it in shares of an Indian bank. Most unfortunately, the bank failed a year or two afterward; and thus the whole of the £6000, which the vessel had cost Livingstone, vanished into air.

His little daughter Anna Mary had a good share of his attention at Bombay:

"*24th December*, 1865.—I went last night to take tea in the house of a Hindoo gentleman who is not a professed Christian. It was a great matter for such to eat with men not of his caste. Most Hindoos would shrink with horror from contact with us. Seven little girls were present, belonging to two Hindoo families. They were from four or five to eight years old. They were very pleasant-looking, of olive complexions. Their hair was tied in a knot behind, with a wreath of flowers round the knot; they had large gold ear-rings and European dresses. One played very nicely on the piano, while the rest sang very nicely a funny song, which shows the native way of thinking about some of our customs. They sang some nice hymns, and repeated some pieces, as the 'Wreck of the Hesperus,' which was given at the examination of Oswell's school. Then all sung, 'There is a happy land, far, far away,' and it, with some of the Christian hymns, was beautiful. They speak English perfectly, but with a little foreign twang. All joined in a metrical prayer before retiring. They have been taught all by their father, and it was very pleasant to see that this teaching had brought out their natural cheerfulness. Native children don't look lively, but these were brimful of fun. One not quite as tall as yourself brought a child's book to me, and with great glee pointed out myself under the lion. She can read fluently, as I suppose you can by this time now. I said that I would like a little girl like her to go with me to Africa to sing these pretty hymns to me there. She said she would like to go, but should not like to have a black husband. This is Christmas season, and to-morrow is held as the day in which our Lord was born, an event which angels made known to men, and it brought great joy, and proclaimed peace on earth and good-will to men. That Saviour must be your friend, and He will be if you ask Him so to be. He will forgive and save you, and take you into his family."

On New Year's Day, 1866, he writes in his Journal: "The Governor told me that he had much pleasure in giving Dr. Kirk an appointment; he would telegraph to him to-day. It is to be at Zanzibar, where he will be of great use in promoting all good works."

It had been arranged that Dr. Livingstone was to cross to Zanzibar in the "Thule," a steamer that had formed part of the squadron of Captain Sherard Osborn in China, and which Livingstone had now the honor of being commissioned to present to the Sultan of Zanzibar, as a present from Sir Bartle Frere and the Bombay Government.

We give a few extracts from his journal at sea:

"*17th January.*—Issued flannel to all the boys from Nassick; the marines have theirs from Government. The boys sing a couple of hymns every evening, and repeat the Lord's Prayer. I mean to keep up this, and make this a Christian Expedition, telling a little about Christ wherever we go. His love in coming down to save men will be our theme. I dislike very much to make my religion distasteful to others. This, with ——'s hypocritical ostentation, made me have fewer religious services on the Zambesi than would have been desirable, perhaps. He made religion itself distasteful by excessive ostentation. . . . Good works gain the approbation of the world, and though there is antipathy in the human heart to the gospel of Christ, yet when Christians make their good works shine all admire them. It is when great disparity exists between profession and practice that we secure the scorn of mankind. The Lord help me to act in all cases in this Expedition as a Christian ought!"

"*23d January.*—My second book has been reviewed very favorably by the *Athenæum* and the *Saturday Review*, and by many newspapers. Old John Crawford gives a snarl in the *Examiner*, but I can afford that it should be so. 4800 copies were sold on first night of Mr. Murray's sale. It is rather a handsome volume. I hope it may do some good."

In a letter to Mr. James Young he writes of his voyage, and discharges a characteristic spurt of humor at a mutual Edinburgh acquaintance who had mistaken an order about a magic lantern:

"*At sea*, 300 *miles from Zanzibar, 26th January,* 1866.—We have enjoyed fair weather in coming across the weary waste of waters. We started on the 5th. The 'Thule,' to be a pleasure yacht, is the most incorrigible roller ever known. The whole 2000 miles has been an everlasting see-saw, shuggy-shoo, and enough to tire the patience of even a chemist, who is the most patient of all animals. I am pretty well gifted in that respect myself, though I say it that shouldn't say it, but that Sandy B——! The world will never get on till we have a few of those instrument-makers hung. I was particular in asking him to get me Scripture slides colored, and put in with the magic lantern, and he has not put in one! The very object for which I wanted it is thus frustrated, and I did not open it till we were at sea. O Sandy! Pity Burk and Hare have no successors in Auld Reekie! . . .

"You will hear that I have the prospect of Kirk being out here. I am very glad of it, as I am sure his services will be found invaluable on the East Coast."

To his daughter Agnes he writes, *à propos* of the rolling of the ship:

"Most of the marine Sepoys were sick. You would have been a victim unless you had tried the new remedy of a bag of pounded ice along the spine, which sounds as hopeful as the old cure for toothache: take a mouthful of cold water, and sit on the fire till it boils, you will suffer no more from toothache. . . . A shark took a bite at the revolving vane of the patent log to-day. He left some pieces of the enamel of his teeth in the brass, and probably has the toothache. You will sympathize with him. . . . If you ask Mr. Murray to send, by Mr. Conyngham, Buckland's *Curiosities of Natural History*, and Mr. Gladstone's *Address to the Edinburgh Students*, it will save me writing to him. When you return home you will be scrutinized to see if you are spoiled. You have only to act naturally and kindly to all your old friends to disarm them of their prejudices. I think you will find the Youngs true friends. Mrs. Williamson, of Widdicombe Hill, near Bath, writes to me that she would like to show you her plans for the benefit of poor orphans. If you thought of going to Bath it might be well to get all the insight you could into that and every other good work. It is well to be able to take a comprehensive view of all benevolent enterprises, and resolve to do our duty in life in some way or other, for we cannot live for ourselves alone. A life of selfishness is one of misery, and it is unlike that of our blessed Saviour, who pleased not Himself. He followed not his own will even, but the will of his Father in heaven. I have read with much pleasure a book called *Rose Douglas*. It is the life of a minister's daughter—with fictitious names, but all true. She was near Lanark, and came through Hamilton. You had better read it if you come in contact with it."

Referring to an alarm, arising from the next house having taken fire, of which she had written him, he adds playfully:

"You did not mention what you considered most precious on the night of the fire; so I dreamed that I saw one young lady hugging a German grammar to her bosom; another with a pair of curling tongs, a tooth-pick, and a pinafore; another with a bunch of used-up postage stamps and autographs in a crinoline turned upside down, and a fourth lifted up Madame Hocédé and insisted on carrying her as her most precious baggage. Her name, which I did not catch, will go down to posterity alongside of the ladies who each carried out her husband from the besieged city, and took care never to let him hear the last on't afterward. I am so penetrated with admiration of her that I enclose the wing of a flying-fish for her. It lighted among us last night, while we were at dinner, coming right through the skylight. You will make use of this fact in the *high-flying* speech which you will deliver to her in French."

Zanzibar is at length reached on the 28th January, after a voyage of twenty-three days, tedious enough, though but half the length of the cruise in the "Nyassa" two years before. To Agnes:

"*29th Jan.*—We went to call to-day on the Sultan. His Highness met us at the bottom of the stair, and as he shook hands a brass band, which he got at Bombay, blared forth 'God save the Queen'! This was excessively ridiculous, but I maintained sufficient official gravity. After coffee and sherbet we came away, and the wretched band now struck up 'The British Grenadier,' as if the fact of my being only 5 feet 8, and Brebner about 2 inches lower, ought not to have suggested 'Wee Willie Winkie' as more appropriate. I was ready to explode, but got out of sight before giving way."

Dr. Livingstone brought a very cordial recommendation to the Sultan from Sir Bartle Frere, and experienced much kindness at his hand. Being ill with toothache, the Sultan could not receive the gift of the "Thule" in person, and it was presented through his commodore.

Livingstone was detained in Zanzibar nearly two months waiting for H.M.S. "Penguin," which was to convey him to the mouth of the Rovuma. Zanzibar life was very monotonous—"It is the old, old way of living—eating, drinking, sleeping; sleeping, drinking, eating. Getting fat; slaving-dhows coming and slaving-dhows going away; bad smells; and kindly looks from English folks to each other." The sight of slaves in the Zanzibar market, and the recognition of some who had been brought from Nyassa, did not enliven his visit, though it undoubtedly confirmed his purpose and quickened his efforts to aim another blow at the accursed trade. Always thinking of what would benefit Africa, he writes to Sir Thomas Maclear urging very strongly the starting of a line of steamers between the Cape, Zanzibar, and Bombay: "It would be a most profitable one, and would do great good, besides, in eating out the trade in slaves."

At last the "Penguin" came for him, and once more, and for the last time, Livingstone left for the mainland continent.

CHAPTER XIX.

FROM ZANZIBAR TO UJIJI.

A.D. 1866–1869.

Dr. Livingstone goes to mouth of Rovuma—His prayer—His company—His herd of animals—Loss of his buffaloes—Good spirits when setting out—Difficulties at Rovuma—Bad conduct of Johanna men—Dismissal of his Sepoys—Fresh horrors of slave-trade—Uninhabited tract—He reaches Lake Nyassa—Letter to his son Thomas—Disappointed hopes—His double aim, to teach natives and rouse horror of slave-trade—Tenor of religious addresses —Wikatami remains behind—Livingstone finds no altogether satisfactory station for commerce and missions—Question of the watershed—Was it worth the trouble?—Overruled for good to Africa—Opinion of Sir Bartle Frere—At Marenga's—The Johanna men leave in a body—Circulate rumor of his murder—Sir Roderick disbelieves it—Mr. E. D. Young sent out with Search Expedition—Finds proof against rumor—Livingstone half-starved— Loss of his goats—Review of 1866—Reflections on Divine Providence— Letter to Thomas—His dog drowned—Loss of his medicine-chest—He feels sentence of death passed on him—First sight of Lake Tanganyika—Detained at Chitimba's—Discovery of Lake Moero— Occupations during detention of 1867—Great privations and difficulties—Illness—Rebellion among his men —Discovery of Lake Bangweolo—Its oozy banks—Detention—Sufferings— He makes for Ujiji—Very severe illness in beginning of 1869—Reaches Ujiji—Finds his goods have been wasted and stolen—Most bitter disappoint-ment—His medicines, etc., at Unyanyembe—Letter to Sultan of Zanzibar— Letters to Dr. Moffat and his daughter.

On the 19th of March, fortified by a firman from the Sultan to all his people, and praying the Most High to prosper him, "by granting him influence in the eyes of the heathen, and blessing his intercourse with them," Livingstone left Zanzibar in H.M.S. "Penguin" for the mouth of the Rovuma. His company consisted of thir-teen Sepoys, ten Johanna men, nine Nassick boys, two Shupanga men, and two Waiyau. Musa, one of the Johanna men, had been a sailor in the "Lady Nyassa";

Susi and Amoda, the Shupanga men, had been wood-cutters for the "Pioneer"; and the two Waiyau lads, Wikatani and Chuma, had been among the slaves rescued in 1861, and had lived for some time at the mission station at Chibisa's. Besides these, he carried with him a sort of menagerie in a dhow—six camels, three buffaloes and a calf, two mules, and four donkeys. What man but Dr. Livingstone would have encumbered himself with such baggage, and for what conceivable purpose except the benefit of Africa? The tame buffaloes of India were taken that he might try whether, like the wild buffaloes of Africa, they would resist the bite of the tsetse-fly; the other animals for the same purpose. There were two words of which Livingstone might have said, as Queen Mary said of Calais, that at his death they would be found engraven on his heart—fever and tsetse; the one the great scourge of man, the other of beast, in South Africa. To help to counteract two such foes to African civilization no trouble or expense would have been judged too great. Already he had lost nine of his buffaloes at Zanzibar. It was a sad pity that owing to the ill-treatment of the remaining animals by his people, who turned out a poor lot, it could never be known conclusively whether the tsetse-bite was fatal to them or not.

In spite of all he had suffered in Africa, and though he was without the company of a single European, he had, in setting out, something of the exhilarating feeling of a young traveler starting on his first tour in Switzerland, deepened by the sense of nobility which there is in every endeavor to do good to others. "The mere animal pleasure of traveling in a wild unexplored country is very great. . . . The sweat of one's brow is no longer a curse when one works for God; it proves a tonic to the system, and is actually a blessing." The Rovuma was found to have changed greatly since his last visit, so that he had to land his goods twenty-five miles to the north at Mikindany harbor, and find his way down to the river farther up. The toil was fitted to wear out the strongest of his men. Nothing could have been more grateful than the Sunday rest. Through his Nassick boys, he tried to teach the Makondé—a tribe that bore a very bad character, but failed; however, the people were wonderfully civil, and, contrary to all previous usage, neither inflicted fines nor made complaints, though the animals had done some damage to their corn. He set this down as an answer to his prayers for influence among the heathen.

His vexations, however, were not long of beginning. Both the Sepoy marines and the Nassick boys were extremely troublesome, and treated the animals abominably. The Johanna men were thieves. The Sepoys became so intolerable that after four months' trial he sent most of them back to the coast. It required an effort to resist the effect of such things, owing to the tendency of the mind to brood over the ills of travel. The natives were not unkindly, but food was very scarce. As they advanced, the horrors of the slave-trade presented themselves in all their hideous aspects. Women were found dead, tied to trees, or lying in the path shot and stabbed, their fault having been inability to keep up with the party, while their amiable owners, to prevent them from becoming the property of any one else, put an end to their lives. In some instances the captives, yet in the slave-sticks, were found not quite dead. Brutality was sometimes seen in another form, as when some natives laughed at a poor boy suffering from a very awkward form of hernia, whose mother was trying to bind up the part. The slave-trade utterly demoralized the people; the Arabs bought whoever was brought to them, and the great extent of forest in the country favored kidnapping; otherwise the people were honest.

Farther on they passed through an immense uninhabited

tract, that had once evidently had a vast population. Then, in the Waiyau country, west of Mataka's, came a splendid district 3400 feet above the sea, as well adapted for a settlement as Magomero, but it had taken them four months to get at it, while Magomero was reached in three weeks. The abandonment of that mission he would never cease to regret. As they neared Lake Nyassa, slave parties became more common. On the 8th August, 1866, they reached the lake, which seemed to Livingstone like an old familiar friend which he never expected to see again. He thanked God, bathed again in the delicious water, and felt quite exhilarated.

Writing to his son Thomas, 28th August, he says:

"The Sepoys were morally unfit for travel, and then we had hard lines, all of us. Food was not to be had for love or money. Our finest cloths only brought miserable morsels of the common grain. I trudged it the whole way, and having no animal food save what turtle-doves and guinea-fowls we occasionally shot, I became like one of Pharaoh's lean kine. The last tramp [to Nyassa] brought us to a land of plenty. It was over a very fine country, but quite depopulated. . . . The principal chief, named Mataka, lives on the watershed overhanging this, but fifty miles or more distant from this; his town contained a thousand houses—many of them square, in imitation of the Arabs. Large patches of English peas in full bearing grew in the moist hollows, or were irrigated. Cattle showed that no tsetse existed. When we arrived, Mataka was just sending back a number of cattle and captives to their own homes. They had been taken by his people without his knowledge from Nyassa. I saw them by accident: there were fifty-four women and children, about a dozen young men and boys, and about twenty-five or thirty head of cattle. As the act was spontaneous, it was the more gratifying to witness. . . .

"I sometimes remember you with some anxiety, as not knowing what opening may be made for you in life. . . . Whatever you feel yourself best fitted for, 'commit thy way to the Lord, trust also in Him, and He will bring it to pass.' One ought to endeavor to devote the peculiarities of his nature to his Redeemer's service, whatever these may be."

Resting at the lake, and working up journal, lunars, and altitudes, he hears of the arrival of an Englishman at

Mataka's, with cattle for him, "who had two eyes behind as well as two in front—news enough for awhile." Zoology, botany, and geology engage his attention as usual. He tries to get across the lake, but cannot, as the slavers own all the dhows, and will neither lend nor sell to him; he has therefore to creep on foot round its southern end. Marks of destruction and desolation again shock the eye —skulls and bones everywhere. At the point where the Shiré leaves Nyassa, he could not but think of disappointed hopes—the death of his dear wife, and of the Bishop, the increasing vigor of the slave-trade, and the abandonment of the Universities Mission. But faith assured him of good times coming, though he might not live to see them. Would only he had seen through the vista of the next ten years! Bishop Tozer done with Africa, and Bishop Steere returning to the old neighborhood, and resuming the old work of the Universities Mission; and his own countrymen planted his name on the promontory on which he gazed so sorrowfully, training the poor natives in the arts of civilization, rearing Christian households among them, and proclaiming the blessed Gospel of the God of love!

Invariably as he goes along, Dr. Livingstone aims at two things: at teaching some of the great truths of Christianity, and rousing consciences on the atrocious guilt of the slave-trade. In connection with the former he discovers that his usual way of conducting divine service— by the reading of prayers—does not give ignorant persons any idea of an unseen Being; kneeling and praying with the eyes shut is better. At the foot of the lake he goes out of his way to remonstrate with Mukaté, one of the chief marauders of the district. The tenor of his addresses is in some degree shaped by the practices he finds so prevalent:

"We mention our relationship to our Father, the guilt of selling any of his children, the consequences:—*e.g.* it begets war, for as they don't like to sell their own, they steal from other villagers, who retaliate. Arabs and Waiyau, invited into the country by their selling, foster feuds,—wars and depopulation ensue. We mention the Bible—future state—prayer; advise union, that they would unite as one family to expel enemies, who came first as slave-traders, and ended by leaving the country a wilderness."

It was about this time that Wikatani, one of the two Waiyau boys who had been rescued from slavery, finding, as he believed or said, some brothers and sisters on the western shore of the lake, left Livingstone and remained with them. There had been an impression in some quarters, that, according to his wont, Livingstone had made him his slave; to show the contrary, he gave him his choice of remaining or going, and, when the boy chose to remain, he acquiesced.

Dr. Livingstone had ere now passed over the ground where, if anywhere, he might have hoped to find a station for a commercial and missionary settlement, independent of the Portuguese. In this hope he was rather disappointed. The only spot he refers to is the district west of Mataka's, which, however, was so difficult of access. Nearer the coast a mission might be established, and to this project his mind turned afterward; but it would not command the Nyassa district. On the whole he preferred the Zambesi and Shiré valley, with all their difficulties. But the Rovuma was not hopeless, and indeed, within the last few years, the Universities Mission has occupied the district successfully.

The geographical question of the watershed had now to be grappled with. It is natural to ask whether this question was of sufficient importance to engage his main energies, and justify the incalculable sacrifices undergone by him during the remaining six years of his life. First of all, we must remember, it was not his own scheme

—it was pressed on him by Sir Roderick Murchison and the Geographical Society; and it may perhaps be doubted whether, had he foreseen the cost of the enterprise, he would have deemed the object worthy of the price. But ever and anon, he seemed to be close on what he was searching for, and certain to secure it by just a little further effort; while as often, like the cup of Tantalus, it was snatched from his grasp. Moreover, during a lifetime of splendid self-discipline, he had been training himself to keep his promises, and to complete his tasks; nor could he in any way see it his duty to break the one or leave the other unfinished. He had undertaken to the Geographical Society to solve that problem, and he would do it if it could be done. Wherever he went he had always some opportunity to make known the fatherhood of God and his love in Christ, although the seed he sowed seemed seldom to take root. Then he was gathering fresh information on the state of the country and the habits of the people. He was especially gathering information on the accursed slave-trade.

This question of the watershed, too, had fascinated his mind, for he had a strong impression that the real sources of the Nile were far higher than any previous traveler had supposed—far higher than Lake Victoria Nyanza, and that it would be a service to religion as well as science to discover the fountains of the stream on whose bosom, in the dawn of Hebrew history, Moses had floated in his ark of bulrushes. A strong impression lurked in his mind that if he should only solve that old problem he would acquire such influence that new weight would be given to his pleadings for Africa; just as, at the beginning of his career, he had wished for a commanding style of composition, to be able to rouse the attention of the world to that ill-treated continent.

He was strongly disposed to think that in the account of the sources given to Herodotus by the Registrar of

Minerva in the temple of Saïs, that individual was not joking, as the father of history supposed. He thought that in the watershed the two conical hills, Crophi and Mophi might be found, and the fountains between them which it was impossible to fathom; and that it might be seen that from that region there was a river flowing north to Egypt, and another flowing south to a country that might have been called Ethiopia. But whatever might be his views or aims, it was ordained that in the wanderings of his last years he should bring within the sympathies of the Christian world many a poor tribe otherwise unknown; that he should witness sights, surpassing all he had ever seen before of the inhumanity and horrors of the slave-traffic—sights that harrowed his inmost soul; and that when his final appeal to his countrymen on behalf of its victims came, not from his living voice but from his tomb, it should gather from a thousand touching associations a thrilling power that would rouse the world, and finally root out the accursed thing.

A very valuable testimony was borne by Sir Bartle Frere to the real aims of Livingstone, and the value of his work, especially in this last journey, in a speech delivered in the Glasgow Chamber of Commerce, 10th November, 1876:

"The object," he said, "of Dr. Livingstone's geographical and scientific explorations was to lead his countrymen to the great work of Christianizing and civilizing the millions of Central Africa. You will recollect how, when first he came back from his wonderful journey, though we were all greatly startled by his achievements and by what he told us, people really did not lay what he said much to heart. They were stimulated to take up the cause of African discovery again, and other travelers went out and did excellent service; but the great fact which was from the very first upon Livingstone's mind, and which he used to impress upon you, did not make the impression he wished, and although a good many people took more and more interest in the civilization of Africa and in the abolition of the slave-trade, which he pointed out was the great obstacle to all progress, still it did not come home to the people generally. It was not until his third and last journey, when he was no more to return among us, that the descriptions which he gave of the horrors of the slave-trade in the interior really took hold upon the mind of the people of this country, and made them determine that what used to be considered the crotchet of a few religious minds and humanitarian sort of persons, should be a phase of the great work which this country had undertaken, to free the African races, and to abolish, in the first place, the slave-trade by sea, and then, as we hope, the slaving by land."

In September an Arab slaver was met at Marenga's, who told Musa, one of the Johanna men, that all the country in front was full of Mazitu, a warlike tribe; that forty-four Arabs and their followers had been killed by them at Kasunga, and that he only had escaped. Musa's heart was filled with consternation. It was in vain that Marenga assured him that there were no Mazitu in the direction in which he was going, and that Livingstone protested to him that he would give them a wide berth. The Johanna men wanted an excuse for going back, but in such a way that, when they reached Zanzibar, they should get their pay. They left him in a body, and when they got to Zanzibar, circulated a circumstantial report that he had been murdered. In December, 1866, Musa appeared at Zanzibar, and told how Livingstone had crossed Lake Nyassa to its western or northwestern shore, and was pushing on west or northwest, when, between Marenga and Maklisoora, a band of savages stopped their way, and rushed on him and his small band of followers, now reduced to twenty. Livingstone fired twice, and killed two; but, in the act of reloading, three Mafite leaped upon him through the smoke, one of them felled him with an axe-cut from behind, and the blow nearly severed his head from his body. The Johanna men fled into the thick jungle, and miraculously escaped. Returning to the scene of the tragedy, they found the body of their master, and in a shallow grave dug with some stakes, they committed his remains to the ground.

Many details were given regarding the Sepoys, and regarding the after fortunes of Musa and his companions. Under cross-examination Musa stood firmly to his story, which was believed both by Dr. Seward and Dr. Kirk, of Zanzibar. But when the tidings reached England, doubt was thrown on them by some of those best qualified to judge. Mr. Edward D. Young, who had had dealings with Musa, and knew him to be a liar, was suspicious of the story; so was Mr. Horace Waller. Sir Roderick Murchison, too, proclaimed himself an unbeliever, notwithstanding all the circumstantiality and apparent conclusiveness of the tale. The country was resounding with lamentations, the newspapers were full of obituary notices, but the strong-minded disbelievers were not to be moved.

Sir Roderick and his friends of the Geographical Society determined to organize a search expedition, and Mr. E. D. Young was requested to undertake the task. In May, 1867, all was ready for the departure of the Expedition; and on the 25th July, Mr. E. D. Young, who was accompanied by Mr. Faulkner, John Reid, and Patrick Buckley, cast anchor at the mouth of the Zambesi. A steel boat named "The Search," and some smaller boats, were speedily launched, and the party were moving up the river. We have no space for an account of Mr. Young's most interesting journey, not even for the detail of that wonderful achievement, the carrying of the pieces of the "Search" past the Murchison Cataracts, and their reconstruction at the top, without a single piece missing. The sum and substance of Mr. Young's story was, that first, quite unexpectedly, he came upon a man near the south end of Lake Nyassa, who had seen Livingstone there, and who described him well, showing that he had not crossed at the north end, as Musa had said, but, for some reason, had come round by the south; then, the chief Marenga not only told him of Livingstone's stay there, but also

34

of the return of Musa, after leaving him, without any story of his murder; also, at Mapunda, they came on traces of the boy Wikatani, and learned his story, though they did not see himself. The most ample proof of the falsehood of Musa's story was thus obtained, and by the end of 1867, Mr. Young, after a most active, gallant, and successful campaign, was approaching the shores of England.[1] No enterprise could have brought more satisfactory results, and all in the incredibly short period of eight months.

Meanwhile, Livingstone, little thinking of all the commotion that the knave Musa had created, was pushing on in the direction of Lake Tanganyika. Though it was not true that he had been murdered, it was true that he was half-starved. The want of other food compelled him to subsist to a large extent on African maize, the most tasteless and unsatisfying of food. It never produced the feeling of sufficiency, and it would set him to dream of dinners he had once eaten, though dreaming was not his habit, except when he was ill. Against his will, the thought of delicious feasts would come upon him, making it all the more difficult to be cheerful, with, probably, the poorest fare on which life could be in any way maintained. To complete his misery, his four goats were lost, so that the one comfort of his table—a little milk along with his maize—was taken from him when most eagerly sought and valued.

In reviewing the year 1866, he finds it less productive of results than he had hoped for: "We now end 1866. It has not been so fruitful or useful as I intended. Will try to do better in 1867, and be better—more gentle and loving; and may the Almighty, to whom I commit my way, bring my desires to pass, and prosper me! Let all the sins of '66 be blotted out, for Jesus' sake. May He who was full of grace and truth impress his character on

---

[1] See *The Search for Livingstone*, by E. D. Young: London, 1868.

mine: grace—eagerness to show favor; truth—truthfulness, sincerity, honor—for his mercy's sake."

Habitually brave and fearless though Livingstone was, it was not without frequent self-stimulation, and acts of faith in unseen truth, that the peace of his mind was maintained. In the midst of his notes of progress, such private thoughts as the following occur from time to time: "It seems to have been a mistake to imagine that the Divine Majesty on high was too exalted to take any notice of our mean affairs. The great minds among men are remarkable for the attention they bestow on minutiæ. An astronomer cannot be great unless his mind can grasp an infinity of very small things, each of which, if unattended to, would throw his work out. A great general attends to the smallest details of his army. The Duke of Wellington's letters show his constant attention to minute details. And so with the Supreme Mind of the universe, as He is revealed to us in his Son. 'The very hairs of your head are all numbered.' 'A sparrow cannot fall to the ground without your Father.' 'He who dwelleth in the light which no man can approach unto' condescends to provide for the minutest of our wants, directing, guarding, and assisting in each hour and moment, with an infinitely more vigilant and excellent care than our own utmost self-love can ever attain to. With the ever-watchful, loving eye constantly upon me, I may surely follow my bent, and go among the heathen in front, bearing the message of peace and good-will. All appreciate the statement that it is offensive to our common Father to sell and kill his children. I will therefore go, and may the Almighty help me to be faithful!"

Writing to his son Thomas, 1st February, 1867, he complains again of his terrible hunger:

"The people have nothing to sell but a little millet-porridge and mushrooms. Woe is me! good enough to produce fine dreams of the roast beef of old England, but nothing else. I have become very thin, though I was so before; but now, if you weighed me, you might calcu-late very easily how much you might get for the bones. But—we got a cow yesterday, and I am to get milk to-morrow. . . . I grieve to write it, poor poodle 'Chitane' was drowned" [15th January, in the Chimbwé]; "he had to cross a marsh a mile wide, and waist-deep. . . . I went over first, and forgot to give directions about the dog— all were too much engaged in keeping their balance to notice that he swam among them till he died. He had more spunk than a hundred country dogs—took charge of the whole line of march, ran to see the first in the line, then back to the last, and barked to haul him up; then, when he knew what hut I occupied, would not let a country cur come in sight of it, and never stole himself. We have not had any difficulties with the people, made many friends, imparted a little knowl-edge sometimes, and raised a protest against slavery very widely."

The year 1867 was signalized by a great calamity, and by two important geographical feats. The calamity was the loss of his medicine-chest. It had been intrusted to one of his most careful people; but, without authority, a carrier hired for the day took it and some other things to carry for the proper bearer, then bolted, and neither carrier nor box could be found. "I felt," says Living-stone, "as if I had now received the sentence of death, like poor Bishop Mackenzie." With the medicine-chest was lost the power of treating himself in fever with the medicine that had proved so effectual. We find him not long after in a state of insensibility, trying to raise him-self from the ground, falling back with all his weight, and knocking his head upon a box. The loss of the medicine-box was probably the beginning of the end; his system lost the wonderful power of recovery which it had hitherto shown; and other ailments—in the lungs, the feet, and the bowels, that might have been kept under in a more vigor-ous state of general health, began hereafter to prevail against him.

The two geographical feats were—his first sight of Lake Tanganyika, and his discovery of Lake Moero. In April he reached Lake Liemba, as the lower part of Tanganyika was called. The scenery was wonderfully beautiful, and the air of the whole region remarkably peaceful. The want of medicine made an illness here very severe; on recovering, he would have gone down the lake, but was dissuaded, in consequence of his hearing that a chief was killing all that came that way. He therefore returns to Chitimba's, and resolves to explore Lake Moero, believing that there the question of the watershed would be decided. At Chitimba's, he is detained upward of three months, in consequence of the disturbed state of the country. At last he gets the escort of some Arab traders, who show him much kindness, but again he is prostrated by illness, and at length he reaches Lake Moero, 8th November, 1867. He hears of another lake, called Bembo or Bangweolo, and to hear of it is to resolve to see it. But he is terribly wearied with two years' traveling without having heard from home, and he thinks he must first go to Ujiji, for letters and stores. Meanwhile, as the traders are going to Casembe's, he accompanies them thither. Casembe he finds to be a fierce chief, who rules his people with great tyranny, cutting off their ears, and even their hands, for the most trivial offenses. Persons so mutilated, seen in his village, excite a feeling of horror. This chief was not one easily got at, but Livingstone believed that he gained an influence with him, only he could not quite overcome his prejudice against him. The year 1867 ended with another severe attack of illness.

"The chief interest in Lake Moero," says Livingstone, "is that it forms one of a chain of lakes, connected by a river some 500 miles in length. First of all, the Chambezé rises in the country of Mambwé, N.E. of Molemba; it then flows southwest and west, till it reaches lat. 11° s., and long. 29° E., where it forms Lake Bemba or Bangweolo; emerging thence, it assumes the name of Luapula, and comes down here to fall into Moero. On going out of this lake it is known by the name of Lualaba, as it flows N.W. in Rua to form another lake with many islands, called Urengé or Ulengé. Beyond this, information is not positive as to whether it enters Lake Tanganyika, or another lake beyond that. . . . Since coming to Casembe's, the testimony of natives and Arabs has been so united and consistent, that I am but ten days from Lake Bemba or Bangweolo, that I cannot doubt its accuracy."

The detentions experienced in 1867 were long and wearisome, and Livingstone disliked them because he was never well when doing nothing. His light reading must have been pretty well exhausted; even *Smith's Dictionary of the Bible*, which accompanied him in these wanderings, and which we have no doubt he read throughout, must have got wearisome sometimes. He occupied himself in writing letters, in the hope that somehow or sometime he might find an opportunity of despatching them. He took the rainfall carefully during the year, and lunars and other observations, when the sky permitted. He had intended to make his observations more perfect on this journey than on any previous one, but alas for his difficulties and disappointments! A letter to Sir Thomas Maclear and Mr. Mann, his assistant, gives a pitiful account of these: "I came this journey with a determination to observe very carefully all your hints as to occultations and observations, east and west, north and south, but I have been so worried by lazy, deceitful Sepoys, and thievish Johanna men, and indifferent instruments, that I fear the results are very poor." He goes on to say that some of his instruments were defective, and others went out of order, and that his time-taker, one of his people, had no conscience, and could not be trusted. The records of his observations, notwithstanding, indicate much care and pains. In April, he had been very unwell, taking fits of total insensibility, but as he had not said anything of this to his people at home, it was to be kept a secret.

His Journal for 1867 ends with a statement of the poverty of his food, and the weakness to which he was reduced. He had hardly anything to eat but the coarsest grain of the country, and no tea, coffee, or sugar. An Arab trader, Mohamad Bogharib, who arrived at Casembe's about the same time, presented him with a meal of vermicelli, oil, and honey, and had some coffee and sugar; Livingstone had had none since he left Nyassa. The Journal for 1868 begins with a prayer that if he should die that year, he might be prepared for it. The year was spent in the same region, and was signalized by the discovery of Lake Bemba, or, as it may more properly be called, Lake Bangweolo. Early in the year he heard accounts of what interested him greatly—certain underground houses in Rua, ranging along a mountain side for twenty miles. In some cases the doorways were level with the country adjacent; in others, ladders were used to climb up to them; inside they were said to be very large, and not the work of men, but of God. He became eagerly desirous to visit these mysterious dwellings.

Circumstances turning out more favorable to his going to Lake Bangweolo, Dr. Livingstone put off his journey to Ujiji, on which his men had been counting, and much against the advice of Mohamad, his trader friend and companion, determined first to see the lake of which he had heard so much. The consequence was a rebellion among his men. With the exception of five, they refused to go with him. They had been considerably demoralized by contact with the Arab trader and his slave-gang. Dr. Livingstone took this rebellion with wonderful placidity, for in his own mind he could not greatly blame them. It was no wonder they were tired of the everlasting tramping, for he was sick of it himself. He reaped the fruit of his mildness by the men coming back to him, on his return from the lake, and offering their services. It cannot be said of him that he was not disposed to make any allowance for human weakness. When recording a fault, and how he dealt with it, he often adds, " consciousness of my own defects makes me lenient." " I also have my weaknesses."

The way to the lake was marked by fresh and lamentable tokens of the sufferings of slaves. "24*th June.—* Six men-slaves were singing as if they did not feel the weight and degradation of the slave-sticks. I asked the cause of their mirth, and was told that they rejoiced at the idea of 'coming back after death, and haunting and killing those who had sold them.' Some of the words I had to inquire about; for instance, the meaning of the words, 'to haunt and kill by spirit power;' then it was, 'Oh, you sent me off to Manga (sea-coast), but the yoke is off when I die, and back I shall come to haunt and to kill you.' Then all joined in the chorus, which was the name of each vendor. It told not of fun, but of the bitterness and tears of such as were oppressed; and on the side of the oppressors there was power. There be higher than they!"

His discovery of Lake Bangweolo is recorded as quietly as if it had been a mill-pond: "On the 18th July, I walked a little way out, and saw the shores of the lake for the first time, thankful that I had come safely hither." The lake had several inhabited islands, which Dr. Livingstone visited, to the great wonder of the natives, who crowded around him in multitudes, never having seen such a curiosity as a white man before. In the middle of the lake the canoe-men whom he had hired to carry him across refused to proceed further, under the influence of some fear, real or pretended, and he was obliged to submit. But the most interesting, though not the most pleasant, thing about the lake, was the ooze or sponge which occurred frequently on its banks. The spongy places were slightly depressed valleys, without trees or bushes, with grass a foot or fifteen inches high; they were usually from two to ten miles long, and from a quarter of a mile to a mile broad. In the course of thirty geographical miles, he crossed twenty-nine, and that, too, at the end of the fourth month of the dry season. It was necessary for him to strip the lower part of his person before fording them, and then the leeches pounced on him, and in a moment had secured such a grip, that even twisting them round the fingers failed to tear them off.

It was Dr. Livingstone's impression at this time that in discovering Lake Bangweolo, with the sponges that fed it, he had made another discovery—that these marshy places might be the real sources of the three great rivers, the Nile, the Congo, and the Zambesi. A link, however, was yet wanting to prove his theory. It had yet to be shown that the waters that flowed from Lake Bangweolo into Lake Moero, and thence northward by the river Lualaba, were connected with the Nile system. Dr. Livingstone was strongly inclined to believe that this connection existed; but toward the close of his life he had more doubts of it, although it was left to others to establish conclusively that the Lualaba was the Congo, and sent no branch to the Nile.

On leaving Lake Bangweolo, detention occurred again as it had occurred before. The country was very disturbed and very miserable, and Dr. Livingstone was in great straits and want. Yet with a grim humor he tells how, when lying in an open shed, with all his men around him, he dreamed of having apartments at Mivart's Hotel. It was after much delay that he found himself at last, under the escort of a slave-party, on the way to Ujiji. Mr. Waller has graphically described the situation. "At last he makes a start on the 11th of December, 1868, with the Arabs, who are bound eastward for Ujiji. It is a motley group, composed of Mohamad and his friends, a gang of Unyamwezi hangers-on, and strings of wretched slaves yoked together in their heavy slave-sticks. Some carry ivory, others copper, or food for the march, while hope and fear, misery and villainy, may be read off on the various faces that pass in line out of this country, like a serpent dragging its accursed folds away from the victim it has paralyzed with its fangs."

New Year's Day, 1869, found Livingstone laboring under a worse attack of illness than any he had ever had before. For ten weeks to come his situation was as painful as can be conceived. A continual cough, night and day, the most distressing weakness, inability to walk, yet the necessity of moving on, or rather of being moved on, in a kind of litter arranged by Mohamad Bogharib,— where, with his face poorly protected from the sun, he was jolted up and down and sideways, without medicine or food for an invalid,—made the situation sufficiently trying. His prayer was that he might hold out to Ujiji, where he expected to find medicines and stores, with the rest and shelter so necessary in his circumstances. So ill was he, that he lost count of the days of the week and the month. "I saw myself lying dead in the way to Ujiji, and all the letters I expected there—useless. When I think of my children, the lines ring through my head perpetually:

> "'I shall look into your faces,
> And listen to what you say;
> And be often very near you
> When you think I'm far away.'"

On the 26th February, 1869, he embarked in a canoe on Tanganyika, and on the 14th March he reached the longed-for Ujiji, on the eastern shore of the lake. To complete his trial, he found that the goods he expected had been made away with in every direction. A few fragments were about all he could find. Medicines, wine, and cheese had been left at Unyanyembe, thirteen days distant. A war was raging on the way, so that they could not be sent for till the communications were restored.

To obviate as far as possible the recurrence of such a disaster to a new store of goods which he was now asking Dr. Kirk to send him, Livingstone wrote a letter to the Sultan of Zanzibar, 20th April, 1869, in which he frankly and cordially acknowledged the benefit he had derived from the letter of recommendation his Highness had given him, and the great kindness of the Arabs, especially

Mohamad Bogharib, who had certainly saved his life. Then he complains of the robbery of his goods, chiefly by one Musa bin Salim, one of the people of the Governor of Unyanyembe, who had bought ivory with the price, and another man who had bought a wife. Livingstone does not expect his cloth and beads to be brought back, or the price of the wife and ivory returned, but he says:

"I beg the assistance of your authority to prevent a fresh stock of goods, for which I now send to Zanzibar, being plundered in the same way. Had it been the loss of ten or twelve pieces of cloth only, I should not have presumed to trouble your Highness about the loss; but 62 pieces or gorahs out of 80, besides beads, is like cutting a man's throat. If one or two guards of good character could be sent by you, no one would plunder the pagasi next time.

"I wish also to hire twelve or fifteen good freemen to act as canoe-men or porters, or in any other capacity that may be required. I shall be greatly obliged if you appoint one of your gentlemen who knows the country to select that number, and give them and their headman a charge as to their behavior. If they know that you wish them to behave well it will have great effect. I wish to go down Tanganyika, through Luanda and Chowambe, and pass the river Karagwe, which falls into Lake Chowambe. Then come back to Ujiji, visit Manyuema and Rua, and then return to Zanzibar, when I hope to see your Highness in the enjoyment of health and happiness."

Livingstone showed only his usual foresight in taking these precautions for the protection of his next cargo of goods. In stating so plainly his intended route, his purpose was doubtless to prevent carelessness in executing his orders, such as might have arisen had it been deemed uncertain where he was going, and whether or not he meant to return by Zanzibar.

Of letters during the latter part of this period very few seem to have reached their destination. A short letter to

Dr. Moffat, bearing date "Near Lake Moero, March, 1868,"
dwells dolefully on his inability to reach Lake Bemba in
consequence of the flooded state of the country, and then
his detention through the strifes of the Arabs and the
natives. The letter, however, is more occupied with reviewing the past than narrating the present. In writing
to Dr. Moffat, he enters more minutely than he would
have done with a less intimate and sympathetic friend
into the difficulties of his lot—difficulties that had been
increased by some from whom he might have expected
other things. He had once seen a map displayed in the
rooms of the Geographical Society, substantially his own,
but with another name in conspicuous letters. On the
Zambesi he had had difficulties, little suspected, of which
in the meantime he would say nothing to the public. A
letter to his daughter Agnes, after he had gone to Bangweolo, dwells also much on his past difficulties—as if he
felt that the slow progress he was making at the moment
needed explanation or apology. Amid such topics, almost
involuntary touches of the old humor occur: "I broke
my teeth tearing at maize and other hard food, and they
are coming out. One front tooth is out, and I have such
an awful mouth. If you expect a kiss from me, you must
take it through a speaking-trumpet." In one respect,
amid all his trials, his heart seems to become more tender
than ever—in affection for his children, and wise and
considerate advice for their guidance. In his letter to
Agnes, he adverts with some regret to a chance he lost of
saying a word for his family when Lord Palmerston sent
Mr. Hayward, Q.C., to ask him what he could do to serve
him. "It never occurred to me that he meant anything
for me or my children till I was out here. I thought only
of my work in Africa, and answered accordingly." It was
only the fear that his family would be in want that occasioned this momentary regret at his disinterested answer
to Lord Palmerston.

CHAPTER XX.

MANYUEMA.

A.D. 1869–1871.

He sets out to explore Manyuema and the river Lualaba—Loss of forty-two letters—His feebleness through illness—He arrives at Bambarré—Becomes acquainted with the soko or gorilla—Reaches the Luama River—Magnificence of the country—Repulsiveness of the people—Cannot get a canoe to explore the Lualaba—Has to return to Bambarré—Letter to Thomas, and retrospect of his life—Letter to Sir Thomas Maclear and Mr. Mann—Miss Tinné—He is worse in health than ever, yet resolves to add to his programme and go round Lake Bangweolo—Letter to Agnes—Review of the past—He sets out anew in a more northerly direction—Overpowered by constant wet—Reaches Nyangwe—Long detention—Letter to his brother John—Sense of difficulties and troubles—Nobility of his spirit—He sets off with only three attendants for the Lualaba—Suspicions of the natives—Influence of Arab traders—Frightful difficulties of the way—Lamed by footsores—Has to return to Bambarré—Long and wearisome detention—Occupations—Meditations and reveries—Death no terror—Unparalleled position and trials—He reads his Bible from beginning to end four times—Letter to Sir Thomas Maclear—To Agnes—His delight at her sentiments about his coming home—Account of the soko—Grief to hear of death of Lady Murchison—Wretched character of men sent from Zanzibar—At last sets out with Mohamad—Difficulties—Slave-trade most horrible—Cannot get canoes for Lualaba—Long waiting—New plan—Frustrated by horrible massacre on banks of Lualaba—Frightful scene—He must return to Ujiji—New illness—Perils of journey to Ujiji—Life three times endangered in one day—Reaches Ujiji—Shereef has sold off his goods—He is almost in despair—Meets Henry M. Stanley and is relieved—His contributions to Natural Science during last journeys—Professor Owen in the *Quarterly Review*.

AFTER resting for a few weeks at Ujiji, Dr. Livingstone set out, 12th July, 1869, to explore the Manyuema country. Ujiji was not a place favorable for making arrangements; it was the resort of the worst scum of Arab traders. Even to send his letters to the coast was a difficult undertaking, for the bearers were afraid he would expose their doings.

On one day he despatched no fewer than forty-two—enough, no doubt, to form a large volume; none of these ever arrived at Zanzibar, so that they must have been purposely destroyed. The slave-traders of Urungu and Itawa, where he had been, were gentlemen compared with those of Ujiji, who resembled the Kilwa and Portuguese, and with whom trading was simply a system of murder. Here lay the cause of Livingstone's unexampled difficulties at this period of his life; he was dependent on men who were not only knaves of the first magnitude, but who had a special animosity against him, and a special motive to deceive, rob, and obstruct him in every possible way.

After considerable deliberation he decided to go to Manyuema, in order to examine the river Lualaba, and determine the direction of its flow. This would settle the question of the watershed, and in four or five months, if he should get guides and canoes, his work would be done. On setting out from Ujiji he first crossed the lake, and then proceeded inland on foot. He was still weak from illness, and his lungs were so feeble that to walk up-hill made him pant. He became stronger, however, as he went on, refreshed doubtless by the interesting country through which he passed, and the aspect of the people, who were very different from the tribes on the coast.

On the 21st September he arrived at Bambarré, in Manyuema, the village of the Chief Moenékuss. He found the people in a state of great isolation from the rest of the world, with nothing to trust to but charms and idols,—both being bits of wood. He made the acquaintance of the soko or gorilla, not a very social animal, for it always tries to bite off the ends of its captor's fingers and toes. Neither is it particularly intellectual, for its nest shows no more contrivance than that of a cushat dove. The curiosity of the people was very great, and sometimes it took an interesting direction. "Do people die with you?" asked two intelligent young men. "Have you no charm against death? Where do people go after death?" Livingstone spoke to them of the great Father, and of their prayers to Him who hears the cry of his children; and they thought this to be natural.

He rested at Bambarré till the 1st of November, and then went westward till he reached the Luamo River, and was within ten miles of its confluence with the Lualaba. He found the country surpassingly beautiful: " Palms crown the highest heights of the mountains, and their gracefully-bent fronds wave beautifully in the wind. Climbers of cable size in great numbers are hung among the gigantic trees; many unknown wild fruits abound, some the size of a child's head, and strange birds and monkeys are everywhere. The soil is excessively rich, and the people, though isolated by old feuds that are never settled, cultivate largely."

The country was very populous, and Livingstone so excited the curiosity of the people that he could hardly get quit of the crowds. It was not so uninteresting to be stared at by the women, but he was wearied with the ugliness of the men. Palm-toddy did not inspire them with any social qualities, but made them low and disagreeable. They had no friendly feeling for him, and could not be inspired with any. They thought that he and his people were like the Arab traders, and they would not do anything for them. It was impossible to procure a canoe for navigating the Lualaba, so that there was nothing for it but to return to Bambarré, which was reached on the 19th December, 1869.

A long letter to his son Thomas (Town of Moenékuss, Manyuema Country, 24th September, 1869) gives a retrospect of this period, and indeed, in a sense, of his life:

"My dear Tom,—I begin a letter, though I have no prospect of being able to send it off for many months to come. It is to have something in readiness when the hurry usual in preparing a mail does arrive. I am in the Manyuema Country, about 150 miles west of Ujiji, and at the town of Moenekoos or Moenékuss, a principal chief among the reputed cannibals. His name means 'Lord of the light-gray parrot with a red tail,' which abounds here, and he points away still further west to the country of the real cannibals. His people laugh, and say, 'Yes, we eat the flesh of men,' and should they see the inquirer to be credulous, enter into particulars. A black stuff smeared on the cheeks is the sign of mourning, and they told one of my people who believes all they say that it is animal charcoal made of the bones of the relatives they have eaten. They showed him the skull of one recently devoured, and he pointed it out to me in triumph. It was the skull of a gorilla, here called 'soko,' and this they do eat. They put a bunch of bananas in his way, and hide till he comes to take them, and spear him. Many of the Arabs believe firmly in the cannibal propensity of the Manyuema. Others who have lived long among them, and are themselves three-fourths African blood, deny it. I suspect that this idea must go into oblivion with those of people who have no knowledge of fire, of the Supreme Being, or of language. The country abounds in food,—goats, sheep, fowls, buffaloes, and elephants: maize, holcuserghum, cassaba, sweet potatoes, and other farinaceous eatables, and with ground-nuts, palm-oil, palms, and other fat-yielding nuts, bananas, plantains, sugar-cane in great plenty. So there is little inducement to eat men, but I wait for further evidence.

"Not knowing how your head has fared, I sometimes feel greatly distressed about you, and if I could be of any use I would leave my work unfinished to aid you. But you will have every medical assistance that can be rendered, and I cease not to beg the Lord who healeth his people to be gracious to your infirmity.

"The object of my Expedition is the discovery of the sources of the Nile. Had I known all the hardships, toil, and time involved, I would of been of the mind of St. Mungo, of Glasgow, of whom the song says that he let the Molendinar Burn 'rin by,' when he could get something stronger. I would have let the sources 'rin by' to Egypt, and never been made 'drumly' by my plashing through them. But I shall make this country and people better known. 'This,' Professor Owen said to me, 'is the first step; the rest will in due time follow.' By different agencies the Great Ruler is bringing all things into a focus. Jesus is gathering all things unto Himself, and He is daily becoming more and more the centre of the world's hopes and of the world's fears. War brought freedom to 4,000,000 of the most hopeless and helpless slaves. The world never saw such fiendishness as that with which the Southern slaveocracy clung to slavery. No power in this world or the next would ever make them relax their iron grasp. The lie had entered into their soul. Their cotton was King. With it they would force England and France to make them independent, because without it the English and French must starve. Instead of being made a nation, they made a nation of the North. War has elevated and purified the Yankees, and now they have the gigantic task laid at their doors to elevate and purify 4,000,000 of slaves. I earnestly hope that the Northerners may not be found wanting in their portion of the superhuman work. The day for Africa is yet to come. Possibly the freed men may be an agency in elevating their fatherland.

"England is in the rear. This affair in Jamaica brought out the fact of a large infusion of bogiephobia in the English. Frightened in early years by their mothers with 'Bogie Blackman,' they were terrified out of their wits by a riot, and the sensation writers, who act the part of the 'dreadful boys' who frightened aunts, yelled out that emancipation was a mistake. 'The Jamaica negroes were as savage as when they left Africa.' They might have put it much stronger by saying, as the rabble that attended Tom Sayers's funeral, or that collects at every execution at Newgate. But our golden age is not in the past. It is in the future—in the good time coming yet for Africa and for the world.

"The task I undertook was to examine the watershed of South Central Africa. This was the way Sir Roderick put it, and though he mentioned it as the wish of the Geographical Council, I suspect it was his own idea; for two members of the Society wrote out 'instructions' for me, and the watershed was not mentioned. But scientific words were used which the writers evidently did not understand.

"The examination of the watershed contained the true scientific mode of procedure, and Sir Roderick said to me: 'You will be the discoverer of the sources of the Nile.' I shaped my course for a path across the north end of Lake Nyassa, but to avoid the certainty of seeing all my attendants bolting at the first sight of the wild tribes there, the Nindi, I changed off to go round the south end, and if not, cross the middle. What I feared for the north took place in the south when the Johanna men heard of the Mazitu, though we were 150 miles from the marauders, and I offered to go due west till past their beat. They were terrified, and ran away as soon as they saw my face turned west. I got carriers from village to village, and got on nicely with people who had never engaged in the slave-trade; but it was slow work. I came very near to the Mazitu three times, but obtained information in time to avoid them. Once we were taken for Mazitu ourselves, and surrounded by a crowd of excited savages. They produced a state of confusion and terror, and men fled hither and thither with the fear of death on them. Casembe would not let me go into his southern district till he had sent men to see that the Mazitu, or, as they are called in Lunda, the Watuta, had left. Where they had been all the food was swept off, and we suffered cruel hunger. We had goods to buy with, but the people had nothing to sell, and were living on herbs and mushrooms. I had to feel every step of the way, and generally was groping in the dark. No one knew anything beyond his own district, and who cared where the rivers ran? Casembe said, when I was going to Lake Bangweolo: 'One piece of water was just like another (it is the Bangweolo water), but as your chief desired you to visit that one, go to it. If you see a traveling party going north, join it. If not, come back to me and I will send you safely along my path by Moero;' and gave me a man's load of a fish like whitebait. I gradually gained more light on the country, and slowly and surely saw the problem of the fountains of the Nile developing before my eyes. The vast volume of water draining away to the north made me conjecture that I had been working at the sources of the Congo too. My present trip to Manyuema proves that all goes to the river of Egypt. In fact, the head-waters of the Nile are gathered into two or three arms, very much as was depicted by Ptolemy in the second century of our era. What we moderns can claim is rediscovery of what had fallen into oblivion, like the circumnavigation of Africa by the Phœnician admiral of one of the Pharaohs, B.C. 600. He was not believed, because 'he had the sun on his right hand in going round from east to west.' Though to us this stamps his tale as genuine, Ptolemy was not believed, because his sources were between 10 and 12 north latitude, and collected into two or three great head branches. In my opinion, his informant must have visited them.

"I cared nothing for money, and contemplated spending my life as a hard-working, poor missionary. By going into the country beyond Kuruman we pleased the Directors, but the praises they bestowed excited envy. Mamma and you all had hard times. The missionaries at Kuruman, and south of it, had comfortable houses and gardens. They could raise wheat, pumpkins, maize, at very small expense, and their gardens yielded besides apples, pears, apricots, peaches, quinces, oranges, grapes, almonds, walnuts, and all vegetables, for little more than the trouble of watering. A series of droughts compelled us to send for nearly all our food 270 miles off. Instead of help we had to pay the uttermost farthing for everything, and got bitter envy besides. Many have thought that I was inflated by the praises I had lavished upon me, but I made it a rule never to read anything of praise. I am thankful that a kind Providence has enabled me to do what will reflect honor on my children, and show myself a stout-hearted servant of Him from whom comes every gift. None of you must become mean, craven-hearted, untruthful, or dishonest, for if you do, you don't inherit it from me. I hope that you have selected a profession that suits your taste. It will make you hold up your head among men, and is your most serious duty. I shall not live long, and it would not be well to rely on my influence. I could help you a little while living, but have little else but what people call a great name to bequeath afterward. I am nearly toothless, and in my second childhood. The green maize was in one part the only food we could get with any taste. I ate the hard fare, and was once horrified by finding most of my teeth loose. They never fastened again, and generally became so loose as to cause pain. I had to extract them, and did so by putting on a strong thread with what sailors call a clove-hitch, tie the other end to a stump above or below, as the tooth was upper or lower, strike the thread with a heavy pistol or stick, and the tooth dangled at the stump, and no pain was felt. Two upper front teeth are thus out, and so many more, I shall need a whole set of artificials. I may here add that the Manyuema stole the bodies of slaves which were buried, till a threat was used. They said the hyenas had exhumed the dead, but a slave was cast out by Banyamwezi, and neither hyenas nor men touched it for seven days. The threat was effectual. I think that they are cannibals, but not ostentatiously so. The disgust expressed by native traders has made them ashamed. Women never partook of human flesh. Eating sokos or gorillas must have been a step in the process of teaching them to eat men. The sight of a soko nauseates me. He is so hideously ugly, I can conceive no other use for him than sitting for a portrait of Satan. I have lost many months by rains, refusal of my attendants to go into a canoe, and irritable eating ulcers on my feet from wading in mud instead of sailing. They are frightfully common, and often kill slaves. I am recovering, and hope to go down Lualaba, which I would call Webb River or Lake; touch then another Lualaba, which I will name Young's River or Lake; and then by the good hand of our Father above turn homeward through Karagwe. As ivory-trading is here like gold-digging, I felt constrained to offer a handsome sum of money and goods to my friend Mohamad Bogharib for men. It was better to do this than go back to Ujiji, and then come over the whole 260 miles. I would have waited there for men from Zanzibar, but the authority at Ujiji behaved so oddly about my letters, I fear they never went to the coast. The worthless slaves I have saw that I was at their mercy, for no Manyuema will go into the next district, and they behaved as low savages who have been made free alone can. Their eagerness to enslave and kill their own countrymen is distressing. . . .

"Give my love to Oswell and Anna Mary and the Aunties. I have received no letter from any of you since I left home. The good Lord bless you all, and be gracious to you.—Affectionately yours,

"DAVID LIVINGSTONE."

Another letter is addressed to Sir Thomas Maclear and Mr. Mann, September, 1869. He enters at considerable length into his reasons for the supposition that he had discovered, on the watershed, the true sources of the Nile. He refers in a generous spirit to the discoveries of other travelers, mistaken though he regarded their views on the sources, and is particularly complimentary to Miss Tinné:

"A Dutch lady whom I never saw, and of whom I know nothing save from scraps in the newspapers, moves my sympathy more than any other. By her wise foresight in providing a steamer, and pushing on up the river after the severest domestic affliction—the loss by fever of her two aunts—till after she was assured by Speke and Grant that they had already discovered in Victoria Nyanza the sources she sought, she proved herself a genuine explorer, and then by trying to go s.w. on land. Had they not, honestly enough of course, given her their mistaken views, she must inevitably, by boat or on land, have reached the head-waters of the Nile. I cannot conceive of her stopping short of Bangweolo. She showed such indomitable pluck she must be a descendant of Van Tromp, who swept the English Channel till killed by our Blake, and whose tomb every Englishman who goes to Holland is sure to visit.

"We great he-beasts say, 'Exploration was not becoming her sex.' Well, considering that at least 1600 years have elapsed since Ptolemy's informants reached this region, and kings, emperors, and all the great men of antiquity longed in vain to know the fountains, exploration does not seem to have become the other sex either. She came much further up than the two centurions sent by Nero Cæsar.

"I have to go down and see where the two arms unite,—the lost city Meroe ought to be there,—then get back to Ujiji to get a supply of goods which I have ordered from Zanzibar, turn bankrupt after I secure them, and let my creditors catch me if they can, as I finish up by going round outside and south of all the sources, so that I may be sure no one will cut me out and say he found other sources south of mine. This is one reason for my concluding trip; another is to visit the underground houses in stone, and the copper mines of Katanga which have been worked for ages (Malachite). I have still a seriously long task before me. My letters have been delayed inexplicably, so I don't know my affairs. If I have a salary I don't know it, though the *Daily Telegraph* abused me for receiving it when I had none. Of this alone I am sure—my friends will all wish me to make a complete work of it before I leave, and in their wish I join. And it is better to go in now than to do it in vain afterward."

"I have still a seriously long task before me." Yet he had lately been worse in health and weaker than he had ever been; he was much poorer than he expected to be, and the difficulties had proved far beyond any he had hitherto experienced. But so far from thinking of taking things more easily than before, he actually enlarges his programme, and resolves to "finish up by going round outside and south of all the sources." His spirit seems only to rise as difficulties are multiplied.

He writes to his daughter Agnes at the same time: "You remark that you think you could have traveled as well as Mrs. Baker, and I think so too. Your mamma was famous for roughing it in the bush, and was never a trouble." The allusion carries him to old days—their travels to Lake 'Ngami, Mrs. Livingstone's death, the Helmores, the Bishop, Thornton. Then he speaks of recent troubles and difficulties, his attack of pneumonia, from which he had not expected to recover, his annoyances with his men, so unlike the old Makololo, the loss of his letters and boxes, with the exception of two from an unknown donor that contained the *Saturday Review* and his old friend *Punch* for 1868. Then he goes over African travelers and their achievements, real and supposed. He returns again to the achievements of ladies, and praises Miss Tinné and other women. "The death-knell of American slavery was rung by a woman's hand. We great he-beasts say Mrs. Stowe exaggerated. From what I have seen of slavery I say exaggeration is a simple impossibility. I go with the sailor who, on seeing slave-traders, said: 'If the devil don't catch these fellows, we might as well have no devil at all.'"

The year 1870 was begun with the prayer that in the course of it he might be able to complete his enterprise, and retire through the Basango before the end of it. In February he hears with gratitude of Mr. E. D. Young's Search Expedition up the Shiré and Nyassa. In setting out anew he takes a more northerly course, proceeding through paths blocked with very rank vegetation, and suffering from choleraic illness caused by constant wettings. In the course of a month the effects of the wet became overpowering, and on 7th February Dr. Livingstone had to go into winter quarters. He remained quiet till 26th June.

In April, 1870, from "Manyuema or Cannibal Country, say 150 miles N.W. of Ujiji," he began a letter to Sir Roderick Murchison, but changed its destination to his brother John in Canada. He notices his immediate object —to ascertain where the Lualaba joined the eastern branch of the Nile, and contrasts the lucid reasonable problem set him by Sir Roderick with the absurd instructions he had received from some members of the Geographical Society. "I was to furnish 'a survey on successive pages of my journal,' 'latitudes every night,' 'hydrography of Central Africa,' and because they voted one-fifth or perhaps one-sixth part of my expenses, give them 'all my notes, copies if not the originals!' For mere board and no lodgings I was to work for years and hand over the results to them." Contrasted with such absurdities, Sir Roderick's proposal had quite fascinated him. He had ascertained that the watershed extended 800 miles from west to east, and had traversed it in every direction, but at a cost which had been wearing out both to mind and body. He drops a tear over the Universities Mission, but becomes merry over Bishop Tozer strutting about with his crosier at

Zanzibar, and in a fine clear day getting a distant view of the continent of which he claimed to be Bishop. He denounces the vile policy of the Portuguese, and laments the indecision of some influential persons who virtually upheld it. He is tickled with the generous offer of a small salary, when he should settle somewhere, that had been make to him by the Government, while men who had risked nothing were getting handsome salaries of far greater amount; but rather than sacrifice the good of Africa, HE WOULD SPEND EVERY PENNY OF HIS PRIVATE MEANS. He seems surrounded by a whole sea of difficulties, but through all, the nobility of his spirit shines undimmed. To persevere in the line of duty is his only conceivable course. He holds as firmly as ever by the old anchor—" All will turn out right at last."

When ready, they set out on 26th June. Most of his people failed him; but nothing daunted, he set off then with only three attendants, Susi, Chuma, and Gardner, to the northwest for the Lualaba. Whenever he comes among Arab traders he finds himself suspected and hated because he is known to condemn their evil deeds.

The difficulties by the way were terrible. Fallen trees and flooded rivers made marching a perpetual struggle. For the first time, Livingstone's feet failed him. Instead of healing as hitherto, when torn by hard travel, irritating sores fastened upon them, and as he had but three attendants, he had to limp back to Bambarré, which he reached in the middle of July.

And here he remained in his hut for eighty days, till 10th October, exercising patience, harrowed by the wickedness he could not stop, extracting information from the natives, thinking about the fountains of the Nile, trying to do some good among the people, listening to accounts of soko-hunting, and last, not least, reading his Bible. He did not leave Bambarré till 16th February, 1871. From what he had seen and what he had heard he was more and more persuaded that he was among the true fountains of the Nile. His reverence for the Bible gave that river a sacred character, and to throw light on its origin seemed a kind of religious act. He admits, however, that he is not quite certain about it, though he does not see how he can be mistaken. He dreams that in his early life Moses may have been in these parts, and if he should only discover any confirmation of sacred history or sacred chronology he would not grudge all the toil and hardship, the pain and hunger, he had undergone. The very spot where the fountains are to be found becomes defined in his mind. He even drafts a despatch which he hopes to write, saying that the fountains are within a quarter of a mile of each other!

Then he bethinks him of his friends who have done noble battle with slavery, and half in fancy, half in earnest, attaches their names to the various waters. The fountain of the Liambai or Upper Zambesi he names Palmerston Fountain, in fond remembrance of that good man's long and unwearied labor for the abolition of the slave-trade. The lake formed by the Lufira is to be Lincoln Lake, in gratitude to him who gave freedom to four millions of slaves. The fountain of Lufira is associated with Sir Bartle Frere, who accomplished the grand work of abolishing slavery in Sindia, in Upper India. The central Lualaba is called the River Webb, after the warm-hearted friend under whose roof he wrote *The Zambesi and its Tributaries;* while the western branch is named the Young River, to commemorate his early instructor in chemistry and life-long friend, James Young. "He has shed pure white light in many lowly cottages and in some rich palaces. I, too, have shed light of another kind, and am fain to believe that I have performed a small part in the grand revolution which our Maker has been for ages carrying on, by multitudes of conscious and many unconscious agents, all over the world."[1]

---

[1] See *Last Journals,* vol. ii. pp. 65, 66.

He is by no means unaware that death may be in the cup. But, fortified as he was by an unalterable conviction that he was in the line of duty, the thought of death had no influence to turn him either to the right hand or to the left. For the first three years he had a strong presentiment that he would fall. But it had passed away as he came near the end, and now he prayed God that when he retired it might be to his native home.

Probably no human being was ever in circumstances parallel to those in which Livingstone now stood. Years had passed since he had heard from home. The sound of his mother-tongue came to him only in the broken sentences of Chuma or Susi or his other attendants, or in the echoes of his own voice as he poured it out in prayer, or in some cry of home-sickness that could not be kept in. In long pain and sickness there had been neither wife nor child nor brother to cheer him with sympathy, or lighten his dull hut with a smile. He had been baffled and tantalized beyond description in his efforts to complete the little bit of exploration which was yet necessary to finish his task. His soul was vexed for the frightful exhibitions of wickedness around him, where "man to man," instead of brothers, were worse than wolves and tigers to each other. During all his past life he had been sowing his seed weeping, but so far was he from bringing back his sheaves rejoicing, that the longer he lived the more cause there seemed for his tears. He had not yet seen of the travail of his soul. In opening Africa he had seemed to open it for brutal slave-traders, and in the only instance in which he had yet brought to it the feet of men "beautiful upon the mountains, publishing peace," disaster had befallen, and an incompetent leader had broken up the enterprise. Yet, apart from his sense of duty, there was no necessity for his remaining there. He was offering himself a freewill-offering, a living sacrifice. What could have sustained his heart and kept him firm to his purpose in such a wilderness of desolation?

"I read the whole Bible through four times whilst I was in Manyuema."

So he wrote in his Diary, not at the time, but the year after, on the 3d October, 1871.[1] The Bible gathers wonderful interest from the circumstances in which it is read. In Livingstone's circumstances it was more the Bible to him than ever. All his loneliness and sorrow, the sickness of hope deferred, the yearnings for home that could neither be repressed nor gratified, threw a new light on the Word. How clearly it was intended for such as him, and how sweetly it came home to him! How faithful, too, were its pictures of human sin and sorrow! How true its testimony against man, who will not retain God in his knowledge, but, leaving Him, becomes vain in his imaginations and hard in his heart, till the bloom of Eden is gone, and a waste, howling wilderness spreads around! How glorious the out-beaming of Divine Love, drawing near to this guilty race, winning and cherishing them with every endearing act, and at last dying on the cross to redeem them! And how bright the closing scene of Revelation —the new heaven and the new earth wherein dwelleth righteousness—yes, he can appreciate *that* attribute— the curse gone, death abolished, and all tears wiped from the mourner's eye!

So the lonely man in his dull hut is riveted to the well-worn book; ever finding it a greater treasure as he goes along; and fain, when he has reached its last page, to turn back to the beginning, and gather up more of the riches which he has left upon the road.

To Sir Thomas Maclear and Mr. Mann he writes during his detention (September, 1870) on a leaf of his cheque-book, his paper being done. He gives his theory

---

[1] See *Last Journals*, vol. ⁚. p. 154.

of the rivers, enlarges on the fertility of the country, bewails his difficulty in getting men, as the Manyuema never go beyond their own country, and the traders, who have only begun to come there, are too busy collecting ivory to be able to spare men. " The tusks were left in the terrible forests, where the animals were killed; the people, if treated civilly, readily go and bring the precious teeth, some half rotten, or gnawed by the teeth of a rodent called dezi. I think that mad naturalists name it Aulocaudatus Swindermanus, or some equally wise agglutination of syllables. . . . My chronometers are all dead; I hope my old watch was sent to Zanzibar; but I have got no letters for years, save some, three years old, at Ujiji. I have an intense and sore longing to finish and retire, and trust that the Almighty may permit me to go home."

In one of his letters to Agnes from Manyuema he quotes some words from a letter of hers that he ever after cherished as a most refreshing cordial:

" I commit myself to the Almighty Disposer of events, and if I fall, will do so doing my duty, like one of his stout-hearted servants. I am delighted to hear you say that, much as you wish me home, you would rather hear of my finishing my work to my own satisfaction than come merely to gratify you. That is a noble sentence, and I felt all along sure that all my friends would wish me to make a complete work of it, and in that wish, in spite of every difficulty, I cordially joined. I hope to present to my young countrymen an example of manly perseverance. I shall not hide from you that I am made by it very old and shaky, my cheeks fallen in, space round the eyes ditto; mouth almost toothless,—a few teeth that remain, out of their line, so that a smile is that of a he-hippopotamus,—a dreadful old fogie, and you must tell Sir Roderick that it is an utter impossibility for me to appear in public till I get new teeth, and even then the less I am seen the better."

Another letter to Agnes from Manyuema gives a curious account of the young soko or gorilla a chief had lately presented to him:

"She sits crouching eighteen inches high, and is the most intelligent and least mischievous of all the monkeys I have seen. She holds out her hand to be lifted and carried, and if refused makes her face as in a bitter human weeping, and wrings her hands quite humanly, sometimes adding a foot or third hand to make the appeal more touching. . . . She knew me at once as a friend, and when plagued by any one always placed her back to me for safety, came and sat down on my mat, decently made a nest of grass and leaves, and covered herself with the mat to sleep. I cannot take her with me, though I fear that she will die before I return, from people plaguing her. Her fine long black hair was beautiful when tended by her mother, who was killed. I am mobbed enough alone; two sokos—she and I—would not have got breath.

"I have to submit to be a gazing-stock. I don't altogether relish it, here or elsewhere, but try to get over it good-naturedly, get into the most shady spot of the village, and leisurely look at all my admirers. When the first crowd begins to go away, I go into my lodgings to take what food may be prepared, as coffee, when I have it, or roasted maize infusion when I have none. The door is shut, all save a space to admit light. It is made of the inner bark of a gigantic tree, not a quarter of an inch thick, and slides in a groove behind a post on each side of the doorway. When partially open it is supported by only one of the posts. Eager heads sometimes crowd the open space, and crash goes the thin door, landing a Manyuema beauty on the floor. 'It was not I,' she gasps out, 'it was Bessie Bell and Jeanie Gray that shoved me in, and—' as she scrambles out of the lion's den, 'see they're laughing'; and, fairly out, she joins in the merry giggle too. To avoid darkness or being half-smothered, I often eat in public, draw a line on the ground, then 'toe the line,' and keep them out of the circle. To see me eating with knife, fork, and spoon is wonderful. 'See!—they don't touch their food!—what oddities, to be sure.' . . .

"Many of the Manyuema women are very pretty; their hands, feet, limbs, and form are perfect. The men are handsome. Compared with them the Zanzibar slaves are like London door-knockers, which some atrocious iron-founder thought were like lions' faces. The way in which these same Zanzibar Mohammedans murder the men and seize the women and children makes me sick at heart. It is not slave-trade. It is murdering free people to make slaves. It is perfectly indescribable. Kirk has been working hard to get this murdersome system put a stop to. Heaven prosper his noble efforts! He says in one of his letters to me, 'It is monstrous injustice to compare the free people in the interior, living under their own chiefs and laws, with what slaves at Zanzibar afterward become by the abominable system which robs them of their manhood. I think it is like comparing the anthropologists with their ancestral sokos.' . . .

"I am grieved to hear of the departure of good Lady Murchison. Had I known that she kindly remembered me in her prayers, it would have been great encouragement. . . .

"The men sent by Dr. Kirk are Mohammedans, that is, unmitigated liars. Musa and his companions are fair specimens of the lower class of Moslems. The two head-men remained at Ujiji, to feast on my goods, and get pay without work. Seven came to Bambarré, and in true Moslem style swore that they were sent by Dr. Kirk to bring me back, not to go with me, if the country were bad or dangerous. Forward they would not go. I read Dr. Kirk's words to them to follow wheresoever I led. 'No, by the old liar Mohamed, they were to force me back to Zanzibar.' After a superabundance of falsehood, it turned out that it all meant only an advance of pay, though they had double the Zanzibar wages. I gave it, but had to threaten on the word of an Englishman to shoot the ringleaders before I got them to go. They all speak of English as men who do not lie. . . . I have traveled more than most people, and with all sorts of followers. The Christians of Kuruman and Kolobeng were out of sight the best I ever had. The Makololo, who were very partially Christianized, were next best— honest, truthful, and brave. Heathen Africans are much superior to the Mohammedans, who are the most worthless one can have."

Toward the end of 1870, before the date of this letter, he had so far recovered that, though feeling the want of medicine as much as of men, he thought of setting out, in order to reach and explore the Lualaba, having made a bargain with Mohamad, for £270, to bring him to his destination. But now he heard that Syde bin Habib, Dugumbé, and others were on the way from Ujiji, perhaps bringing letters and medicines for him. He cannot move till they arrive; another weary time. "Sorely am I perplexed, and grieve and mourn."

The New Year 1871 passes while he is at Bambarré, with its prayer that he might be permitted to finish his task. At last, on 4th February, ten of the men despatched to him from the coast arrive, but only to bring a fresh disappointment. They were slaves, the property of Banians, who were British subjects! and they brought only one letter! Forty had been lost. There had been cholera at Zanzibar, and many of the porters sent by Dr. Kirk had died of it. The ten men came with a lie in their mouth; they would not help him, swearing that the Consul told them not to go forward, but to force Livingstone back. On the 10th they mutinied, and had to receive an advance of pay. It was apparent that they had been instructed by their Banian masters to baffle him in every way, so that their slave-trading should not be injured by his disclosures. Their two head-men, Shereef and Awathe, had refused to come farther than Ujiji, and were reveling in his goods there. Dr. Livingstone never ceased to lament and deplore that the men who had been sent to him were so utterly unsuitable. One of them actually formed a plot for his destruction, which was only frustrated through his being overheard by one whom Livingstone could trust. Livingstone wrote to his friends that owing to the inefficiency of the men, he lost two years of time, about a thousand pounds in money, had some 2000 miles of useless traveling, and was four several times subjected to the risk of a violent death.

At length, having arranged with the men, he sets out on 16th February over a most beautiful country, but woefully difficult to pass through. Perhaps it was hardly a less bitter disappointment to be told, on the 25th, that the Lualaba flowed west-southwest, so that after all it might be the Congo.

On the 29th March Livingstone arrived at Nyangwe, on the banks of the Lualaba. This was the farthest point westward that he reached in his last Expedition.

The slave-trade here he finds to be as horrible as in any other part of Africa. He is heart-sore for human blood. He is threatened, bullied, and almost attacked. In some places, however, the rumor spreads that he makes no slaves, and he is called "the good one." His men are a ceaseless trouble, and for ever mutinying, or otherwise harassing him. And yet he perseveres in his old kind way, hoping by kindness to gain influence with them. Mohamad's people, he finds, have passed him on the west, and thus he loses a number of serviceable articles he was to get from them, and all the notes made for him of the rivers they had passed. The difficulties and discouragements are so great that he wonders whether, after all, God is smiling on his work.

His own men circulate such calumnious reports against him that he is unable to get canoes for the navigation of the Lualaba. This leads to weeks and months of weary waiting, and yet all in vain; but afterward he finds some consolation on discovering that the navigation was perilous, that a canoe had been lost from the inexperience of her crew in the rapids, so that had he been there, he should very likely have perished, as his canoe would probably have been foremost.

A change of plan was necessary. On 5th July he offered to Dugumbé £400, with all the goods he had at Ujiji besides, for men to replace the Banian slaves, and for the other means of going up the Lomamé to Katanga, then returning and going up Tanganyika to Ujiji. Dugumbé took a little time to consult his friends before replying to the offer.

Meanwhile an event occurred of unprecedented horror, that showed Livingstone that he could not go to Lomamé in the company of Dugumbé. Between Dugumbé's people and another chief a frightful system of pillage, murder, and burning of villages was going on with horrible activity. One bright summer morning, 15th July, when fifteen hundred people, chiefly women, were engaged peacefully in marketing in a village on the banks of the Lualaba, and while Dr. Livingstone was sauntering about, a murderous fire was opened on the people, and a massacre ensued of such measureless atrocity that he could describe it only by saying that it gave him the impression of being in hell. The event was so superlatively horrible, and had such an overwhelming influence on Livingstone, that we copy at full length the description of it given in the *Last Journals:*

"Before I had got thirty yards out, the discharge of two guns in the middle of the crowd told me that slaughter had begun : crowds dashed off from the place, and threw down their wares in confusion, and ran. At the same time that the three opened fire on the mass of people near the upper end of the market-place, volleys were discharged from a party down near the creek on the panic-stricken women, who dashed at the canoes. These, some fifty or more, were jammed in the creek, and the men forgot their paddles in the terror that seized all. The canoes were not to be got out, for the creek was too small for so many; men and women, wounded by the balls, poured into them, and leaped and scrambled into the water, shrieking. A long line of heads in the river showed that great numbers struck out for an island a full mile off; in going toward it they had to put the left shoulder to a current of about two miles an hour; if they had struck away diagonally to the opposite bank, the current would have aided them, and, though nearly three miles off, some would have gained land; as it was, the heads above water showed the long line of those that would inevitably perish.

"Shot after shot continued to be fired on the helpless and perishing. Some of the long line of heads disappeared quietly; whilst other poor creatures threw their arms high, as if appealing to the great Father above, and sank. One canoe took in as many as it could hold, and all paddled with hands and arms; three canoes, got out in haste, picked up sinking friends, till all went down together, and disappeared. One man in a long canoe, which could have held forty or fifty, had clearly lost his head; he had been out in the stream before the massacre began, and now paddled up the river nowhere, and never looked to the drowning. By and by all the heads disappeared; some had turned down stream toward the bank, and escaped. Dugumbé put people into one of the deserted vessls to save those in the water, and saved twenty-one; but one woman refused to be taken on board, from thinking that she was to be made a slave of; she preferred the chance of life by swimming, to the lot of a slave. The Bagenya women are expert in the water, as they are accustomed to dive for oysters, and those who went down stream may have escaped, but the Arabs themselves estimated the loss of life at between 330 and 400 souls. The shooting-party near the canoes were so reckless, they killed two of their own people; and a Banyamwezi follower, who got into a deserted canoe to plunder, fell into the water, went down, then came up again, and down to rise no more.

"After the terrible affair in the water, the party of Tagamoio, who was the chief perpetrator, continued to fire on the people there, and fire their villages. As I write I hear the loud wails on the left bank over those who are there slain, ignorant of their many friends now in the depths of Lualaba. Oh, let Thy kingdom come! No one will ever know the exact loss on this bright sultry summer morning; it gave me the impression of being in Hell. All the slaves in the camp rushed at the fugitives on land, and plundered them; women were for hours collecting and carrying loads of what had been thrown down in terror."

The remembrance of this awful scene was never effaced from Livingstone's heart. The accounts of it published in the newspapers at home sent a thrill of horror through the country. It was recorded at great length in a despatch to the Foreign Secretary, and indeed, it became one of the chief causes of the appointment of a Royal Commission to investigate the subject of the African slave-trade, and of the mission of Sir Bartle Frere to Africa to concert measures for bringing it to an end.

Dugumbé had not been the active perpetrator of the massacre, but he was mixed up with the atrocities that had been committed, and Livingstone could have nothing to do with him. It was a great trial, for, as the Banian men were impracticable, there was nothing for it now but to go back to Ujiji, and try to get other men there with whom he would repeat the attempt to explore the river. For twenty-one months, counting from the period of their engagement, he had fed and clothed these men, all in vain, and now he had to trudge back forty-five days, a journey equal, with all its turnings and windings, to six hundred miles. Livingstone was ill, and after such an exciting time he would probably have had an attack of fever, but for another ailment to which he had become more especially subject. The intestinal canal had given way, and he was subject to attacks of severe internal hæmorrhage, one of which came on him now.[1] It appeared afterward that had he gone with Dugumbé, he would have been exposed to an assault in force by the Bakuss, as they made an attack on the party and routed them, killing two hundred. If Livingstone had been among them, he might have fallen in this engagement. So again, he saw how present disappointments work for good.

The journey back to Ujiji, begun 20th July, 1871, was a very wretched one. Amid the universal desolation caused by the very wantonness of the marauders, it was impossible for Livingstone to persuade the natives that he did not belong to the same set. Ambushes were set for him and his company in the forest. On the 8th August they came to an ambushment all prepared, but it had been abandoned for some unknown reason. By and by, on the same day, a large spear flew past Livingstone, grazing his neck; the native who flung it was but ten yards off; the hand of God alone saved his life.[2] Farther on, another spear was thrown, which missed him by a foot. On the same day a large tree, to which fire had been applied to fell it, came down within a yard of him. Thus on one day he was delivered three times from impending death. He went on through the forest, expecting every minute to be attacked, having no fear, but perfectly indifferent whether he should be killed or not. He lost all his remaining calico that day, a telescope, umbrella, and five spears. By and by he was prostrated with grievous illness. As soon as he could move he went onward, but he felt as if dying on his feet. And he was ill-rigged for the road, for the light French shoes to which he was reduced, and which had been cut to ease his feet till they would hardly hang together, failed

---

[1] His friends say that for a considerable time before he had been subject to the most grievous pain from hæmorrhoids. His sufferings were often excruciating.

[2] The head of this spear is among the Livingstone relics at Newstead Abbey.

to protect him from the sharp fragments of quartz with which the road was strewed. He was getting near to Ujiji, however, where abundance of goods and comforts were no doubt safely stowed away for him, and the hope of relief sustained him under all his trials.

At last, on the 23d October, reduced to a living skeleton, he reached Ujiji. What was his misery, instead of finding the abundance of goods he had expected, to learn that the wretch Shereef, to whom they had been consigned, had sold off the whole, not leaving one yard of calico out of 3000, or one string of beads out of 700 pounds! The scoundrel had divined on the Koran, found that Livingstone was dead, and would need the goods no more. Livingstone had intended, if he could not get men at Ujiji to go with him to the Lualaba, to wait there till suitable men should be sent up from the coast; but he had never thought of having to wait in beggary. If anything could have aggravated the annoyance, it was to see Shereef come, without shame, to salute him, and tell him on leaving, that he was going to pray; or to see his slaves passing from the market with all the good things his property had bought! Livingstone applied a term to him which he reserved for men—black or white—whose wickedness made them alike shameless and stupid—he was a "moral idiot."

It was the old story of the traveler who fell among thieves that robbed him of all he had; but where was the good Samaritan? The Government and the Geographical Society appeared to have passed by on the other side. But the good Samaritan was not as far off as might have been thought. One morning Syed bin Majid, an Arab trader, came to him with a generous offer to sell some ivory and get goods for him; but Livingstone had the old feeling of independence, and having still a few barter goods left, which he had deposited with Mohamad bin Saleh before going to Manyuema, he declined for the present Syed's generous offer. But the kindness of Syed was not the only proof that he was not forsaken. Five days after he reached Ujiji the good Samaritan appeared from another quarter. As Livingstone had been approaching Ujiji from the southwest, another white man had been approaching it from the east. On 28th October, 1871, Henry Moreland Stanley, who had been sent to look for him by Mr. James Gordon Bennett, Jr., of the *New York Herald* newspaper, grasped the hand of David Livingstone. An angel from heaven could hardly have been more welcome. In a moment the sky brightened. Stanley was provided with ample stores, and was delighted to supply the wants of the traveler. The sense of sympathy, the feeling of brotherhood, the blessing of fellowship, acted like a charm. Four good meals a day, instead of the spare and tasteless food of the country, made a wonderful change on the outer man; and in a few days Livingstone was himself again— hearty and happy and hopeful as before.

Before closing this chapter and entering on the last two years of Livingstone's life, which have so lively an interest of their own, it will be convenient to glance at the contributions to natural science which he continued to make to the very end. In doing this, we avail ourselves of a very tender and Christian tribute to the memory of his early friend, which Professor Owen contributed to the *Quarterly Review*, April, 1875, after the publication of Livingstone's *Last Journals.*

Mr. Owen appears to have been convinced by Livingstone's reasoning and observations, that the Nile sources were in the Bangweolo watershed—a supposition now ascertained to have been erroneous. But what chiefly attracted and delighted the great naturalist was the many interesting notices of plants and animals scattered over the *Last Journals.* These Journals contain important contributions both to economic and physiological botany. In the former department, Livingstone makes valuable observations on plants useful in the arts, such as gum-copal, papyrus, cotton, india-rubber, and the palm-oil tree; while in the latter, his notices of "carnivorous plants," which catch insects that probably yield nourishment to the plant, of silicified wood and the like, show how carefully he watched all that throws light on the life and changes of plants. In zoölogy he was never weary of observing, especially when he found a strange-looking animal with strange habits. Spiders, ants, and bees of unknown varieties were brought to light, but the strangest of his new acquaintances were among the fishy tribes. He found fish that made long excursions on land, thanks to the wet grass through which they would wander for miles, thus proving that "a fish out of water." is not always the best symbol for a man out of his element. There were fish, too, that burrowed in the earth; but most remarkable at first sight were the fish that appeared to bring forth their young by ejecting them from their mouths. If Bruce or Du Chaillu had made such a statement, remarks Professor Owen, what ridicule would they not have encountered! But Livingstone was not the man to make a statement of what he had not ascertained, or to be content until he had found a scientific explanation of it. He found that in the branchial openings of the fish, there occur bags or pouches, on the same principle as the pouch of the opossum, where the young may be lodged for a time for protection or nourishment, and that when the creatures are discharged through the mouth into the water, it is only from a temporary cradle where they were probably enjoying repose, beyond the reach of enemies.

Perhaps the greatest of Livingstone's scientific discoveries during this journey was that "of a physical condition of the earth's surface in elevated tracts of the great continent, unknown before." The bogs or earth-sponges, that from his first acquaintance with them gave him so much trouble, and at last proved the occasion of his death.

were not only remarkable in themselves, but interesting as probably explaining the annual inundations of most of the rivers. Wherever there was a plain sloping toward a narrow opening in hills or higher ground, there were the conditions for an African sponge. The vegetation falls down and rots, and forms a rich black loam, resting often, two or three feet thick, on a bed of pure river sand. The early rains turn the vegetation into slush, and fill the pools. The later rains, finding the pools already full, run off to the rivers, and form the inundation. The first rains occur south of the equator when the sun goes vertically over any spot, and the second or greater rains happen in his course north again. This, certainly, was the case as observed on the Zambesi and Shiré, and taking the different times for the sun's passage north of the equator, it explained the inundations of the Nile.

Such notices show that in his love of nature, and in his careful observation of all her agencies and processes, Livingstone, in his last journeys, was the same as ever. He looked reverently on all plants and animals, and on the solid earth in all its aspects and forms, as the creatures of that same God whose love in Christ it was his heart's delight to proclaim. His whole life, so varied in its outward employments, yet so simple and transparent in its one great object, was ruled by the conviction that the God of nature and the God of revelation were one. While thoroughly enjoying his work as a naturalist, Professor Owen frankly admits that it was but a secondary object of his life. "Of his primary work the record is on high, and its imperishable fruits remain on earth. The seeds of the Word of Life implanted lovingly, with pains and labor, and above all with faith; the out-door scenes of the simple Sabbath service; the testimony of Him to whom the worship was paid, given in terms of such simplicity as were fitted to the comprehension of the dark-skinned listeners,—these seeds will not have been scattered by him in vain. Nor have they been sown in words alone, but in deeds, of which some part of the honor will redound to his successors. The teaching by forgiveness of injuries,—by trust, however unworthy the trusted,—by that confidence which imputed his own noble nature to those whom he would win,—by the practical enforcement of the fact that a man might promise and perform—might say the thing he meant,—of this teaching by good deeds, as well as by the words of truth and love, the successor who treads in the steps of LIVINGSTONE, and accomplishes the discovery he aimed at, and pointed the way to, will assuredly reap the benefit."[1]

[1] *Quarterly Review,* April, 1875, pp. 498, 499.

CHAPTER XXI.

LIVINGSTONE AND STANLEY.

A.D. 1871–1872.

Mr. Gordon Bennett sends Stanley in search of Livingstone—Stanley at
Zanzibar—Starts for Ujiji—Reaches Unyanyembe—Dangerous illness—War
between Arabs and natives—Narrow escape of Stanley—Approach to Ujiji
—Meeting with Livingstone—Livingstone's story—Stanley's news—Livingstone's goods and men at Bagamoio—Stanley's accounts of Livingstone—
Refutation of foolish and calumnious charges—They go to the north of the
lake—Livingstone resolves not to go home, but to get fresh men and return
to the sources—Letter to Agnes—to Sir Thomas Maclear—The travelers go
to Unyanyembe—More plundering of stores—Stanley leaves for Zanzibar—
Stanley's bitterness of heart at parting—Livingstone's intense gratitude to
Stanley—He intrusts his Journal to him, and commissions him to send
servants and stores from Zanzibar—Stanley's journey to the coast—Finds
Search Expedition at Bagamoio—Proceeds to England—Stanley's reception
—Unpleasant feelings—Éclaircissement—England grateful to Stanley.

THE meeting of Stanley and Livingstone at Ujiji was
as unlikely an occurrence as could have happened, and,
along with many of the earlier events in Livingstone's life,
serves to show how wonderfully an Unseen Hand shaped
and guarded his path. Neither Stanley nor the gentleman
who sent him had any personal interest in Livingstone.
Mr. Bennett admitted frankly that he was moved neither
by friendship nor philanthropy, but by regard to his
business and interest as a journalist. The object of a
journal was to furnish its readers with the news which
they desired to know; the readers of the *New York Herald*
desired to know about Livingstone; as a journalist, it was
his business to find out and tell them. Mr. Bennett
determined that, cost what it might, he would find out,
and give the news to his readers. These were the very

unromantic notions, with an under-current probably of better quality, that were passing through his mind at Paris, on the 16th October, 1869, when he sent a telegram to Madrid, summoning Henry M. Stanley, one of the "own correspondents" of his paper, to "come to Paris on important business." On his arrival, Mr. Bennett asked him bluntly, "Where do you think Livingstone is?" The correspondent could not tell—could not even tell whether he was alive. "Well," said Mr. Bennett, "I think he is alive, and that he may be found, and I am going to send you to find him." Mr. Stanley was to have whatever money should be found necessary; only he was to find Livingstone. It is very mysterious that he was not to go straight to Africa—he was to visit Constantinople, Palestine, and Egypt first. Then, from India, he was to go to Zanzibar; get into the interior, and find him if alive; obtain all possible news of his discoveries; and if he were dead, get the fact fully verified, find out the place of his burial, and try to obtain possession of his bones, that they might find a resting-place at home.

It was not till January, 1871, that Stanley reached Zanzibar. To organize an expedition into the interior was no easy task for one who had never before set foot in Africa. To lay all his plans without divulging his object would, perhaps, have been more difficult if it had ever entered into any man's head to connect the *New York Herald* with a search for Livingstone. But indomitable vigor and perseverance succeeded, and by the end of February and beginning of March, one hundred and ninety-two persons in all had started in five caravans at short intervals from Bagamoio for Lake Tanganyika, two white men being of the party besides Stanley, with horses, donkeys, bales, boats, boxes, rifles, etc., to an amount that made the leader of the expedition ask himself how such an enormous weight of material could ever be carried into the heart of Africa.

The ordinary and extraordinary risks and troubles of travel in these parts fell to Mr. Stanley's lot in unstinted abundance. But when Unyanyembe was reached, the half-way station to Ujiji, troubles more than extraordinary befell. First, a terrible attack of fever that deprived him of his senses for a fortnight. Then came a worse trouble. The Arabs were at war with a chief Mirambo, and Stanley and his men, believing they would help to restore peace more speedily, sided with the Arabs. At first they were apparently victorious, but immediately after, part of the Arabs were attacked on their way home by Mirambo, who lay in ambush for them, and were defeated. Great consternation prevailed. The Arabs retreated in panic, leaving Stanley, who was ill, to the tender mercies of the foe. Stanley, however, managed to escape. After this experience of the Arabs in war, he resolved to discontinue his alliance with them. As the usual way to Ujiji was blocked, he determined to try a route more to the south. But his people had forsaken him. One of his two English companions was dead, the other was sick and had to be sent back. Mirambo was still threatening. It was not till the 20th September that new men were engaged by Stanley, and his party were ready to move.

They marched slowly, with various adventures and difficulties, until, by Mr. Stanley's reckoning, on the 10th November (but by Livingstone's earlier), they were close on Ujiji. Their approach created an extraordinary excitement. First one voice saluted them in English, then another; these were the salutations of Livingstone's servants, Susi and Chuma. By and by the Doctor himself appeared. "As I advanced slowly toward him," says Mr. Stanley, "I noticed he was pale, looked wearied, had a gray beard, wore a bluish cap with a faded gold band round it, had on a red-sleeved waistcoat and a pair of gray tweed trousers. I would have run to him, only I was a coward in the presence of such a mob,—would have embraced him, only he, being an Englishman, I did not know how he would receive me; so I did what cowardice and false pride suggested was the best thing—walked deliberately to him, took off my hat and said, 'Dr. Livingstone, I presume?' 'Yes,' said he, with a kind smile, lifting his cap slightly. I replace my hat on my head, and he puts on his cap, and we both grasp hands, and then I say aloud—'I thank God, Doctor, I have been permitted to see you.' He answered, 'I feel thankful that I am here to welcome you.'"

The conversation began—but Stanley could not remember what it was. "I found myself gazing at him, conning the wonderful man at whose side I now sat in Central Africa. Every hair of his head and beard, every wrinkle of his face, the wanness of his features, and the slightly wearied look he bore, were all imparting intelligence to me—the knowledge I craved for so much ever since I heard the words, 'Take what you want, but find Livingstone.' What I saw was deeply interesting intelligence to me and unvarnished truth. I was listening and reading at the same time. What did these dumb witnesses relate to me?

"Oh, reader, had you been at my side on this day in Ujiji, how eloquently could be told the nature of this man's work? Had you been there but to see and hear! His lips gave me the details; lips that never lie. I cannot repeat what he said; I was too much engrossed to take my note-book out, and begin to stenograph his story. He had so much to say that he began at the end, seemingly oblivious of the fact that five or six years had to be accounted for. But his account was oozing out; it was growing fast into grand proportions—into a most marvelous history of deeds."

And Stanley, too, had wonderful things to tell the Doctor. "The news," says Livingstone, "he had to tell one who had been two full years without any tidings from

Europe made my whole frame thrill. The terrible fate
that had befallen France, the telegraphic cables successfully laid in the Atlantic, the election of General Grant,
the death of good Lord Clarendon, my constant friend;
the proof that Her Majesty's Government had not forgotten me in voting £1000 for supplies, and many other
points of interest, revived emotions that had lain dormant
in Manyuema." As Stanley went on, Livingstone kept
saying, "You have brought me new life—you have
brought me new life."

There was one piece of news brought by Stanley to
Livingstone that was far from satisfactory. At Bagamoio,
on the coast, Stanley had found a caravan with supplies
for Livingstone that had been despatched from Zanzibar
three or four months before, the men in charge of which
had been lying idle there all that time on the pretext that
they were waiting for carriers. A letter-bag was also lying
at Bagamoio, although several caravans for Ujiji had left
in the meantime. On hearing that the Consul at Zanzibar,
Dr. Kirk, was coming to the neighborhood to hunt, the
party at last made off. Overtaking them at Unyanyembe,
Stanley took charge of Livingstone's stores, but was not
able to bring them on; only he compelled the letter-carrier
to come on to Ujiji with his bag. At what time, but for
Stanley, Livingstone would have got his letters, which after
all were a year on the way, he could not have told. For
his stores, or such fragments of them as might remain, he
had afterward to trudge all the way to Unyanyembe. His
letters conveyed the news that Government had voted a
thousand pounds for his relief, and were besides to pay
him a salary.[1] The unpleasant feeling he had had so long
as to his treatment by Government was thus at last somewhat relieved. But the goods that had lain in neglect at
Bagamoio, and were now out of reach at Unyanyembe,

---

[1] The intimation of salary was premature. Livingstone got a pension of
£300 afterward, which lasted only for a year and a half.

represented one-half the Government grant, and would probably be squandered, like his other goods, before he could reach them. The impression made on Stanley by Livingstone was remarkably vivid, and the portrait drawn by the American will be recognized as genuine by every one who knows what manner of man Livingstone was:

" I defy any one to be in his society long without thoroughly fathoming him, for in him there is no guile, and what is apparent on the surface is the thing that is in him. . . . Dr. Livingstone is about sixty years old, though after he was restored to health he looked like a man who had not passed his fiftieth year. His hair has a brownish color yet, but is here and there streaked with gray lines over the temples; his beard and moustaches are very gray. His eyes, which are hazel, are remarkably bright; he has a sight keen as a hawk's. His teeth alone indicate the weakness of age; the hard fare of Lunda has made havoc in their lines. His form, which soon assumed a stoutish appearance, is a little over the ordinary height, with the slightest possible bow in the shoulders. When walking he has a firm but heavy tread, like that of an overworked or fatigued man. He is accustomed to wear a naval cap with a semicircular peak, by which he has been identified throughout Africa. His dress, when first I saw him, exhibited traces of patching and repairing, but was scrupulously clean.

"I was led to believe that Livingstone possessed a splenetic, misanthropic temper; some have said that he is garrulous; that he is demented; that he is utterly changed from the David Livingstone whom people knew as the reverend missionary ; that he takes no notes or observations but such as those which no other person could read but himself, and it was reported, before I proceeded to Africa, that he was married to an African princess.

" I respectfully beg to differ with all and each of the above statements. I grant he is not an angel; but he approaches to that being as near as the nature of a living man will allow. I never saw any spleen or misanthropy in him : as for being garrulous, Dr. Livingstone is quite the reverse; he is reserved, if anything; and to the man who says Dr. Livingstone is changed, all I can say is, that he never could have known him, for it is notorious that the Doctor has a fund of quiet humor, which he exhibits at all times when he is among friends." [After repudiating the charge as to his notes and observations, Mr. Stanley continues:] " As to the report of his African marriage, it is unnecessary to say more than that it is untrue, and it is utterly beneath a gentleman even to hint at such a thing in connection with the name of Dr. Livingstone.

"You may take any point in Dr. Livingstone's character, and analyze it carefully, and I would challenge any man to find a fault in it. . . . . His gentleness never forsakes him; his hopefulness never deserts him. No harassing anxieties, distraction of mind, long separation from home and kindred, can make him complain. He thinks 'all will come out right at last'; he has such faith in the goodness of Providence. The sport of adverse circumstances, the plaything of the miserable beings sent to him from Zanzibar—he has been baffled and worried, even almost to the grave, yet he will not desert the charge imposed upon him by his friend Sir Roderick Murchison. To the stern dictates of duty, alone, has he sacrificed his home and ease, the pleasures, refinements, and luxuries of civilized life. His is the Spartan heroism, the inflexibility of the Roman, the enduring resolution of the Anglo-Saxon—never to relinquish his work, though his heart yearns for home; never to surrender his obligations until he can write FINIS to his work.

"There is a good-natured *abandon* about Livingstone which was not lost on me. Whenever he began to laugh, there was a contagion about it that compelled me to imitate him. It was such a laugh as Teufelsdröckh's—a laugh of the whole man from head to heel. If he told a story, he related it in such a way as to convince one of its truthfulness; his face was so lit up by the sly fun it contained, that I was sure the story was worth relating, and worth listening to.

"Another thing that especially attracted my attention was his wonderfully retentive memory. If we remember the many years he has spent in Africa, deprived of books, we may well think it an uncommon memory that can recite whole poems from Byron, Burns, Tennyson, Longfellow, Whittier, and Lowell. . . .

"His religion is not of the theoretical kind, but it is a constant, earnest, sincere practice. It is neither demonstrative nor loud, but manifests itself in a quiet, practical way, and is always at work. It is not aggressive, which sometimes is troublesome if not impertinent. In him religion exhibits its loveliest features; it governs his conduct not only toward his servants but toward the natives, the bigoted Mohammedans, and all who come in contact with him. Without it, Livingstone, with his ardent temperament, his enthusiasm, his high spirit and courage, must have become uncompanionable, and a hard master. Religion has tamed him and made him a Christian gentleman; the crude and willful have been refined and subdued; religion has made him the most companionable of men and indulgent of masters—a man whose society is pleasurable to a degree. · · ●

"From being thwarted and hated in every possible way by the Arabs and half-castes upon his first arrival at Ujiji, he has, through his uniform kindness and mild, pleasant temper, won all hearts. I observed that universal respect was paid to him. Even the Mohammedans never passed his house without calling to pay their compliments, and to say, 'The blessing of God rest on you!' Each Sunday morning he gathers his little flock around him, and reads prayers and a chapter from the Bible, in a natural, unaffected, and sincere tone; and afterward delivers a short address in the Kisawahili language, about the subject read to them, which is listened to with evident interest and attention."

It was agreed that the two travelers should make a short excursion to the north end of Lake Tanganyika, to ascertain whether the lake had an outlet there. This was done, but it was found that instead of flowing out, the river Lu-'izé flowed into the lake, so that the notion that the lake discharged itself northward turned out to be an error. Meanwhile, the future arrangements of Dr. Livingstone were matter of anxious consideration. One thing was fixed and certain from the beginning: Livingstone would not go home with Stanley. Much though his heart yearned for home and family—all the more that he had just learned that his son Thomas had had a dangerous accident,—and much though he needed to recruit his strength and nurse his ailments, he would not think of it while his work remained unfinished. To turn back to those dreary sponges, sleep in those flooded plains, encounter anew that terrible pneumonia which was "worse than ten fevers," or that distressing hæmorrhage which added extreme weakness to extreme agony—might have turned any heart; Livingstone never flinched from it. What a reception awaited him if he had gone home to England! What welcome from friends and children, what triumphal cheers from all the great Societies and *savants*, what honors from all who had honors to confer, what opportunity of renewing efforts to establish missions and commerce, and to suppress the slave traffic! Then he might return to Africa in a year, and finish his work. If Livingstone had taken this course, no whisper would have been heard against it. The nobility of his soul never rose higher, his utter abandonment of self, his entire devotion to duty, his right honorable determination to work while it was called to-day never shone more brightly than when he declined all Stanley's entreaties to return home, and set his face steadfastly to go back to the bogs of the watershed. He writes in his journal: "My daughter Agnes says, 'Much as I wish you to come home, I had rather that you finished your work to your own satisfaction, than return merely to gratify me.' Rightly and nobly said, my darling Nannie; vanity whispers pretty loudly, 'She is a chip of the old block.' My blessing on her and all the rest."

After careful consideration of various plans, it was agreed that he should go to Unyanyembe, accompanied by Stanley, who would supply him there with abundance of goods, and who would then hurry down to the coast, organize a new expedition composed of fifty or sixty faithful men to be sent on to Unyanyembe, by whom Livingstone would be accompanied back to Bangweolo and the sources, and then to Rua, until his work should be completed, and he might go home in peace.

A few extracts from Livingstone's letters will show us how he felt at this remarkable crisis. To Agnes:

" *Tanganyika, 18th November,* 1871.—[After detailing his troubles in Manyuema, the loss of all his goods at Ujiji, and the generous offer of Syed bin Majid, he continues:] "Next I heard of an Englishman being at Unyamyembe with boats, etc., but who he was, none could tell. At last, one of my people came running out of breath and shouted, 'An Englishman coming!' and off he darted back again to meet him. An American flag at the head of a large caravan showed the nationality of the stranger. Baths, tents, saddles, big kettles, showed that he was not a poor Lazarus like me. He turned out to be Henry M. Stanley, traveling correspondent of the *New York Herald*, sent specially to find out if I were really alive, and, if dead, to bring home my bones. He had brought abundance of goods at great expense, but the fighting referred to delayed him, and he had to leave a great part at Unyamyembe. To all he had I was made free. [In a later letter, Livingstone says: 'He

laid all he had at my service, divided his clothes into two heaps, and
pressed one heap upon me; then his medicine-chest; then his goods
and everything he had, and to coax my appetite, often cooked dainty
dishes with his own hand.'] He came with the true American characteristic generosity. The tears often started into my eyes on every fresh
proof of kindness. My appetite returned, and I ate three or four times
a day, instead of scanty meals morning and evening. I soon felt strong,
and never wearied with the strange news of Europe and America he told.
The tumble down of the French Empire was like a dream. . . ."

A long letter to his friend Sir Thomas Maclear and Mr.
Mann, of the same date, goes over his travels in Manyuema,
his many disasters, and then his wonderful meeting with
Mr. Stanley at Ujiji. Speaking of the unwillingness of the
natives to believe in the true purpose of his journey, he
says: "They all treat me with respect, and are very much
afraid of being written against; but they consider the
sources of the Nile to be a sham; the true object of my
being sent is to see their odious system of slaving, and *if
indeed my disclosures should lead to the suppression of the East
Coast slave-trade, I would esteem that as a far greater feat than
the discovery of all the sources together.* It is awful, but I
cannot speak of the slaving for fear of appearing guilty of
exaggerating. It is not trading; it is murdering for captives to be made into slaves." His account of himself in
the journey from Nyangwe is dreadful: "I was near a
fourth lake on this central line, and only eighty miles from
Lake Lincoln on our west, in fact almost in sight of the geographical end of my mission, when I was forced to return
[through the misconduct of his men] between 400 and 500
miles. A sore heart, made still sorer by the sad scenes I had
seen of man's inhumanity to man, made this march a terrible
tramp—the sun vertical, and the sore heat reacting on the
physical frame. I was in pain nearly every step of the way,
and arrived a mere ruckle of bones to find myself destitute."
In speaking of the impression made by Mr. Stanley's
kindness: "I am as cold and non-demonstrative as we
islanders are reputed to be, but this kindness was over-

38

whelming. Here was the good Samaritan and no mistake. Never was I more hard pressed; never was help more welcome."

During thirteen months Stanley received no fewer than ten parcels of letters and papers sent up by Mr. Webb, American Consul at Zanzibar, while Livingstone received but one. This was an additional ground for faith in the efficiency of Stanley's arrangements.

The journey to Unyanyembe was somewhat delayed by an attack of fever which Stanley had at Ujiji, and it was not till the 27th December that the travelers set out. On the way Stanley heard of the death of his English attendant Shaw, whom he had left unwell. On the 18th of February, 1872, they reached Unyanyembe, where a new chapter of the old history unfolded itself. The survivor of two head-men employed by Ludha Damji had been plundering Livingstone's stores, and had broken open the lock of Mr. Stanley's store-room and plundered him likewise. Notwithstanding, Mr. Stanley was able to give Livingstone a large amount of calico, beads, brass wire, copper sheets, a tent, boat, bath, cooking-pots, medicine-chest, tools, books, paper, medicines, cartridges, and shot. This, with four flannel shirts that had come from Agnes, and two pairs of boots, gave him the feeling of being quite set up.

On the 14th of March Mr. Stanley left Livingstone for Zanzibar, having received from him a commission to send him up fifty trusty men, and some additional stores. Mr. Stanley had authority to draw from Dr. Kirk the remaining half of the Government grant, but lest it should have been expended, he was furnished with a cheque for 5000 rupees on Dr. Livingstone's agents at Bombay. He was likewise intrusted with a large folio MS. volume containing his journals from his arrival at Zanzibar, 28th January, 1866, to February 20, 1872, written out with all his characteristic care and beauty. Another instruction had been laid upon him. If he should find another set of slaves on the way to him, he was to send them back, for Livingstone would on no account expose himself anew to the misery, risk, and disappointment he had experienced from the kind of men that had compelled him to turn back at Nyangwe.

Dr. Livingstone's last act before Mr. Stanley left him was to write his letters—twenty for Great Britain, six for Bombay, two for New York, and one for Zanzibar. The two for New York were for Mr. Bennett of the *New York Herald*, by whom Stanley had been sent to Africa.

Mr. Stanley has freely unfolded to us the bitterness of his heart in parting from Livingstone. "My days seem to have been spent in an Elysian field; otherwise, why should I so keenly regret the near approach of the parting hour? Have I not been battered by successive fevers, prostrate with agony day after day lately? Have I not raved and stormed in madness? Have I not clenched my fists in fury, and fought with the wild strength of despair when in delirium? Yet, I regret to surrender the pleasure I have felt in this man's society, though so dearly purchased. . . . *March 14th.*—We had a sad breakfast together. I could not eat, my heart was too full; neither did my companion seem to have an appetite. We found something to do which kept us longer together. At eight o'clock I was not gone, and I had thought to have been off at five A.M. . . . We walked side by side; the men lifted their voices in a song. I took long looks at Livingstone, to impress his features thoroughly on my memory. . . . 'Now, my dear Doctor, the best friends must part. You have come far enough; let me beg of you to turn back.' 'Well,' Livingstone replied, 'I will say this to you: You have done what few men could do,—far better than some great travelers I know. And I am grateful to you for what you have done for me. God guide you safe home, and bless you, my friend.'—'And may God bring

you safe back to us all, my dear friend. Farewell!'—
'Farewell!' . . . My friendly reader, I wrote the above
extracts in my Diary on the evening of each day. I look
at them now after six months have passed away ; yet I am
not ashamed of them ; my eyes feel somewhat dimmed at
the recollection of the parting. I dared not erase, nor
modify what I had penned, while my feelings were strong.
God grant that if ever you take to traveling in Africa you
will get as noble and true a man for your companion as
David Livingstone! For four months and four days I lived
with him in the same house, or in the same boat, or in the
same tent, and I never found a fault in him. I am a man
of a quick temper, and often without sufficient cause, I daresay, have broken the ties of friendship ; but with Livingstone I never had cause for resentment, but each day's life
with him added to my admiration for him."

If Stanley's feeling for Livingstone was thus at the
warmest temperature, Livingstone's sense of the service
done to him by Stanley was equally unqualified. Whatever else he might be or might not be, he had proved a
true friend to him. He had risked his life in the attempt
to reach him, had been delighted to share with him every
comfort he possessed, and to leave with him ample stores
of all that might be useful to him in his effort to finish his
work. Whoever may have been to blame for it, it is certain that Livingstone had been afflicted for years, and
latterly worried almost to death, by the inefficency and
worthlessness of the men sent to serve him. In Stanley
he found one whom he could trust implicitly to do everything that zeal and energy could contrive in order to find
him efficient men and otherwise carry out his plans. It
was Stanley therefore whom he commissioned to send him
up men from Zanzibar. It was Stanley to whom he intrusted his Journal and other documents. Stanley had
been his confidental friend for four months—the only
white man to whom he talked for six years. It was matter

of life and death to Livingstone to be supplied for this concluding piece of work far better than he had been for years back. What man in his senses would have failed in these circumstances to avail himself to the utmost of the services of one who had shown himself so efficient; would have put him aside to fall back on others, albeit his own countrymen, who, with all their good-will, had not been able to save him from robbery, beggary, and a half-broken heart.

Stanley's journey from Unyanyembe to Bagamoio was a perpetual struggle against hostile natives, flooded roads, slush, mire, and water, roaring torrents, ants and mosquitos, or, as he described it, the ten plagues of Egypt. On his reaching Bagamoio, on the 6th May, he found a new surprise. A white man dressed in flannels and helmet appeared, and as he met Stanley congratulated him on his splendid success. It was Lieutenant Henn, R.N., a member of the Search Expedition which the Royal Geographical Society and others had sent out to look for Livingstone. The resolution to organize such an Expedition was taken after news had come to England of the war between the Arabs and the natives at Unyanyembe, stopping the communication with Ujiji, and rendering it impossible, as it was thought, for Mr. Stanley to get to Livingstone's relief. The Expedition had been placed under command of Lieutenant Dawson, R.N., with Lieutenant Henn as second, and was joined by the Rev. Charles New, a Missionary from Mombasa, and Mr. W. Oswell Livingstone, youngest son of the Doctor. Stanley's arrival at Bagamoio had been preceded by that of some of his men, who brought the news that Livingstone had been found and relieved. On hearing this, Lieutenant Dawson hurried to Zanzibar to see Dr. Kirk, and resigned his command. Lieutenant Henn soon after followed his example by resigning too. They thought that as Dr. Livingstone had been relieved there was no need for their

going on. Mr. New likewise declined to proceed. Mr. W.
Oswell Livingstone was thus left alone, at first full of the
determination to go on to his father with the men whom
Stanley was providing; but owing to the state of his
health, and under the advice of Dr. Kirk, he, too, declined
to accompany the Expedition, so that the men from Zanzibar proceeded to Unyanyembe alone.

On the 29th of May, Stanley, with Messrs. Henn, Livingstone, New, and Morgan, departed in the "Africa" from
Zanzibar, and in due time reached Europe.

It was deeply to be regretted that an enterprise so
beautiful and so entirely successful as Mr. Stanley's
should have been in some degree marred by ebullitions
of feeling little in harmony with the very joyous event.
The leaders of the English Search Expedition and their
friends felt, as they expressed it, that the wind had been
taken out of their sails. They could not but rejoice
that Livingstone had been found and relieved, but it
was a bitter thought that they had had no hand in the
process. It was galling to their feelings as Englishmen
that the brilliant service had been done by a stranger, a
newspaper correspondent, a citizen of another country.
On a small scale that spirit of national jealousy showed
itself, which on a wider arena has sometimes endangered
the relations of England and America.

When Stanley reached England, it was not to be overwhelmed with gratitude. At first the Royal Geographical
Society received him coldly. Instead of his finding Livingstone, it was surmised that Livingstone had found him.
Strange things were said of him at the British Association
at Brighton. The daily press actually challenged his
truthfulness; some of the newspapers affected to treat his
whole story as a myth. Stanley says frankly that this reception gave a tone of bitterness to his book—*How I Found
Livingstone*—which it would not have had if he had understood the real state of things. But the heart of the nation

was sound; the people believed in Stanley, and appreciated his service. At last the mists cleared away, and England acknowledged its debt to the American. The Geographical Society gave him the right hand of fellowship "with a warmth and generosity never to be forgotten." The President apologized for the words of suspicion he had previously used. Her Majesty the Queen presented Stanley with a special token of her regard. Unhappily, in the earlier stages of the affair, wounds had been inflicted which are not likely ever to be wholly healed. Words were spoken on both sides which cannot be recalled. But the great fact remains, and will be written on the page of history, that Stanley did a noble service to Livingstone, earning thereby the gratitude of England and of the civilized world.

CHAPTER XXII.

FROM UNYANYEMBE TO BANGWEOLO.

A.D. 1872–73.

Livingstone's long wait at Unyanyembe—His plan of operations—His fifty-
ninth birthday—Renewal of self-dedication—Letters to Agnes—to *New
York Herald*—Hardness of the African battle—Waverings of judgment,
whether Lualaba was the Nile or the Congo—Extracts from Journal—Gleams
of humor—Natural history—His distress on hearing of the death of Sir
Roderick Murchison—Thoughts on mission-work—Arrival of his escort—His
happiness in his new men—He starts from Unyanyembe—Illness—Great
amount of rain—Near Bangweolo—Incessant moisture—Flowers of the
forest—Taking of observations regularly prosecuted—Dreadful state of the
country from rain—Hunger—Furious attack of ants—Greatness of Livingstone's sufferings—Letters to Sir Thomas Maclear, Mr. Young, his brother,
and Agnes—His sixtieth birthday—Great weakness in April—Sunday services and observations continued—Increasing illness—The end approaching—
Last written words—Last day of his travels—He reaches Chitambo's village,
in Ilala—Is found on his knees dead, on morning of 1st May—Courage and
affection of his attendants—His body embalmed—Carried toward shore—
Dangers and sufferings during the march—The party meet Lieutenant Cameron at Unyanyembe—Determine to go on—*Ruse* at Kasekéra—Death of
Dr. Dillon—The party reach Bagamoio, and the remains are placed on board
a cruiser—The Search Expeditions from England—to East Coast under Cameron—to West Coast under Grandy—Explanation of Expeditions by Sir
Henry Rawlinson—Livingstone's remains brought to England—Examined
by Sir W. Fergusson and others—Buried in Westminster Abbey—Inscription
on slab—Livingstone's wish for a forest grave—Lines from *Punch*—Tributes
to his memory—Sir Bartle Frere—The *Lancet*—Lord Polwarth—Florence
Nightingale.

WHEN Stanley left Livingstone at Unyanyembe there
was nothing for the latter but to wait there until the men
should come to him who were to be sent up from Zanzibar.
Stanley left on the 14th March; Livingstone calculated
that he would reach Zanzibar on the 1st May, that his men

would be ready to start about the 22d May, and that they ought to arrive at Unyanyembe on the 10th or 15th July. In reality, Stanley did not reach Bagamoio till the 6th May, the men were sent off about the 25th, and they reached Unyanyembe about the 9th August. A month more than had been counted on had to be spent at Unyanyembe, and this delay was all the more trying because it brought the traveler nearer to the rainy season.

The intention of Dr. Livingstone, when the men should come, was to strike south by Ufipa, go round Tanganyika, then cross the Chambeze, and bear away along the southern shore of Bangweolo, straight west to the ancient fountains; from them in eight days to Katanga copper mines; from Katanga, in ten days, northeast to the great underground excavations, and back again to Katanga; from which N.N.W. twelve days to the head of Lake Lincoln. "There I hope devoutly," he writes to his daughter, "to thank the Lord of all, and turn my face along Lake Kamolondo, and over Lualaba, Tanganyika, Ujiji, and home."

His stay at Unyanyembe was a somewhat dreary one; there was little to do and little to interest him. Five days after Stanley left him occurred his fifty-ninth birthday. How his soul was exercised appears from the renewal of his self-dedication recorded in his Journal:

"*19th March, Birthday.*—My Jesus, my King, my Life, my All; I again dedicate my whole self to Thee. Accept me, and grant, O gracious Father, that ere this year is gone I may finish my task. In Jesus' name I ask it. Amen. So let it be.   DAVID LIVINGSTONE."

Frequent letters were written to his daughter from Unyanyembe, and they dwelt a good deal upon his difficulties, the treacherous way in which he had been treated, and the indescribable toil and suffering which had been the result. He said that in complaining to Dr. Kirk of the men whom he had employed, and the disgraceful use they had made of his (Kirk's) name, he never meant to charge him with being the author of their crimes, and it never occurred to him to say to Kirk, " I don't believe you to be the traitor they imply;" but Kirk took his complaint in high dudgeon as a covert attack upon himself, and did not act toward him as he ought to have done, considering what he owed him. His cordial and uniform testimony of Stanley was, " altogether he has behaved right nobly."

On the 1st May he finished a letter for the *New York Herald,* and asked God's blessing on it. It contained the memorable words afterward inscribed on the stone to his memory in Westminster Abbey: " All I can add in my loneliness is, may Heaven's rich blessing come down on every one—American, English, or Turk—who will help to heal the open sore of the world." It happened that the words were written precisely a year before his death.

Amid the universal darkness around him, the universal ignorance of God and of the grace and love of Jesus Christ, it was hard to believe that Africa should ever be won. He had to strengthen his faith amid this universal desolation. We read in his Journal:

" 13*th* *May.*—He will keep his word—the gracious One, full of grace and truth ; no doubt of it. He said: " Him that cometh unto me, I will in no wise cast out ;' and ' Whatsoever ye shall ask in my name, I will give it.' He WILL keep his word : then I can come and humbly present my petition, and it will be all right. Doubt is here inadmissible, surely.   D. L."

His mind ruminates on the river system of the country and the probability of his being in error:

" 21*st* *May.*—I wish I had some of the assurance possessed by others, but I am oppressed with the apprehension that, after all, it may turn out that I have been following the Congo; and who would risk being put into a cannibal pot, and converted into black man for *it ?*"

" 31*st* *May.*—In reference to this Nile source, I have been kept in perpetual doubt and perplexity. I know too much to be positive. Great Lualaba, or Lualubba, as Manyuema say, may turn out to be the Congo, and Nile a shorter river after all.[1] The fountains flowing north and

---

[1] From false punctuation, this passage is unintelligible in the *Last Journals,* vol. ii. p. 193.

south seem in favor of its being the Nile. Great westing is in favor of the Congo."

"*24th June.*—The medical education has led me to a continual tendency to suspend the judgment. What a state of blessedness it would have been, had I possessed the dead certainty of the homœopathic persuasion, and as soon as I found the Lakes Bangweolo, Moero, and Kamolondo pouring out their waters down the great central valley, bellowed out, 'Hurrah! Eureka!' and gone home in firm and honest belief that I had settled it, and no mistake. Instead of that, I am even now not at all 'cock-sure' that I have not been following down what may after all be the Congo.

We now know that this was just what he had been doing. But we honor him all the more for the diffidence that would not adopt a conclusion while any part of the evidence was wanting, and that led him to encounter unexampled risks and hardships before he would affirm his favorite view as a fact. The moral lesson thus enforced is invaluable. We are almost thankful that Livingstone never got his doubts solved, it would have been such a disappointment; even had he known that in all time coming the great stream which had cast on him such a resistless spell would be known as the Livingstone River, and would perpetuate the memory of his life and his efforts for the good of Africa.

Occasionally his Journal gives a gleam of humor: "*18th June.*—The Ptolemaic map defines people according to their food,—the Elephantophagi, the Struthiophagi, the Ichthiophagi, and the Anthropophagi. If we followed the same sort of classification, our definition would be by the drink, thus: the tribe of stout-guzzlers, the roaring potheen-fuddlers, the whisky-fishoid-drinkers, the vin-ordinaire bibbers, the lager-beer-swillers, and an outlying tribe of the brandy cocktail persuasion."

Natural History furnishes an unfailing interest: "*19th June.*—Whydahs, though full-fledged, still gladly take a feed from their dam, putting down the breast to the ground, and cocking up the bill and chirruping in the most engaging manner and winning way they know. She still gives them a little, but administers a friendly shove-off too. They all pick up feathers or grass, and hop from side to side of their mates, as if saying, 'Come, let us play at making little houses.' The wagtail has shaken her young quite off, and has a new nest. She warbles prettily, very much like a canary, and is extremely active in catching flies, but eats crumbs of bread-and-milk too. Sun-birds visit the pomegranate flowers, and eat insects therein too, as well as nectar. The young whydah birds crouch closely together at night for heat. They look like a woolly ball on a branch. By day they engage in pairing and coaxing each other. They come to the same twig every night. Like children, they try and lift heavy weights of feathers above their strength."

On 3d July a very sad entry occurs: "Received a note from Oswell, written in April last, containing the sad intelligence of Sir Roderick's departure from among us. Alas! alas! this is the only time in my life I ever felt inclined to use the word, and it bespeaks a sore heart; the best friend I ever had,—true, warm, and abiding,—he loved me more than I deserved; he looks down on me still." This entry indicates extraordinary depth of emotion. Sir Roderick exercised a kind of spell on Livingstone. Respect for him was one of the subordinate motives that induced him to undertake this journey. The hope of giving him satisfaction was one of the subordinate rewards to which he looked forward. His death was to Livingstone a kind of scientific widowhood, and must have deprived him of a great spring to exertion in this last wandering. On Sir Roderick's part the affection for him was very great. "Looking back," says his biographer, Professor Geikie, "upon his scientific career when not far from its close, Murchison found no part of it which brought more pleasing recollections than the support he had given to African explorers—Speke, Grant, and notably Livingstone. 'I rejoice,' he said, 'in the steadfast tenacity with which I have upheld my confidence in the ultimate success of the last-named of these brave men. In fact, it was the confidence I placed in the undying vigor of my dear friend Livingstone which has sustained me in the hope that I might live to enjoy the supreme delight of welcoming him back to his own country.' But that consummation was not to be. He himself was gathered to his rest just six days before Stanley brought news and relief to the forlorn traveler on Lake Tanganyika. And Livingstone, while still in pursuit of his quest, and within ten months of his death, learned in the heart of Africa the tidings which he chronicled in his journal." [1]

At other times he is ruminating on mission-work:

"10*th July.*—No great difficulty would be encountered in establishing a Christian mission a hundred miles or so from the East Coast. . . . To the natives the chief attention of the mission should be directed. It would not be desirable or advisable to refuse explanation to others; but I have avoided giving offense to intelligent Arabs, who, having pressed me, asking if I believed in Mohamed, by saying, 'No, I do not; I am a child of Jesus bin Miriam,' avoiding anything offensive in my tone, and often adding that Mohamed found their forefathers bowing down to trees and stones, and did good to them by forbidding idolatry, and teaching the worship of the only One God. This they all know, and it pleases them to have it recognized. It might be good policy to hire a respectable Arab to engage free porters, and conduct the mission to the country chosen, and obtain permission from the chief to build temporary houses. . . . A couple of Europeans beginning and carrying on a mission without a staff of foreign attendants, implies coarse country fare, it is true; but this would be nothing to those who at home amuse themselves with vigils, fasting, etc. A great deal of power is thus lost in the Church. Fastings and vigils, without a special object in view, are time run to waste. They are made to minister to a sort of self-gratification, instead of being turned to account for the good of others. They are like groaning in sickness: some people amuse themselves when ill with continuous moaning. The forty days of Lent might be annually spent in visiting adjacent tribes, and bearing unavoidable hunger and thirst with a good

---

[1] *Life of Sir R. I. Murchison,* vol. ii. pp. 297-8.

39

grace. Considering the greatness of the object to be attained, men might go without sugar, coffee, tea, as I went from September, 1866, to December, 1868, without either."

On the subject of Missions he says, at a later period, 8th November: "The spirit of missions is the spirit of our Master; the very genius of his religion. A diffusive philanthropy is Christianity itself. It requires perpetual propagation to attest its genuineness."

Thanks to Mr. Stanley and the American Consul, who made arrangements in a way that drew Livingstone's warmest gratitude, his escort arrived at last, consisting of fifty-seven men and boys. Several of these had gone with Mr. Stanley from Unyanyembe to Zanzibar; among the new men were some Nassick pupils who had been sent from Bombay to join Lieutenant Dawson. John and Jacob Wainwright were among these. To Jacob Wainwright, who was well-educated, we owe the earliest narrative that appeared of the last eight months of Livingstone's career. How happy he was with the men now sent to him appears from a letter to Mr. Stanley, written very near his death: "I am perpetually reminded that I owe a great deal to you for the men you sent. With one exception, the party is working like a machine. I give my orders to Manwa Sera, and never have to repeat them." Would that he had had such a company before!

On the 25th August the party started. On the 8th October they reached Tanganyika, and rested, for they were tired, and several were sick, including Livingstone, who had been ill with his bowel disorder. The march went on slowly, and with few incidents. As the season advanced, rain, mist, swollen streams, and swampy ground became familiar. At the end of the year they were approaching the river Chambeze. Christmas had its thanksgiving: "I thank the good Lord for the good gift of his Son, Jesus Christ our Lord."

In the second week of January they came near Bangweolo, and the reign of Neptune became incessant. We are told of cold rainy weather; sometimes a drizzle, sometimes an incessant pour; swollen streams and increasing sponges,—making progress a continual struggle. Yet, as he passes through a forest, he has an eye to its flowers, which are numerous and beautiful:

" There are many flowers in the forest; marigolds, a white jonquil-looking flower without smell, many orchids, white, yellow, and pink asclepias, with bunches of French-white flowers, clematis—*Methonica gloriosa*, gladiolus, and blue and deep purple polygalas, grasses with white starry seed-vessels, and spikelets of brownish red and yellow. Besides these, there are beautiful blue flowering bulbs, and new flowers of pretty, delicate form and but little scent. To this list may be added balsams, compositæ of blood-red color and of purple; other flowers of liver color, bright canary yellow, pink orchids on spikes thickly covered all round, and of three inches in length; spiderworts of fine blue or yellow or even pink. Different colored asclepiadeæ; beautiful yellow and red umbelliferous flowering plants; dill and wild parsnips; pretty flowering aloes, yellow and red, in one whorl of blossoms; peas and many other flowering plants which I do not know."

Observations were taken with unremitting diligence, except when, as was now common, nothing could be seen in the heavens. As they advanced, the weather became worse. It rained as if nothing but rain were ever known in the watershed. The path lay across flooded rivers, which were distinguished by their currents only from the flooded country along their banks. Dr. Livingstone had to be carried over the rivers on the back of one of his men, in the fashion so graphically depicted on the cover of the *Last Journals*. The stretches of sponge that came before and after the rivers, with their long grass and elephant-holes, were scarcely less trying. The inhabitants were, commonly, most unfriendly to the party; they refused them food, and, whenever they could, deceived them as to the way. Hunger bore down on the party with its bitter gnawing. Once a mass of furious ants attacked the Doctor by night, driving him in despair from hut to hut. Any frame but one of iron must have succumbed to a single month of such a life, and before a week was out, any body of men, not held together by a power of discipline and a charm of affection unexampled in the history of difficult expeditions, would have been scattered to the four winds. Livingstone's own sufferings were beyond all previous example.

About this time he began an undated letter—his last —to his old friends Sir Thomas Maclear and Mr. Mann. It was never finished, and never despatched; but as one of the latest things he ever wrote, it is deeply interesting, as showing how clear, vigorous, and independent his mind was to the very last:

"LAKE BANGWEOLO, SOUTH CENTRAL AFRICA.

"MY DEAR FRIENDS MACLEAR AND MANN,— . . . My work at present is mainly retracing my steps to take up the thread of my exploration. It counts in my lost time, but I try to make the most of it by going round outside this lake and all the sources, so that no one may come afterward and cut me out. I have a party of good men, selected by H. M. Stanley, who, at the instance of James Gordon Bennett, of the *New York Herald*, acted the part of a good Samaritan truly, and relieved my sore necessities. A dutiful son could not have done more than he generously did. I bless him. The men, fifty-six in number, have behaved as well as Makololo. I cannot award them higher praise, though they have not the courage of that brave kind-hearted people. From Unyanyembe we went due south to avoid an Arab war which had been going on for eighteen months. It is like one of our Caffre wars, with this difference—no one is enriched thereby, for all trade is stopped, and the Home Government pays nothing. We then went westward to Tanganyika, and along its eastern excessively mountainous bank to the end. The heat was really broiling among the rocks. No rain had fallen, and the grass being generally burned off, the heat rose off the black ashes as if out of an oven, yet the flowers persisted in coming out of the burning soil, and generally without leaves, as if it had been a custom that they must observe by a law of the Medes and Persians. This part detained us long; the men's limbs were affected with a sort of subcutaneous inflammation,—black rose or erysipelas,— and when I proposed mildly and medically to relieve the tension it was too horrible to be thought of, but they willingly carried the helpless. Then we mounted up at once into the high, cold region Urungu, south of Tanganyika, and into the middle of the rainy season, with well-grown grass and everything oppressively green; rain so often that no observations could be made, except at wide intervals. I could form no opinion as to our longitude, and but little of our latitudes. Three of the Baurungu chiefs, one a great friend of mine, Nasonso, had died, and the population all turned topsy-turvy, so I could make no use of previous observations. They elect sisters' or brothers' sons to the chieftainship, instead of the heir-apparent. Food was not to be had for either love or money.

"I was at the mercy of guides who did not know their own country, and when I insisted on following the compass, they threatened, ' no food for five or ten days in that line.' They brought us down to the back or north side of Bangweolo, while I wanted to cross the Chambeze and go round its southern side. So back again southeastward we had to bend. The Portuguese crossed this Chambeze a long time ago, and are therefore the first European discoverers. We were not black men with Portuguese names like those for whom the feat of crossing the continent was eagerly claimed by Lisbon statesmen. Dr. Lacerda was a man of scientific attainments, and Governor of Tette, but finding Cazembe at the rivulet called Chungu, he unfortunately succumbed to fever ten days after his arrival. He seemed anxious to make his way across to Angola. Misled by the similarity of Chambeze to Zambesi, they all thought it to be a branch of the river that flows past Tette, Senna, and Shupanga, by Luabo and Kongoné to the sea.

"I rather stupidly took up the same idea from a map saying 'Zambesi' (eastern branch), believing that the map printer had some authority for his assertion. My first crossing was thus as fruitless as theirs, and I was less excusable, for I ought to have remembered that while Chambeze is the true native name of the northern river, Zambesi is not the name of the southern river at all. It is a Portugese corruption of Dombazi, which we adopted rather than introduce confusion by new names, in the same way that we adopted Nyassa instead of Nyanza ia Nyinyesi = Lake of the Stars, which the Portuguese, from hearsay, corrupted into Nyassa   The English have been worse propagators of nonsense than Portuguese. 'Geography of Nyassa' was thought to be a learned way of writing the name, though ' Nyassi' means long grass and nothing else. It took me twenty-two months to eliminate the error into which I was led, and then it was not by my own acuteness, but by the chief Cazembe, who was lately routed and slain by a party of Banyamwezi. He gave me the first hint of the truth, and that rather in a bantering strain : ' One piece of water is just like another; Bangweolo water is just like Moero water, Chambeze water like Luapula water; they are all the same; but your chief ordered you to go to the Bangweolo, therefore by all means go, but wait a few days, till I have looked out for good men as guides, and good food for you to eat,' etc. etc.

"I was not sure but that it was all royal chaff, till I made my way back south to the head-waters again, and had the natives of the islet Mpabala slowly moving the hands all around the great expanse, with 183° of sea horizon, and saying that is Chambeze, forming the great Bangweolo, and disappearing behind that western headland to change its name to Luapula, and run down past Cazembe to Moero. That was the moment of discovery, and not my passage or the Portuguese passage of the river. If, however, any one chooses to claim for them the discovery of Chambeze as one line of drainage of the Nile Valley, I shall not fight with him; Culpepper's astrology was in the same way the forerunner of the Herschels' and the other astronomers that followed."

To another old friend, Mr James Young, he wrote about the same time: "*Opere peracto ludemus*—the work being finished, we will play—you remember in your Latin Rudiments lang syne. It is true for you, and I rejoice to think it is now your portion, after working nobly, to play. May you have a long spell of it! I am differently situated; I shall never be able to play. . . . To me it seems to be said, 'If thou forbear to deliver them that are drawn unto death, and those that be ready to be slain; if thou sayest, Behold we knew it not, doth not He that pondereth the heart consider, and He that keepeth thy soul doth He not know, and shall He not give to every one according to his works?' I have been led, unwittingly, into the slaving field of the Banians and Arabs in Central Africa. I have seen the woes inflicted, and I must still work and do all I can to expose and mitigate the evils. Though hard work is still to be my lot, I look genially on others more favored in their lot. I would not be a member of the 'International,' for I love to see and think of others enjoying life.

"During a large part of this journey I had a strong presentiment that I should never live to finish it. It is weakened now, as I seem to see the end toward which I have been striving looming in the distance. This presentiment did not interfere with the performance of any duty; it only made me think a great deal more of the future state of being."

In his latest letters there is abundant evidence that the great desire of his heart was to expose the slave-trade, rouse public feeling, and get that great hindrance to all good for ever swept away. "Spare no pains," he wrote to Dr. Kirk in 1871, "in attempting to persuade your superior to this end, and the Divine blessing will descend on you and yours."

To his daughter Agnes he wrote (15th August, 1872): "No one can estimate the amount of God-pleasing good that will be done, if, by Divine favor, this awful slave-trade, into the midst of which I have come, be abolished. This will be something to have lived for, and the conviction has grown in my mind that it was *for this end* I have been detained so long."

To his brother in Canada he says (December, 1872): "If the good Lord permits me to put a stop to the enormous evils of the inland slave-trade, I shall not grudge my hunger and toils. I shall bless his name with all my heart. The Nile sources are valuable to me only as a means of enabling me to open my mouth with power among men. It is this power I hope to apply to remedy an enormous evil, and join my poor little helping hand in the enormous revolution that in his all-embracing Providence He has been carrying on for ages, and is now actually helping forward. Men may think I covet fame, but I make it a rule never to read aught written in my praise."

Livingstone's last birthday (19th March, 1873) found him in much the same circumstances as before. "Thanks to the Almighty Preserver of men for sparing me thus far on the journey of life. Can I hope for ultimate success? So many obstacles have arisen. Let not Satan prevail over me, O my good Lord Jesus." A few days after (24th

March): "Nothing earthly will make me give up my work in despair. I encourage myself in the Lord my God, and go forward."

In the beginning of April, the bleeding from the bowels, from which he had been suffering, became more copious, and his weakness was pitiful; still he longed for strength to finish his work. Even yet the old passion for natural history was strong; the aqueous plants that abounded everywhere, the caterpillars that after eating the plants ate one another, and were such clumsy swimmers; the fish with the hook-shaped lower jaw that enabled them to feed as they skimmed past the plants; the morning summons of the cocks and turtle-doves; the weird scream of the fish eagle—all engaged his interest. Observations continued to be taken, and the Sunday services were always held.

But on the 21st April a change occurred. In a shaky hand he wrote: "Tried to ride, but was forced to lie down, and they carried me back to vil. exhausted." A kitanda or palanquin had to be made for carrying him. It was sorry work, for his pains were excruciating and his weakness excessive. On the 27th April[1] he was apparently at the lowest ebb, and wrote in his Journal the last words he ever penned—"Knocked up quite, and remain = recover sent to buy milch goats. We are on the banks of R. Molilamo."

The word "recover" seems to show that he had no presentiment of death, but cherished the hope of recovery; and Mr. Waller has pointed out, from his own sad observation of numerous cases in connection with the Universities Mission, that malarial poisoning is usually unattended with the apprehension of death, and that in none of these instances, any more than in the case of Livingstone, were there any such messages, or instructions, or expressions of trust and hope as are usual on the part of Christian men when death is near.

---

[1] This was the eleventh anniversary of his wife's death.

The 29th of April was the last day of his travels. In the morning he directed Susi to take down the side of the hut that the kitanda might be brought along, as the door would not admit it, and he was quite unable to walk to it. Then came the crossing of a river; then progress through swamps and plashes; and when they got to anything like a dry plain, he would ever and anon beg of them to lay him down. At last they got him to Chitambo's village, in Ilala, where they had to put him under the eaves of a house during a drizzling rain, until the hut they were building should be got ready.

Then they laid him on a rough bed in the hut, where he spent the night. Next day he lay undisturbed. He asked a few wandering questions about the country— especially about the Luapula. His people knew that the end could not be far off. Nothing occurred to attract notice during the early part of the night, but at four in the morning, the boy who lay at his door called in alarm for Susi, fearing that their master was dead. By the candle still burning they saw him, not in bed, but kneeling at the bedside with his head buried in his hands upon the pillow. The sad yet not unexpected truth soon became evident: he had passed away on the furthest of all his journeys, and without a single attendant. But he had died in the act of prayer—prayer offered in that rever- ential attitude about which he was always so particular; commending his own spirit, with all his dear ones, as was his wont, into the hands of his Saviour; and commending AFRICA—his own dear Africa—with all her woes and sins and wrongs, to the Avenger of the oppressed and the Redeemer of the lost.

If anything were needed to commend the African race, and prove them possessed of qualities fitted to make a noble nation, the courage, affection, and persevering loyalty shown by his attendants after his death might well have this effect. When the sad event became known among the men, it was cordially resolved that every effort should be made to carry their master's remains to Zanzibar. Such an undertaking was extremely perilous, for there were not merely the ordinary risks of travel to a small body of natives, but there was also the superstitious horror everywhere prevalent connected with the dead. Chitambo must be kept in ignorance of what had happened, otherwise a ruinous fine would be sure to be inflicted on them. The secret, however, oozed out, but happily the chief was reasonable. Susi and Chuma, the old attendants of Livingstone, became now the leaders of the company, and they fulfilled their task right nobly. The interesting narrative of Mr. Waller at the end of the *Last Journals* tells us how calmly yet efficiently they set to work. Arrangements were made for drying and embalming the body, after removing and burying the heart and other viscera. For fourteen days the body was dried in the sun. After being wrapped in calico, and the legs bent inward at the knees, it was enclosed in a large piece of bark from a Myonga-tree in the form of a cylinder; over this a piece of sail-cloth was sewed; and the package was lashed to a pole, so as to be carried by two men. Jacob Wainwright carved an inscription on the Mvula tree under which the body had rested, and where the heart was buried, and Chitambo was charged to keep the grass cleared away, and to protect two posts and a cross-piece which they erected to mark the spot.

They then set out on their homeward march. It was a serious journey, for the terrible exposure had affected the health of most of them, and many had to lie down through sickness. The tribes through which they passed were generally friendly, but not always. At one place they had a regular fight. On the whole, their progress was wonderfully quiet and regular. Everywhere they found that the news of the Doctor's death had got before them. At one place they heard that a party of Englishmen, headed by

Dr. Livingstone's son, on their way to relieve his father, had been seen at Bagamoio some months previously. As they approached Unyanyembe, they learned that the party was there, but when Chuma ran on before, he was disappointed to find that Oswell Livingstone was not among them. Lieutenant Cameron, Dr. Dillon, and Lieutenant Murphy were there, and heard the tidings of the men with deep emotion. Cameron wished them to bury the remains where they were, and not run the risk of conveying them through the Ugogo country; but the men were inflexible, determined to carry out their first intention. This was not the only interference with these devoted and faithful men. Considering how carefully they had gathered all Livingstone's property, and how conscientiously, at the risk of their lives, they were carrying it to the coast, to transfer it to the British Consul there, it was not warrantable in the new-comers to take the boxes from them, examine their contents, and carry off a part of them. Nor do we think Lieutenant Cameron was entitled to take away the instruments with which all Livingstone's observations had been made for a series of seven years, and use them, though only temporarily, for the purpose of his Expedition, inasmuch as he thereby made it impossible so to reduce Livingstone's observations as that correct results should be obtained from them. Sir Henry Rawlinson seems not to have adverted to this result of Mr. Cameron's act, in his reference to the matter from the chair of the Geographical Society.

On leaving Unyanyembe the party were joined by Lieutenant Murphy, not much to the promotion of unity of action or harmonious feeling. At Kasekéra a spirit of opposition was shown by the inhabitants, and a *ruse* was resorted to so as to throw them off their guard. It was resolved to pack the remains in such form that when wrapped in calico they should appear like an ordinary bale of merchandise. A fagot of mapira stalks, cut into lengths of about six feet, was then swathed in cloth, to imitate a dead body about to be buried. This was sent back along the way to Unyanyembe, as if the party had changed their minds and resolved to bury the remains there. The bearers, at nightfall, began to throw away the mapira rods, and then the wrappings, and when they had thus disposed of them they returned to their companions. The villagers of Kasekéra had now no suspicion, and allowed the party to pass unmolested. But though one tragedy was averted, another was enacted at Kasekéra— the dreadful suicide of Dr. Dillon while suffering from dysentery and fever.

The cortége now passed on without further incident, and arrived at Bagamoio in February, 1874. Soon after they reached Bagamoio a cruiser arrived from Zanzibar, with the acting Consul, Captain Prideaux, on board, and the remains were conveyed to that island previous to their being sent to England.

The men that for nine long months remained steadfast to their purpose to pay honor to the remains of their master, in the midst of innumerable trials and dangers and without hope of reward, have established a strong claim to the gratitude and admiration of the world. Would that the debt were promptly repaid in efforts to free Africa from her oppressors, and send throughout all her borders the Divine proclamation, "Glory to God in the highest, on earth peace, good-will to men."

In regard to the Search party to which reference has been made, it may be stated that when Livingstone's purpose to go back to the barbarous regions where he had suffered so much before became known in England it excited a feeling of profound concern. Two Expeditions were arranged. That to the East Coast, organized by the Royal Geographical Society, was placed under Lieutenant Cameron, and included in its ranks Robert Moffat, a grandson of Dr. Moffat's, who (as has been already stated) fell early a sacrifice to fever. The members of the Expedition suffered much from sickness; it was broken up at Unyanyembe, when the party bearing the remains of Dr. Livingstone was met. The other party, under command of Lieutenant Grandy, was to go to the West Coast, start from Loanda, strike the Congo, and move on to Lake Lincoln. This Expedition was fitted out solely at the cost of Mr. Young. He was deeply concerned for the safety of his friend, knowing how he was hated by the slave-traders whose iniquities he had exposed, and thinking it likely that if he once reached Lake Lincoln he would make for the west coast along the Congo. The purpose of these Expeditions is carefully explained in a letter addressed to Dr. Livingstone by Sir Henry Rawlinson, then President of the Royal Geographical Society:

"LONDON, *November* 20, 1872.

"DEAR DR. LIVINGSTONE,—You will no doubt have heard of Sir Bartle Frere's deputation to Zanzibar long before you receive this, and you will have learnt with heartfelt satisfaction that there is now a definite prospect of the infamous East African slave-trade being suppressed. For this great end, if it be achieved, we shall be mainly indebted to your recent letters, which have had a powerful effect on the public mind in England, and have thus stimulated the action of the Government. Sir Bartle will keep you informed of his arrangements, if there are any means of communicating with the interior, and I am sure you will assist him to the utmost of your power in carrying out the good work in which he is engaged.

"It was a great disappointment to us that Lieutenant Dawson's Expedition, which we fitted out in the beginning of the year with such completeness, did not join you at Unyanyembe, for it could not have failed to be of service to you in many ways. We are now trying to aid you with a second Expedition under Lieutenant Cameron, whom we have sent out under Sir Bartle's orders, to join you if possible in the vicinity of Lake Tanganyika, and attend to your wishes in respect to his further movements. We leave it entirely to your discretion whether you like to keep Mr. Cameron with you or to send him on to the Victoria Nyanza, or any other points that you are unable to visit yourself. Of course the great point of interest connected with your present exploration is the determination of the lower course of the Lualaba. Mr. Stanley still adheres to the view, which you formerly held, that it

40

drains into the Nile; but if the levels which you give are correct, this is impossible. At any rate, the opinion of the identity of the Congo and Lualaba is now becoming so universal that Mr. Young has come forward with a donation of £2000 to enable us to send another Expedition to your assistance up that river, and Lieutenant Grandy, with a crew of twenty Kroomen, will accordingly be pulling up the Congo before many months are over. Whether he will really be able to penetrate to your unvisited lake, or beyond it to Lake Lincoln, is, of course, a matter of great doubt; but it will at any rate be gratifying to you to know that support is approaching you both from the west and east. We all highly admire and appreciate your indomitable energy and perseverance, and the Geographical Society will do everything in its power to support you, so as to compensate in some measure for the loss you have sustained in the death of your old friend Sir Roderick Murchison. My own tenure of office expires in May, and it is not yet decided who is to succeed me, but whoever may be our President, our interest in your proceedings will not slacken. Mr. Waller will, I daresay, have told you that we have just sent a memorial to Mr. Gladstone, praying that a pension may be at once conferred upon your daughters, and I have every hope that our prayer may be successful. You will see by the papers, now sent to you, that there has been much acrimonious discussion of late on African affairs. I have tried myself in every possible way to throw oil on the troubled waters, and begin to hope now for something like peace. I shall be very glad to hear from you if you can spare time to send me a line, and will always keep a watchful eye over your interests.—I remain, yours very truly,

"H. C. RAWLINSON."

The remains were brought to Aden on board the "Calcutta," and thereafter transferred to the P. and O. steamer "Malwa," which arrived at Southampton on the 15th of April. Mr. Thomas Livingstone, eldest surviving son of the Doctor, being then in Egypt on account of his health,[1] had gone on board at Alexandria. The body was conveyed to London by special train and deposited in the rooms of the Geographical Society in Saville Row.

In the course of the evening the remains were examined by Sir William Fergusson and several other medical gentleman, including Dr. Loudon, of Hamilton,

[1] Thomas never regained robust health. He died at Alexandria, 15th March, 1876.

whose professional skill and great kindness to his family had gained for him a high place in the esteem and love of Livingstone. To many persons it had appeared so incredible that the remains should have been brought from the heart of Africa to London, that some conclusive identification of the body seemed to be necessary to set all doubt at rest. The state of the arm, the one that had been broken by the lion, supplied the crucial evidence. "Exactly in the region of the attachment of the deltoid to the humerus" (said Sir William Fergusson in a contribution to the *Lancet*, April 18, 1874), "there were the indications of an oblique fracture. On moving the arm there were the indications of an ununited fracture. A closer identification and dissection displayed the false joint that had so long ago been so well recognized by those who had examined the arm in former days. . . . The first glance set my mind at rest, and that, with the further examination, made me as positive as to the identification of these remains as that there has been among us in modern times one of the greatest men of the human race—David Livingstone."

On Saturday, April 18, 1874, the remains of the great traveier were committed to their resting-place near the centre of the nave of Westminster Abbey. Many old friends of Livingstone came to be present, and many of his admirers, who could not but avail themselves of the opportunity to pay a last tribute of respect to his memory. The Abbey was crowded in every part from which the spectacle might be seen. The pall-bearers were Mr. H. M. Stanley, Jacob Wainwright, Sir T. Steele, Dr. Kirk, Mr. W. F. Webb, Rev. Horace Waller, Mr. Oswell, and Mr. E. D. Young. Two of these, Mr. Waller and Dr. Kirk, along with Dr. Stewart, who was also present, had assisted twelve years before at the funeral of Mrs. Livingstone at Shupanga. Dr. Moffat, too, was there, full of sorrowful admiration. Amid a service which was emphatically impressive throughout, the simple words of the hymn, sung to the tune of Tallis, were peculiarly touching:

"O God of Bethel! by whose hand
Thy people still are fed,
Who through this weary pilgrimage
Hast all our fathers led."

The black slab that now marks the resting-place of Livingstone bears this inscription:

BROUGHT BY FAITHFUL HANDS

OVER LAND AND SEA,

HERE RESTS

DAVID LIVINGSTONE,

MISSIONARY, TRAVELER, PHILANTHROPIST,

BORN MARCH 19, 1813,

AT BLANTYRE, LANARKSHIRE.

DIED MAY 4,[1] 1873,

AT CHITAMBO'S VILLAGE, ILALA.

For thirty years his life was spent in an unwearied effort to evangelize
the native races, to explore the undiscovered secrets,
and abolish the desolating slave-trade of Central Africa,
and where, with his last words he wrote:
"All I can say in my solitude is, may Heaven's rich blessing
come down on every one—American, English, Turk—
who will help to heal this open sore of the world."

Along the right border of the stone are the words:

TANTUS AMOR VERI, NIHIL EST QUOD NOSCERE MALIM
QUAM FLUVII CAUSAS PER SÆCULA TANTA LATENTES.

And along the left border:

OTHER SHEEP I HAVE WHICH ARE NOT OF THIS FOLD,
THEM ALSO I MUST BRING, AND THEY SHALL HEAR MY VOICE.

---

[1] In the *Last Journals* the date is 1st May; on the stone, 4th May. The attendants could not quite determine the day.

On the 25th June, 1868, not far from the northern border of that lake Bangweolo on whose southern shore he passed away, Dr. Livingstone came on a grave in a forest. He says of it:

"It was a little rounded mound, as if the occupant sat in it in the usual native way; it was strewed over with flour, and a number of the large blue beads put on it; a little path showed that it had visitors. This is the sort of grave I should prefer: to be in the still, still forest, and no hand ever disturb my bones. The graves at home always seemed to me to be miserable, especially those in the cold, damp clay, and without elbow-room; but I have nothing to do but wait till He who is over all decides where I have to lay me down and die. Poor Mary lies on Shupanga brae, 'and beeks fornent the sun.'"

"He who is over all" decreed that while his heart should lie in a leafy forest, in such a spot as he loved, his bones should repose in a great Christian temple, where many, day by day, as they read his name, would recall his noble Christian life, and feel how like he was to Him of whom it is written: "The Spirit of the Lord God is upon me; because the Lord hath anointed me to preach good tidings to the meek: He hath sent me to bind up the broken-hearted, to proclaim liberty to the captives, and the opening of the prison to them that are bound; to proclaim the acceptable year of the Lord, and the day of vengeance of our God; to comfort all that mourn; to appoint unto them that mourn in Zion, to give unto them beauty for ashes, the oil of joy for mourning, the garment of praise for the spirit of heaviness; that they might be called trees of righteousness, the planting of the Lord, that He might be glorified."

> "Droop half-mast colors, bow, bareheaded crowds,
>   As this plain coffin o'er the side is slung,
> To pass by woods of masts and ratlined shrouds,
>   As erst by Afric's trunks, liana-hung.

'Tis the last mile of many thousands trod
  With failing strength but never-failing will,
By the worn frame, now at its rest with God,
  That never rested from its fight with ill.

Or if the ache of travel and of toil
  Would sometimes wring a short, sharp cry of pain
From agony of fever, blain, and boil,
  'Twas but to crush it down and on again!

He knew not that the trumpet he had blown
  Out of the darkness of that dismal land,
Had reached and roused an army of its own
  To strike the chains from the slave's fettered hand.

Now we believe, he knows, sees all is well;
  How God had stayed his will and shaped his way,
To bring the light to those that darkling dwell
  With gains that life's devotion well repay.

Open the Abbey doors and bear him in
  To sleep with king and statesman, chief and sage,
The missionary come of weaver-kin,
  But great by work that brooks no lower wage.

He needs no epitaph to guard a name
  Which men shall prize while worthy work is known;
He lived and died for good—be that his fame:
  Let marble crumble: this is Living—stone."—*Punch.*

Eulogiums on the dead are often attempts, sometimes
sufficiently clumsy, to conceal one-half of the truth and
fill the eye with the other. In the case of Livingstone
there is really nothing to conceal. In tracing his life
in these pages we have found no need for the brilliant
colors of the rhetorician, the ingenuity of the partisan,
or the enthusiasm of the hero-worshiper. We have felt,
from first to last, that a plain, honest statement of the
truth regarding him would be a higher panegyric than
any ideal picture that could be drawn. The best tributes
paid to his memory by distinguished countrymen were

the most literal—we might almost say the most prosaic. It is but a few leaves we can reproduce of the many wreaths that were laid on his tomb.

Sir Bartle Frere, as President of the Royal Geographical Society, after a copious notice of his life, summed it up in these words: "As a whole, the work of his life will surely be held up in ages to come as one of singular nobleness of design, and of unflinching energy and self-sacrifice in execution. It will be long ere any one man will be able to open so large an extent of unknown land to civilized mankind. Yet longer, perhaps, ere we find a brighter example of a life of such continued and useful self-devotion to a noble cause."

In a recent letter to Dr. Livingstone's eldest daughter, Sir Bartle Frere (after saying that he was first introduced to Dr. Livingstone by Mr. Phillip, the painter, as "one of the noblest men he had ever met," and rehearsing the history of his early acquaintance) remarks:

"I could hardly venture to describe my estimate of his character as a Christian further than by saying that I never met a man who fulfilled more completely my idea of a perfect Christian gentleman,—actuated in what he thought and said and did by the highest and most chivalrous spirit, modeled on the precepts of his great Master and Exemplar.

"As a man of science, I am less competent to judge, for my knowledge of his work is to a great extent secondhand; but derived, as it is, from observers like Sir Thomas Maclear, and geographers like Arrowsmith, I believe him to be quite unequaled as a scientific traveler, in the care and accuracy with which he observed. In other branches of science I had more opportunities of satisfying myself, and of knowing how keen and accurate was his observation, and how extensive his knowledge of everything connected with natural science; but every page of his journals, to the last week of his life, testified to his wonderful natural powers and accurate observation. Thirdly, as a missionary and explorer I have always put him in the very first rank. He seemed to me to possess in the most wonderful degree that union of opposite qualities which were required for such a work as opening out heathen Africa to Christianity and civilization. No man had a keener sympathy with even the most barbarous and unenlightened; none had a more ardent desire to benefit and improve the most abject. In his aims, no man attempted, on a grander or more thorough scale, to benefit and improve those of his race who most needed improvement and light. In the execution of what he undertook, I never met his equal for energy and sagacity, and I feel sure that future ages will place him among the very first of those missionaries, who, following the apostles, have continued to carry the light of the gospel to the darkest regions of the world, throughout the last 1800 years. As regards the value of the work he accomplished, it might be premature to speak,—not that I think it possible I can over-estimate it, but because I feel sure that every year will add fresh evidence to show how well-considered were the plans he took in hand, and how vast have been the results of the movements he set in motion."

The generous and hearty appreciation of Livingstone by the medical profession was well expressed in the words of the *Lancet:* "Few men have disappeared from our ranks more universally deplored, as few have served in them with a higher purpose, or shed upon them the lustre of a purer devotion."

Lord Polwarth, in acknowledging a letter from Dr. Livingstone's daughter, thanking him for some words on her father, wrote thus: "I have long cherished the memory of his example, and feel that the truest beauty was his essentially Christian spirit. Many admire in him the great explorer and the noble-hearted philanthropist; but I like to think of him, not only thus, but as a man who was a servant of God, loved his Word intensely, and while he spoke to men of God, spoke more to God of men.

"His memory will never perish, though the first freshness, and the impulse it gives just now, may fade; but his prayers will be had in everlasting remembrance, and unspeakable blessings will yet flow to that vast continent he opened up at the expense of his life. God called and qualified him for a noble work, which, by grace, he nobly fulfilled, and we can love the honored servant, and adore the gracious Master."

Lastly, we give the beautiful wreath of Florence Nightingale, also in the form of a letter to Dr. Livingstone's daughter:

"LONDON, *Feb.* 18*th*, 1874.

"DEAR MISS LIVINGSTONE,—I am only one of all England which is feeling with you and for you at this moment.

"But Sir Bartle Frere encourages me to write to you.

"We cannot help still yearning to hear of some hope that your great father may be still alive.

"God knows; and in knowing that He knows who is all wisdom, goodness, and power, we must find our rest.

"He has taken away, if at last it be as we fear, the greatest man of his generation, for Dr. Livingstone stood alone.

"There are few enough, but a few statesmen. There are few enough, but a few great in medicine, or in art, or in poetry. There are a few great travelers. But Dr. Livingstone stood alone as the great Missionary Traveler, the bringer-in of civilization; or rather the pioneer of civilization—he that cometh before—to races lying in darkness.

"I always think of him as what John the Baptist, had he been living in the nineteenth century, would have been.

"Dr. Livingstone's fame was so world-wide that there

were other nations who understood him even better than we did.

"Learned philologists from Germany, not at all orthodox in their opinions, have yet told me that Dr. Livingstone was the only man who understood races, and how to deal with them for good; that he was the one true missionary. We cannot console ourselves for our loss. He is irreplaceable.

"It is not sad that he should have died out there. Perhaps it was the thing, much as he yearned for home, that was the fitting end for him. He may have felt it so himself.

"But would that he could have completed that which he offered his life to God to do!

"If God took him, however, it was that his life was completed in God's sight; his work finished, the most glorious work of our generation.

"He has opened those countries for God to enter in. He struck the first blow to abolish a hideous slave-trade.

"He, like Stephen, was the first martyr.

> " 'He climbed the steep ascent of heaven,
>   Through peril, toil, and pain;
>   O God! to us may grace be given
>   To follow in his train!'

"To us it is very dreary, not to have seen him again, that he should have had none of us by him at the last; no last word or message.

"I feel this with regard to my dear father and one who was more than mother to me, Mrs. Bracebridge, who went with me to the Crimean war, both of whom were taken from me last month.

"How much more must we feel it, with regard to our great discoverer and hero, dying so far off!

"But does he regret it? How much he must know now! how much he must have enjoyed!

"Though how much we would give to know *his* thoughts, *alone with God*, during the latter days of his life.

"May we not say, with old Baxter (something altered from that verse)?

> "'My knowledge of that life is small,
> The eye of faith is dim;
> But 'tis enough that *Christ knows all,*
> And he will be with *Him.*'

"Let us think only of him and of his present happiness, his eternal happiness, and may God say to us: 'Let not your heart be troubled.' Let us exchange a 'God bless you,' and fetch a real blessing from God in saying so.

"FLORENCE NIGHTINGALE."

CHAPTER XXIII.

POSTHUMOUS INFLUENCE.

History of his life not completed at his death—Thrilling effect of the tragedy of Ilala—Livingstone's influence on the slave-trade—His letters from Manyuema—Sir Bartle Frere's mission to Zanzibar—Successful efforts of Dr. Kirk with Sultan of Zanzibar—The land route—The sea route—Slave-trade declared illegal—Egypt—The Soudan—Colonel Gordon—Conventions with Turkey—King Mtesa of Uganda—Nyassa district—Introduction of lawful commerce—Various commercial enterprises in progress—Influence of Livingstone on exploration — Enterprise of newspapers — Exploring undertakings of various nations—Livingstone's personal service to science—His hard work in science the cause of respect—His influence on missionary enterprise—Livingstonia—Dr. Stewart.—Mr. E. D. Young—Blantyre—The Universities Mission under Bishop Steere—Its return to the mainland and to Nyassa district—Church Missionary Society at Nyanza—London Missionary Society at Tanganyika—French, Inland, Baptist, and American missions—Medical missions—The Fisk Livingstone hall—Livingstone's great legacy to Africa, a spotless Christian name and character—Honors of the future.

THE heart of David Livingstone was laid under the mvula-tree in Ilala, and his bones in Westminster Abbey; but his spirit marched on. The history of his life is not completed with the record of his death. The continual cry of his heart to be permitted to finish his work was answered, answered thoroughly, though not in the way he thought of. The thrill that went through the civilized world when his death and all its touching circumstances became known, did more for Africa than he could have done had he completed his task and spent years in this country following it up. From the worn-out figure kneeling at the bedside in the hut in Ilala an electric spark seemed to fly, quickening hearts on every side. The statesman felt it; it put new vigor into the despatches